Immunochemical Protocols

Methods in Molecular Biology

John M. Walker, Series Editor

1. **Proteins,** edited by ***John M. Walker,*** *1984*
2. **Nucleic Acids,** edited by ***John M. Walker,*** *1984*
3. **New Protein Techniques,** edited by ***John M. Walker,*** *1988*
4. **New Nucleic Acid Techniques,** edited by ***John M. Walker,*** *1988*
5. **Animal Cell Culture,** edited by ***Jeffrey W. Pollard and John M. Walker,*** *1990*
6. **Plant Cell and Tissue Culture,** edited by ***Jeffrey W. Pollard and John M. Walker,*** *1990*
7. **Gene Transfer and Expression Protocols,** edited by ***E. J. Murray,*** *1991*
8. **Practical Molecular Virology,** edited by ***Mary K. L. Collins,*** *1991*
9. **Protocols in Human Molecular Genetics,** edited by ***Christopher G. Mathew,*** *1991*
10. **Immunochemical Protocols** , edited by ***Margaret M. Manson,*** *1992*

Methods in Molecular Biology • 10

Immunochemical Protocols

Edited by

Margaret M. Manson

Medical Research Council Laboratories, Carshalton, Surrey, UK

Humana Press ✷ Totowa, New Jersey

999 Riverview Drive, Suite 208
Totowa, New Jersey 07512

Printed in the United States of America. 10 9 8 7 6 5 4 3 2

Library of Congress Cataloging in Publication Data
Main entry under title:

Methods in molecular biology.

Immunochemical Protocols / edited by Margaret M. Manson.
p. cm. – (Methods in molecular biology ; 10)
Includes index.
ISBN 0-89603-204-3 (hard cover)
ISBN 0-89603-270-1 (comb bound)
1. Immunochemistry–Methodology. I. Manson, Margaret M.
II. Series: Methods in molecular biology ; v. 10.
QR183.6.I43 1992
616.07'56–dc20 91-41416
CIP

Preface

Molecular biologists can no longer afford to be experts in only one particular area. They need to be equally competent at handling DNA, RNA, and proteins, moving from one area to another as required by the problem. Advances in the field have been greatly helped by numerous applications of antibody–antigen interactions, so that a basic knowledge of immunology has now become useful.

The emphasis of *Immunochemical Protocols* is not on the production of antibodies, since many people will beg or buy these, but rather on the enormous range of applications to which these powerful reagents can be put. For completeness, however, methods are included for the production of both polyclonal and monoclonal antibodies. Techniques that may be helpful in obtaining antibodies to elusive antigens and to synthetic peptides are also provided. Both qualitative and quantitative methods are described for localizing, characterizing, and quantifying various antigens. These range from the use of antibodies in screening DNA libraries to their application in ELISA assays for quantifying adducts in alkylated DNA.

Each method is described by an author who has regularly used the technique in his or her own laboratory. Not all the techniques necessarily represent the state-of-the-art—they are, however, all dependable methods that regularly achieve the desired result.

Each chapter starts with a brief description of the basic theory behind the technique being described. The main aim of this book, however, is to outline the practical steps necessary for carrying out the method successfully. The Methods section therefore contains a detailed step-by-step description of each protocol. The Notes section compliments the Methods section by indicating any major problems or difficulties that might arise in using the technique, as well as those adaptations, modifications, or alterations that might prove helpful in specific applications.

Immunochemical Protocols thus should be particularly useful to those with no previous experience of a technique, appealing to postgraduates and research workers who wish to try a technique for the first time, as well as to undergraduates (especially project students).

Margaret M. Manson

Contents

Contributors

MICHAEL G. BAINES • *National Institute for Biological Standards and Control, Potters Bar, UK*
DAVID L. BATES • *Novo Nordisk Diagnostics Ltd., Cambridge, UK*
EDWARD A. BAYER • *Department of Biophysics, The Weizmann Institute of Science, Rehovot, Israel*
JULIAN E. BEESLEY • *Wellcome Research Laboratories, Beckenham, UK*
BHUPENDRA BHATT • *Department of Haematology, King's College School of Medicine and Dentistry, London, UK*
KURT BIENZ • *Institute for Medical Microbiology, University of Basel, Switzerland*
LESZEK K. BORYSIEWICZ • *Department of Medicine, Addenbrooke's Hospital, Cambridge, UK*
M. DANNY BURKE • *Scotgen Limited, University of Aberdeen, Scotland*
FRANCIS J. CARR • *Scotgen Limited, University of Aberdeen, Scotland*
VICTOR T.-W. CHAN • *Oncology Division, Beth Israel Hospital, Boston, MA*
JOHN A. CROLLA • *Wessex Regional Genetics Laboratory, General Hospital, Salisbury, UK*
ALAN J. CUMBER • *Drug Targeting Laboratory, Institute of Cancer Research, Sutton, Surrey, UK*
REGINALD DAVIES • *Medical Research Council Toxicology Unit, Carshalton, Surrey, UK*
CHRISTOPHER J. DEAN • *Institute of Cancer Research, The Haddow Laboratories, Sutton, Surrey, UK*
MONIQUE DIANO • *Biologie de la Differenciation Cellulaire, Faculte des Sciences de Luminy, Marseilles, France*
RICHARD E. EDWARDS • *Medical Research Council Toxicology Unit, Carshalton, Surrey, UK*
DENISE EGGER • *Institute for Medical Microbiology, University of Basel, Switzerland*
GERARD I. EVAN • *Imperial Cancer Research Fund Laboratories, London, UK*
SIMON J. FORSTER • *Scotgen Limited, University of Aberdeen, Scotland*
YVELINE FROBERT • *Section de Pharmacologie et d'Immunologie, Department de Biologie, Gif-sur-Yvette, France*

JACQUES GRASSI • *Section de Pharmacologie et d'Immunologie, Department de Biologie, Gif-sur-Yvette, France*
JONATHAN A. GREEN • *Medical Research Council Toxicology Unit, Carshalton, Surrey, UK*
MICHAEL J. HAAS • *Eastern Regional Research Center, U. S. Department of Agriculture, Philadelphia, PA*
ANITA A. HAMILTON • *Scotgen Limited, University of Aberdeen, Scotland*
DAVID C. HANCOCK • *Imperial Cancer Research Fund Laboratories, London, UK*
WILLIAM J. HARRIS • *Scotgen Limited, University of Aberdeen, Scotland*
COLIN J. HENDERSON • *Imperial Cancer Research Fund, Molecular Pharmacology and Drug Metabolism Group, Edinburgh, Scotland*
C. SIMON HERRINGTON • *University of Oxford, Nuffield Department of Pathology and Bacteriology, John Radcliffe Hospital, Oxford, UK*
MICHEL HIRN • *Faculte des Sciences de Luminy, Marseilles, France*
SUSANNE F. IMRIE • *Institute of Cancer Research, The Haddow Laboratories, Sutton, Surrey, UK*
SCOTT H. KAUFMANN • *Oncology Center, Johns Hopkins Hospital and Department of Pharmacology and Molecular Science, Johns Hopkins University School of Medicine, Baltimore, MD*
ANNE LAINE • *INSERM, Lille Cedex, France*
ANDRÉ LE BIVIC • *Biologie de la Differenciation Cellulaire, Faculte des Sciences de Luminy, Marseilles, France*
BARBARA I. LUDEKE • *Department of Neuropathology, University Hospital, Zurich, Switzerland*
KITTI MAKARANANDA • *Department of Pharmacology, Mahidol University, Bangkok, Thailand*
MARGARET M. MANSON • *Medical Research Council Toxicology Unit, Carshalton, Surrey, UK*
JANET MARTIN • *Medical Research Council Toxicology Unit, Carshalton, Surrey, UK*
JAMES O'D. MCGEE • *University of Oxford, Nuffield Department of Pathology and Bacteriology, John Radcliffe Hospital, Oxford, UK*
WILLIAM T. MELVIN • *Department of Molecular and Cell Biology, University of Aberdeen, Scotland*
SARA E. MOLE • *CRC Human Cancer Genetics Research Group, Department of Pathology, University of Cambridge, UK*
JOHN P. MOORE • *Aaron Diamond AIDS Research Center, New York*
GORDON E. NEAL • *Medical Research Council Laboratories, Toxicology Unit, Carshalton, Surrey, UK*
MICHAEL G. ORMEROD • *Institute of Cancer Research, The Haddow Laboratories, Sutton, Surrey, UK*

ANTHONY H. V. SCHAPIRA • *Department of Neurological Science, Royal Free Medical School, London, UK and Institute of Neurology, London, UK*

JOEL H. SHAPER • *Oncology Center, Johns Hopkins Hospital and Department of Pharmacology and Molecular Science, Johns Hopkins University School of Medicine, Baltimore, MD*

DONALD F. SUMMERS • *Department of Cellular, Viral, and Molecular Biology, University of Utah School of Medicine, Salt Lake City, UT*

BOGUSLAW SZEWCZYK • *Department of Biochemistry, University of Gdansk, Poland*

T. J. THOMAS • *Department of Medicine, Division of Rheumatology, University of Medicine and Dentistry of New Jersey, Robert Wood Johnson Medical School, New Brunswick, NJ*

ROBIN THORPE • *National Institute for Biological Standards and Control, Potters Bar, England*

MICHAEL J. TILBY • *Leukaemia Research Fund Unit, Newcastle University, Newcastle-upon-Tyne, UK*

EDDIE C. Y. WANG • *Department of Medicine, Addenbrooke's Hospital, Cambridge, UK*

EDWARD J. WAWRZYNCZAK • *Drug Targeting Laboratory, Institute of Cancer Research, Sutton, Surrey, UK*

DUNCAN F. WEBSTER • *Department of Molecular and Cell Biology, University of Aberdeen, Scotland*

ANTHONY P. WEETMAN • *Department of Medicine, Addenbrooke's Hospital, Cambridge, UK*

IAN N. H. WHITE • *Medical Research Council Toxicology Unit, Carshalton, Surrey, UK*

MEIR WILCHEK • *Department of Biophysics, The Weizmann Institute of Science, Rehovot, Israel*

GEORGE D. WILSON • *Cancer Research Campaign, Gray Laboratory, Mount Vernon Hospital, Northwood, Middlesex, UK*

C. ROLAND WOLF • *Imperial Cancer Research Fund, Molecular Pharmacology and Drug Metabolism Group, Edinburgh, Scotland*

CHAPTER 1

Production of Polyclonal Antisera

Jonathan A. Green and Margaret M. Manson

1. Introduction

All immunochemical procedures require a suitable antiserum or monoclonal antibody raised against the antigen of interest. Polyclonal antibodies are raised by injecting an immunogen into an animal and, after an appropriate time, collecting the blood fraction containing the antibodies of interest. In producing antibodies, several parameters must be considered with respect to the final use to which the antibody will be put. These include (1) the specificity of the antibody, i.e., the ability to distinguish between different antigens, (2) the avidity of the antibody, i.e., the strength of binding, and (3) the titer of the antibody, which determines the optimal dilution of the antibody in the assay system. A highly specific antibody with high avidity may be suitable for immunohistochemistry, where it is essential that the antibody remains attached during the extensive washing procedures, but may be less useful for immunoaffinity chromatography, as it may prove impossible to elute the antigen from the column without extensive denaturation.

To produce an antiserum, the antigen for the first immunization is often prepared in an adjuvant (usually a water in oil emulsion containing heat-killed bacteria), which allows it to be released slowly and to stimulate the animal's immune system. Subsequent injections of antigen are done with incomplete adjuvant that does not contain the bacteria. The species used to raise the antibodies depends on animal facilities, amount of antigen available, and the amount of antiserum required. Another consideration is the phylogenetic relationship between antigen and immunized species. A highly conserved mammalian protein may require an avian species in order to raise

From: *Methods in Molecular Biology, Vol. 10: Immunochemical Protocols*
Ed.: M. Manson ©1992 The Humana Press, Inc., Totowa, NJ

an antibody. Production of antibodies is still not an exact science and what may work for one antigen may not work for another.

A simple, generally applicable protocol for raising polyclonal antiserum to a purified protein of greater than 10,000 mol wt is described. This method has been used to raise antibodies against a cytosolic protein, glutathione-*S*-transferase, and a membrane-bound glycosylated protein, gammaglutamyl transpeptidase. The latter was first solubilized by cleavage from the membrane with papain *(1)*. Variations to this basic procedure are discussed in Chapters 2, 3, and 5 of this vol. For proteins or peptides of low molecular weight (<5–10 kDa) conjugation to a carrier protein is required for them to elicit antigenicity (*see* this vol., Chapter 4). Variations on this basic technique can be found in selected references *(2–5)*.

2. Materials

1. Phosphate buffered saline (PBS), pH 7.4: 8 g of NaCl, 0.2 g of KH_2PO_4, 2.8 g of $Na_2HPO_4 \cdot 12H_2O$ and 0.2 g of KCl dissolved and made up to 1 L in distilled water.
2. Antigen: Purified protein diluted to about 100 µg/mL in PBS.
3. Complete and incomplete Freund's adjuvant.
4. Two glass luer lock syringes: 2 mL is the best size.
5. Three-way luer fitting plastic stopcock.
6. 19-g Needles, 0.7 mm × 22 mm Argyle medicut cannula.
7. Xylene.
8. Sterile glass universal tubes.
9. Up to four rabbits about 4–6 months old. Various strains can be used, including half sandy lops or New Zealand whites (*see* Note 1).

3. Method

1. Take up 1 mL of complete adjuvant in one of the syringes and 1 mL of antigen solution containing approx 100 µg of the antigen in another. Attach both to the plastic connector (Fig. 1), making sure that the tap on the connector is open in such a way that only the two ports connecting the two syringes are open. Repeatedly push the mixture from syringe to syringe until it becomes thick and creamy (at least 5–10 min). Push all the mixture into one syringe, disconnect this and attach it to a 19-g needle (*see* Notes 2 and 3).
2. Ensure that the rabbit to be injected is held firmly, but comfortably. For the primary immunization, inject 500 µL deeply into each thigh muscle and also inject 500 µL into each of two sites through the skin on the shoulders.

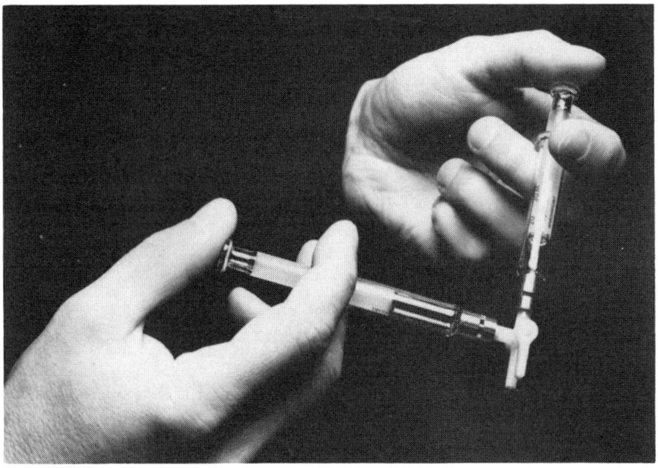

Fig. 1. Preparation of emulsion for immunization. Two luer lock glass syringes connected by a three-way plastic stopcock are used to form a stable emulsion of antigen and adjuvant.

3. Repeat these injections biweekly for a further four weeks, but make the emulsion with incomplete adjuvant.
4. Ten days after the last injection, test-bleed the rabbits from the marginal ear vein. Hold the animal firmly and gently swab the rear marginal vein with xylene to dilate the vein. Then cannulate the vein with an Argyle medicut cannula and withdraw the needle, leaving the plastic cannula in place. Draw blood out of the cannula with a syringe until the required amount has been collected. Transfer the collected blood into a sterile glass universal container.
5. Remove the cannula and stem the blood flow by sustained pressure on the puncture site with a tissue.
6. Allow the collected blood to clot by letting it stand at room temperature for 2 h and then at 4°C overnight. Separate the serum from the blood by detaching the clot carefully with a spatula from the walls of the container and pouring the liquid into a centrifuge tube. Then centrifuge the clot at 2500*g* for 30 min at 4°C and remove any expressed liquid. Add this liquid to the clot-free liquid collected previously and centrifuge the whole pooled liquid as described above. Finally, remove the serum from the cell pellet with a Pasteur pipet (*see* Note 4).
7. At this stage, test the antiserum using an appropriate assay (*see* Note 5). If the antibody has the requirements for the use to which it will be put, up to three further bleeds on successive days may be performed. If the

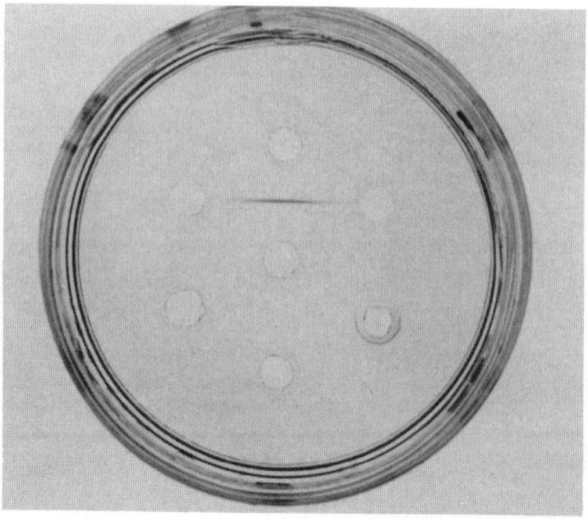

Fig. 2. Ouchterlony double-diffusion technique. The antigen is placed in the center well, cut in an agarose gel, and different antisera in a range of dilutions are placed in the surrounding wells. Antigen and antiserum diffuse toward each other and form a white precipitin line where an antibody recognizes the antigen.

antiserum is unsatisfactory, i.e., the reaction is very weak, inject the rabbit again one month after the test bleed, and again test-bleed 10 days after this injection.

8. Store antibodies in small, preferably sterile, aliquots at a minimum of –20°C. Repeated freezing and thawing should be avoided. For long-term storage, aliquots may be freeze-dried and reconstituted when needed (*see* Note 6).

4. Notes

1. The production of antibodies in animals must be carried out in strict accordance with the legislation of the country concerned.
2. Emulsions containing antigens are just as immunogenic to humans as to the experimental animal. Great care should be exercised during all the procedures.
3. A stable emulsion has been produced when a drop of the preparation does not disperse when placed on water.
4. Serum should be straw colored; a pink coloration shows that hemolysis has taken place. This should not affect the performance of the antibodies during most assay procedures.
5. This can be done by the Ouchterlony diffusion technique (*see* Fig. 2 and ref. *6*), by ELISA (*see* this vol., Chapters 29, 30, and 35), or by Western blot, either using the purified protein or a more complex mixture of

proteins containing the antigen of interest separated on an SDS/PAGE gel (*see* this vol., Chapters 24–28).

6. Some freeze-dried antisera are difficult to reconstitute, or occasionally may lose activity. Test a small sample before drying the whole batch. Any cloudiness after reconstitution is denatured lipoprotein and can be clarified by centrifugation and does not affect antibody binding.

References

1. Cook N.D. and Peters T. J. (1985) Purification of γ-glutamyl transferase by phenyl boronate affinity chromatography. *Biochim. Biophys. Acta* **828,** 205–212.
2. Catty, D. and Raykundalia, C. (1988) Production and quality control of polyclonal antibodies, in *Antibodies vol. 1—A Practical Approach* (Catty, D., ed.), IRL, Oxford.
3. Mayer, R. J. and Walker, J. H. (eds.) (1987) *Immunochemical Methods in Cell and Molecular Biology.* Academic, London.
4. Harlow, E. and Lane, D. (1988) *Antibodies. A Laboratory Manual.* Cold Spring Harbor Laboratory, New York.
5. Langone, J. J. and Van Vunakis, H. (1983) *Methods in Enzymology,* vol. 93. Academic, New York.
6. Ouchterlony, O. and Nilsson, L. A. (1978) Immunodiffusion and immunoelectrophoresis, in *Handbook of Experimental Immunology,* 3rd Ed. (Weir, D. H., ed.). Blackwell, Oxford, UK, pp. 19.1–19.44.

Chapter 2

Efficient Elution of Purified Proteins from Polyvinylidene Difluoride Membranes (Immobilon) After Transfer from SDS-PAGE and Their Use as Immunogens

Boguslaw Szewczyk and Donald F. Summers

1. Introduction

The great analytical power of sodium dodecyl sulfate-polyacrylamide gel electrophoresis (SDS-PAGE) makes it one of the most effective tools of protein chemistry and molecular biology. In the past, there have been many attempts to convert the technique from analytical to preparative scale because, by SDS-PAGE, one can resolve more than one hundred protein species in 5–6 h. The number of papers that describe preparative elution from polyacrylamide gels is immense (for example, *see* refs. *1–5*). In spite of the numerous variations in the procedure of elution, none of the available methods is entirely satisfactory. Some of the methods are very laborious, and others lead to loss of resolution or poor recovery.

In general, the elution of proteins above 100 kDa from polyacrylamide gels always presents considerable problems. Another of the serious limitations of elution from gels is owing to the elastic nature of preparative polyacrylamide gels. The precise excision of a protein band from a complex mixture is difficult and the slice may contain portions of other protein bands located close to the band of interest. To overcome some of the limitations of elution from gels, Parekh et al. *(6)* and Anderson *(7)* attempted to elute proteins

From: *Methods in Molecular Biology, Vol. 10: Immunochemical Protocols*
Ed.: M. Manson ©1992 The Humana Press, Inc., Totowa, NJ

from nitrocellulose replicas of SDS-PAGE gels. Binding of proteins to nitrocellulose is, however, so strong that the dissociating reagents (acetonitrile, pyridine) partly or completely dissolve the membrane. When such preparations are used for immunization, they may cause adverse effects in animals. We have found that when a polyacrylamide gel replica is made on Immobilon membrane and not on nitrocellulose, then the conditions for elution are much milder. Often, there is no need for concentration of the sample or for the removal of elution agents prior to immunization. Furthermore, elution from Immobilon is nearly independent of protein mol wt and recoveries of 70–90% are routinely obtained. We have also shown that, following elution using the technique described herein plus the use of *E. coli* thioredoxin to catalyze protein renaturation, one can recover significant enzymatic activity for some large complex enzymes such as *E. coli* RNA polymerase *(8)* and influenza A virus RNA polymerase *(9)*. Proteins are first separated by SDS-PAGE, and then are electroblotted to Immobilon membranes and stained with amido black or Ponceau S. The protein bands of interest are excised and are then eluted from the membrane with detergent-containing buffers at pH 9.5.

2. Materials

1. Transfer buffer: Tris-glycine (25 m*M* Tris/192 m*M* glycine), pH 8.3.
2. Methanol.
3. Protein stains:
   a. 0.01% Amido black in water.
   b. 0.5% Ponceau S in 1% acetic acid.
4. Elution buffers:
   a. 1% Triton X-100 in 50 m*M* Tris-HCl, pH 9.5.
   b. 1% Triton X-100/2% SDS in 50 m*M* Tris-HCl, pH 9.5.
5. Immobilon (polyvinylidene fluoride) membrane from Millipore Corp., Bedford, MA.
6. Whatman 3MM filter paper.
7. Scotch Brite pads.
8. SDS-PAGE apparatus.
9. Transfer apparatus (e.g., Trans Blot Cell from Bio-Rad Laboratories, Richmond, CA).
10. Glass vessels with flat bottom (e.g., Pyrex baking dishes)
11. Rocker platform.
12. Microfuge.
13. Small dissecting scissors.

3. Method

1. Apply a mixture of proteins containing immunogen to be purified (*see* Notes 1 and 2) to an SDS-PAGE gel and run the gel (*see* vol. 1, Chapter 6 and this vol., Chapter 24).
2. Prepare transfer buffer (about 4 L for Bio-Rad Trans Blot Cell) and five glass dishes, one of them large enough to accommodate the gel holder.
3. Pour methanol (around 50 mL) into one of the dishes, and 100–200 mL of transfer buffer into the other dishes.
4. Place the gel holder and Scotch Brite pads in the biggest dish, and six sheets of Whatman paper in another dish. The size of the Whatman sheets should be slightly smaller than the size of the Scotch Brite pads.
5. Using gloves, cut some Immobilon to a size slightly bigger than the size of the resolving gel. Put the sheet into a dish with methanol for 1 min and then place it in one of the dishes containing transfer buffer.
6. Put a wetted Scotch Brite pad on one side of the gel holder and then three sheets of Whatman paper saturated with transfer buffer on top of the pad.
7. On completion of electrophoresis, carefully remove the upper stacking gel because it may stick to the membrane.
8. Place the lower resolving gel on Whatman paper in the gel holder. Pour a few milliliters of transfer buffer on top of the gel.
9. Place a sheet of Immobilon membrane on the gel. Roll over the membrane with a glass rod to remove air bubbles from between the gel and the membrane.
10. Next place three Whatman sheets prewetted with Tris-glycine buffer on top of the Immobilon and finally a prewetted second Scotch Brite pad. Close the holder.
11. Place the holder in the transfer tank bearing in mind that the membrane should face the anode.
12. Begin electroblotting. Apply 20V for overnight runs. It is not necessary to use methanol in the transfer buffer as it does not improve the binding of proteins to this membrane. (*see* Note 3)
13. After transfer, stain the membrane with amido black solution for 20–30 min or with Ponceau S for 5 min (*see* Notes 4 and 5).
14. Destain the membrane with distilled water.
15. Excise the band(s) of interest with small dissecting scissors and place it in an Eppendorf tube (*see* Note 6).

16. Add 0.2–0.5 mL of elution buffer/cm^2 of Immobilon strip. Two buffers that we used are:
    a. 1% Triton X-100 in 50 m*M* Tris-HCl, pH 9.5.
    b. 2% SDS/1% Triton in 50 m*M* Tris-HCl, pH 9.5.

    The first buffer is less effective (50–75% of total protein eluted) than the second one, but the eluted protein can be injected into animals without the necessity of Triton X-100 removal. On the other hand, the 2% SDS/1% Triton X-100 mixture leads to the complete elution of bound protein from Immobilon, but SDS has to be removed before injections (*see* Notes 7–9).
17. Mix well by vortexing the Immobilon in eluant for 10 min. Spin down (5 min) the Immobilon. Use the supernatant directly for injections (elution with Triton X-100 only) or after protein precipitation with acetone (if SDS and Triton were included in the elution buffer). Protein precipitation is carried out in a dry ice bath. Add 4 vol. of cold acetone to 1 vol. of protein solution. After 2 h at –20°C, pellet the protein, solubilize, and inject into animals by standard procedures (*see* Note 10 and this vol., Chapters 1, 3, 5, and 6).

4. Notes

1. The method was used to obtain a variety of sera against bacterial, viral, and eukaryotic proteins. The amount of immunogen needed to stimulate high levels of antibodies varies for different proteins, but generally, 50–500 µg of protein is sufficient to induce the formation of high levels of specific antibodies.
2. The Immobilon matrix should not be overloaded with protein to prevent its deep penetration into the membrane. The protein band excised from a single electrophoretic lane (about 1 cm in length) should not contain more than 10–20 µg of protein.
3. Transfer of proteins from the gel to Immobilon should not be done at elevated temperatures (above 30°C), as the force of protein binding to Immobilon apparently increases with temperature. Therefore, it is advisable to make transfers in a cold room at 4°C or use precooled transfer buffer.
4. Depending on the supplier and batch of amido black, the sensitivity of protein detection with this reagent may vary. If the sensitivity of staining is not satisfactory, it is advisable to dilute the amido black solution 5–10 times with water rather than to increase its concentration.
5. Staining with Ponceau is done in 1% acetic acid. This may lead to partial denaturation of proteins bound to Immobilon. In this case, 2% SDS/

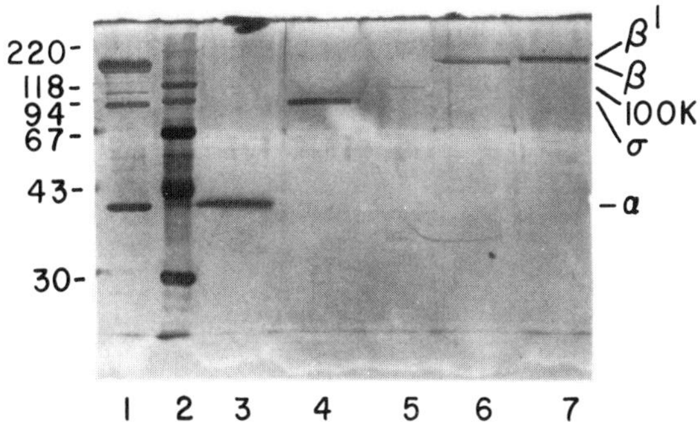

Fig. 1. SDS-PAGE pattern of *Escherichia coli* RNA polymerase subunits eluted from Immobilon membrane. *E. coli* RNA polymerase (1 μg) was resolved by SDS-PAGE and transferred to an Immobilon membrane. The proteins on the membrane were stained with amido black in water; the polymerase subunits were excised and eluted from the membrane with Triton X-100 at pH 9.0. After precipitation with acetone, the subunits were subjected to analytical SDS-PAGE and silver-stained to ascertain their purity. Lane 1, original preparation of *E. coli* RNA polymerase. Lane 2, mixture of Sigma high-mol-wt protein calibration standards; their molecular weights $\times 10^{-3}$ are given at the left-hand side of the figure. Lanes 3–7, individual subunits of the polymerase eluted from the Immobilon membrane; their designations are given at the right-hand side of the figure (100 K protein is not a constituent of the polymerase complex but, it is present in commercial preparations of the enzyme).

1% Triton X-100 in 50 m*M* Tris-HCl, pH 9.5 should be used as the elution buffer.

6. The method described here allows for very precise excision of protein bands from the Immobilon matrix. As an example, the elution of *E. coli* RNA polymerase subunits is shown in Fig. 1. Such precise excision of protein bands is much more difficult when they are cut out of polyacrylamide gels.
7. The elution from Immobilon is strictly pH-dependent. At pH 7.0, there is practically no elution; the maximum efficiency of elution is reached at pH 8.5–9.5 *(8)*.
8. The efficiency of elution is only slightly dependent on the mol wt of the protein. In our hands, a protein of 200 kDa was eluted with only 10% lower efficiency than a protein of 70 kDa when elution was performed in the buffer with Triton X-100 as the eluting agent.
9. Proteins that are insoluble in standard aqueous solutions and are solubilized with SDS before subjecting to electrophoresis may sometimes require special treatment. For example, *E. coli* β-galactosidase (mol wt about 120 K) can be readily eluted from Immobilon membranes under mild conditions. However, β-galactosidase fusion proteins with short segments

of some viral proteins are insoluble in aqueous salt solutions, and must be solubilized with sample buffer for SDS-PAGE. In this case, the proteins can be eluted from the Immobilon only by 2% SDS/1% Triton X-100 in 50 mM Tris-HCl, pH 9.5.

10. Probably the easiest method to obtain emulsions of protein in Freund's adjuvant is by subjecting the mixture placed in an Eppendorf microfuge tube to a short ultrasonic treatment (3–4 pulses 10–20 s each time) in an ultrasonic disintegrator equipped with a microprobe (end diameter of the probe around 1/8 inch).

References

1. Tuszynski, N. Y., Damsky, C. H., Fuhrer, J. P., and Warren, L. (1977) Recovery of concentrated protein samples from sodium dodecyl sulfate-polyacrylamide gels. *Anal. Biochem.* **83,** 119–129.
2. Nguyen, N. Y., DiFonzo, J., and Chrambach, A. (1980) Protein recovery from gel slices by steady-state stacking: An apparatus for the simultaneous extraction and concentration of ten samples. *Anal. Biochem.* **106,** 78–91.
3. Hager, D. A. and Burgess, R. R. (1980) Elution of proteins from sodium dodecyl sulfate-polyacrylamide gels, removal of sodium dodecyl sulfate, and renaturation of enzymatic activity: Results with sigma subunit of *Escherichia coli* RNA polymerase, wheat germ DNA topoisomerase, and other enzymes. *Anal. Biochem.* **109,** 76–86.
4. Stralfors, P. and Belfrage, P. (1983) Electrophoretic elution of proteins from polyacrylamide gel slices. *Anal. Biochem.* **128,** 7–10.
5. Hunkapiller, M. W., Lujan, E., Ostrander, F., and Hood, L. E. (1983) Isolation of microgram quantities of proteins from polyacrylamide gels for amino acid sequence analysis. *Methods Enzymol.* **91,** 227–236.
6. Parekh, B. S., Mehta, H. B., West, M. D., and Montelaro, R. C. (1985) Preparative elution of proteins from nitrocellulose membranes after separation by sodium dodecyl sulfate-polyacrylamide gel electrophoresis. *Anal. Biochem.* **148,** 87–92.
7. Anderson, P. J. (1985) The recovery of nitrocellulose-bound protein. *Anal. Biochem.* **148,** 105–110.
8. Szewczyk, B. and Summers, D. F. (1988) Preparative elution of proteins blotted to immobilon membranes. *Anal. Biochem.* **168,** 48–53.
9. Szewczyk, B., Laver, W. G., and Summers, D. F. (1988) Purification, thioredoxin renaturation, and reconstituted activity of the three subunits of the influenza A virus RNA polymerase. *Proc. Natl. Acad. Sci. USA* **85,** 7907–7911.

CHAPTER 3

Raising Polyclonal Antibodies Using Nitrocellulose-Bound Antigen

Monique Diano, André Le Bivic, and Michel Hirn

1. Introduction

Highly specific antibodies directed against minor proteins, present in small amounts in biological fluids, or against nonsoluble cytoplasmic or membraneous proteins, are often difficult to obtain. The main reasons for this are the small amounts of protein available after the various classical purification processes and the low purity of the proteins.

In general, a crude or partially purified extract is electrophoresed on an SDS polyacrylamide (SDS-PAGE) gel; then the protein band is lightly stained and cut out. In the simplest method, the acrylamide gel band is reduced to a pulp, mixed with Freund' s adjuvant, and injected. Unfortunately, this technique is not always successful. Its failure can probably be attributed to factors such as the difficulty of disaggregating the acrylamide, the difficulty with which the protein diffuses from the gel, the presence of SDS in large quantities resulting in extensive tissue and cell damage, and finally, the toxicity of the acrylamide.

An alternative technique is to extract and concentrate the proteins from the gel by electroelution (*see* Chapter 19 in vol. 1, this series), but this leads to considerable loss of material and low amounts of purified protein.

From: *Methods in Molecular Biology, Vol. 10: Immunochemical Protocols*
Ed.: M. Manson ©1992 The Humana Press, Inc., Totowa, NJ

Another technique is to transfer the separated protein from an SDS-PAGE gel to nitrocellulose. The protein-bearing nitrocellulose can be solubilized with dimethyl sulfoxide (DMSO), mixed with Freund's adjuvant, and injected into a rabbit. However, although rabbits readily tolerate DMSO, mice do not, thus making this method unsuitable for raising monoclonal antibodies.

The monoclonal approach has been considered as the best technique for raising highly specific antibodies, starting from a crude or partially purified immunogen. However, experiments have regularly demonstrated that the use of highly heterogenous material for immunization never results in the isolation of clones producing antibodies directed against all the components of the mixture. Moreover, the restricted specificity of a monoclonal antibody that usually binds to a single epitope of the antigenic molecule is not always an advantage. For example, if the epitope is altered or modified (i.e., by fixative, Lowicryl embedding, or detergent), the binding of the monoclonal antibody might be compromised, or even abolished.

Because conventional polyclonal antisera are complex mixtures of a considerable number of clonal products, they are capable of binding to multiple antigenic determinants. Thus, the binding of polyclonal antisera is usually not altered by slight denaturation, structural changes, or microheterogeneity, making them suitable for a wide range of applications. However, to be effective, a polyclonal antiserum must be of the highest specificity and free of irrelevant antibodies directed against contaminating proteins, copurified with the protein of interest and/or the proteins of the bacterial cell wall present in the Freund's adjuvant. In some cases, the background generated by such irrelevant antibodies severely limits the use of polyclonal antibodies.

A simple technique for raising highly specific polyclonal antisera against minor or insoluble proteins would be of considerable value.

Here, we describe a method for producing polyclonal antibodies, which avoids both prolonged purification of antigenic proteins (with possible proteolytic degradation) and the addition of Freund's adjuvant and DMSO. Two-dimensional gel electrophoresis leads to the purification of the chosen protein in one single, short step. The resolution of this technique results in a very pure antigen, and consequently, in a very high specificity of the antibody obtained. It is a simple, rapid, and reproducible technique for proteins present in sufficiently large quantities to be detected by Coomassie blue staining.

A polyclonal antibody, which by nature cannot be monospecific, can, if its titer is very high, behave like a monospecific antibody in comparison with the low titers of irrelevant antibodies in the same serum. Thus, this method is faster and performs better than other polyclonal antibody techniques while retaining all the advantages of polyclonal antibodies.

2. Materials

1. For isoelectric focusing (IEF) and SDS-PAGE gels, materials are those described by O'Farrell *(2,3)* and Laemmli *(4)*. It should be noted that for IEF, acrylamide and *bis*-acrylamide must be of the highest level of purity, and urea must be ultrapure (enzyme grade) (*see* vol. 1, Chapter 6 and vol. 3, Chapters 15–21).
2. Ampholines with an appropriate pH range, e.g., 5–8 or 3–9.
3. Transfer membranes: 0.45-µm BA 85 nitrocellulose membrane filters (from Schleicher and Schüll GmBH, Kassel, Germany) ; 0.22-µm membranes can be used for low-mol-wt antigenic proteins.
4. Transfer buffer: 20% Methanol, 150 m*M* glycine, and 20 m*M* Tris base, pH 8.3.
5. Phosphate buffered saline (PBS), sterilized by passage through a 0.22-µm filter.
6. Ponceau Red: 0.2% in 3% trichloroacetic acid.
7. Small scissors.
8. Sterile blood-collecting tubes, with 0.1*M* sodium citrate, pH 6, at a final concentration of 3.2%.
9. Ultrasonication apparatus, with 100 W minimum output. We used a 100-W ultrasonic disintegrator with a titanium exponential microprobe with a tip diameter of 3 mm (1/8 in.). The nominal frequency of the system is 20 kc/s, and the amplitude used is 12 µ.

3. Method

This is an immunization method in which nitrocellulcse-bound protein is employed and in which *neither DMSO nor Freund's adjuvant* are used, in contrast to the method described by Knudsen *(1)*. It is equally applicable for soluble and membrane proteins.

3.1. Purification of Antigen

Briefly, subcellular fractionation of the tissue is carried out to obtain the fraction containing the protein of interest. This is then subjected to separation in the first dimension by IEF using OFarrell's technique *(2)*, or as described in vol. 3, Chapter 19. At this stage, it is important to obtain complete solubilization of the protein (*see* Note 1).

Separation in the second dimension is achieved by using an SDS polyacrylamide gradient gel (*see* vol. 1, Chapter 7 and refs. *2* and *4*; *see also* Note 2).

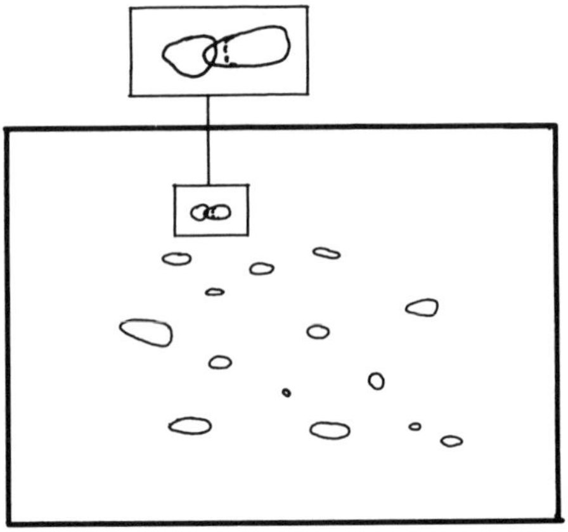

Fig. 1. Excision of the spot containing the antigen. Cut inside the circumference, for instance, along the dotted line for the right spot.

The proteins are then transferred from the gel to nitrocellulose (this Volume, Chapters 24–26 and ref. 5). It is important to work with gloves when handling nitrocellulose to avoid contamination with skin keratins.

3.2. Preparation of Antigen for Immunization

1. Immerse the nitrocellulose sheet in Ponceau red solution for 1–2 min, until deep staining is obtained, then destain the sheet slightly in running distilled water for easier detection of the spots. *Never let the nitrocellulose dry out.*
2. Carefully excise the spot corresponding to the antigenic protein. Excise inside the circumference of the spot to avoid contamination by contiguous proteins (*see* Fig. 1)
3. Immerse the nitrocellulose spot in PBS in an Eppendorf tube (1-mL size). The PBS bath should be repeated several times until the nitrocellulose is thoroughly destained. The last bath should have a volume of about 0.5 mL, adequate for the next step.
4. Cut the nitrocellulose into very small pieces with scissors. Then rinse the scissors into the tube with PBS to avoid any loss (*see* Fig. 2).
5. Macerate the nitrocellulose suspension by sonication. The volume of PBS must be proportional to the surface of nitrocellulose to be sonicated. For example, 70–80 µL of PBS is adequate for about 0.4 cm^2 of nitrocellulose (*see* Notes 3 and 4).
6. After sonication, add about 1 mL of PBS to the nitrocellulose powder to dilute the mixture, and aliquot it in 500, 350, and 250-µL fractions and

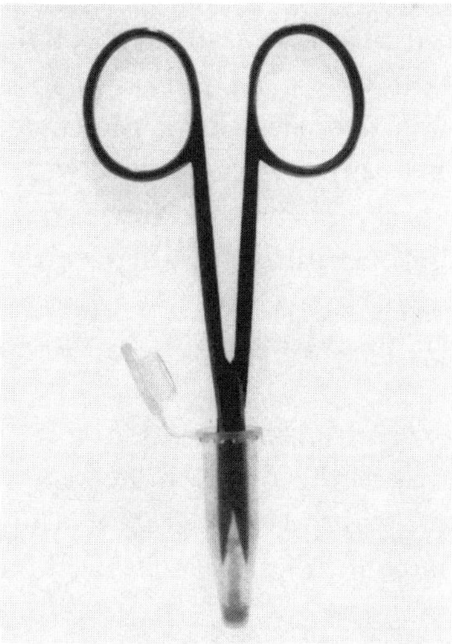

Fig. 2. Maceration of nitrocellulose.

freeze these fractions at –80°C until use. Under these storage conditions, the aliquots may be used for immunization for up to one year or, may be longer. Never store the nitrocellulose without buffer. Never use sodium azide because of its toxicity.

3.3. Immunization

1. Shave the backs of the rabbits. Routinely inject two rabbits with the same antigen.
2. Thaw the 500-µL fraction for the first immunization and add 1.5–2 mL of PBS to reduce the concentration of nitrocellulose powder.
3. Inoculate the antigen, according to Vaitukaitis *(6)*, into 20 or more sites (Vaitukaitis injects at up to 40 sites). Inject subcutaneously, holding the skin between the thumb and forefinger. Inject between 50 and 100 µL—a light swelling appears at the site of injection. As the needle is withdrawn, compress the skin gently. An 18-g hypodermic needle is routinely used, though a finer needle (e.g., 20- or 22-g) may also be used (*see* Note 5). Care should be taken over the last injection; generally, a little powder remains in the head of the needle. Thus, after the last injection, draw up 1 mL of PBS to rinse the needle, resuspend the remaining powder in the syringe, and position the syringe vertically to inject.
4. Three or four weeks after the first immunization, the first booster inoculation is given in the same way. The amount of protein injected is generally less, corresponding to two-thirds of that of the first immunization.

5. Ten days after the second immunization, bleed the rabbit (*see* Note 6). A few milliliters of blood suffice, i.e., enough to check the immune response against a crude preparation of the injected antigenic protein. The antigen is revealed on a Western blot with the specific serum diluted at 1:500 and a horseradish peroxidase-conjugated second antibody. We used 3,3'-Diaminobenzidine tetrahydrochloride (DAB) for color development of peroxidase activity. (*See* Chapter 10 in this vol.) If the protein is highly antigenic, the beginning of the immunological response is detectable.
6. Two weeks after the second immunization, administer a third immunization in the same way as the first two, even if a positive response has been detected. If there was a positive response after the second immunization, one-half of the amount of protein used for the original immunization is sufficient.
7. Bleed the rabbits 10 days after the third immunization and every week thereafter. At each bleeding, check the serum as after the second immunization; but the serum should be diluted at 1:4000 or 1:6000. Bleeding can be continued for as long as the antibody titer remains high (*see* Note 7).

   Another booster should be given when the antibody titer begins to decrease if it is necessary to obtain a very large volume of antiserum (*see* Note 7).
8. After bleeding, keep the blood at room temperature until it clots. Then collect the serum and centrifuge for 10 min at 3000*g* to eliminate microclots and lipids. Store aliquots at –22°C

4. Notes

1. To ensure solubilization, the following techniques are useful:
   a. The concentration of urea in the mixture should be 9–9.5*M*, i.e., close to saturation.
   b. The protein mixture should be frozen and thawed at least six times. Ampholines should be added only after the last thawing because freezing renders them inoperative.
   c. If the antigenic protein is very basic and outside the pH range of the ampholines, it is always possible to carry out NEPHGE (nonequilibrium pH gradient electrophoresis) for the first dimension (Section 3.1., and *see also* vol. 3, Chapter 17).
2. If the antigenic protein is present in small amounts in the homogenate, it is possible to save time by cutting out the part of the IEF gel where the

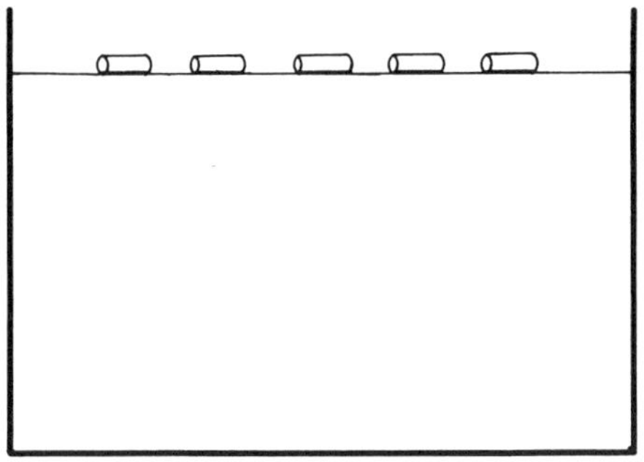

Fig. 3. Second dimension with several IEF gels. Several IEF gels are cut, 0.5 cm above and 0.5 cm below the isoelectric point of the protein of interest. They are placed side by side at the top of the second dimension slab gel. Thus, only one SDS gel is needed to collect several spots of interest.

protein is located and depositing several pieces of the first-dimension gel side by side on the second-dimension gel slab (*see* Fig.3).

3. Careful attention should be paid to temperature during preparation of the antigen; always work between 2 and 4°C. Be particularly careful during sonication; wait 2–3 min between consecutive sonications.
4. This is a crucial point in the procedure. If too much PBS is added, the pieces of nitrocellulose will swirl around the probe and disintegration does not occur. In this case, the nitrocellulose pieces should be allowed to settle to the bottom of the tube before sonication and the excess buffer drawn off with a syringe or other suitable instrument (70–80 µL of PBS is sufficient for about 0.4 cm^2 of nitrocellulose). For these quantities, one or two 10-s cycles suffice to get powdered nitrocellulose. We mention the volume as a reference since the surface of nitrocellulose-bound antigen may vary. In every case the volume of PBS must be adjusted.
5. What is an appropriate amount of antigenic protein to inject? There is no absolute answer to this question. It depends both on the molecular weight of the protein, and also on its antigenicity. It is well known that if the amount of antigen is very low (0.5–1 *µg* for the classic method with Freund's adjuvant), there is no antibody production; if the amount of antigen is very high (e.g., several hundred micrograms of highly antigenic protein), antibody production might also be blocked.

   It would appear that in our method, a lower amount of antigen is required for immunization; the nitrocellulose acts as if it progressively releases the bound protein, and thus, the entire amount of protein is used progressively by the cellular machinery.

Our experiments show that a range of 10–40 µg for the first immunization generally gives good results, although, in some cases, 5 µg of material is sufficient. The nitrocellulose powder has the additional effect of triggering an inflammatory process at the sites of injection, thus enhancing the immune response, as does Freund's adjuvant by means of the emulsion of the antigenic protein with the tubercular bacillus; macrophages abound in the inflamed areas.

6. It is perhaps worth noting that careful attention should be paid to the condition of the rabbit at time of bleeding. We bleed the rabbits at the lateral ear artery. When the rabbit is calm, 80–100 mL of blood may be taken. The best time for bleeding is early in the morning, and after the rabbit has drunk. Under these conditions, the serum is very clear. It is essential to operate in a quiet atmosphere. If the rabbit is nervous or under stress, the arteries are constricted so strongly that only a few drops of blood can be obtained. Note that to avoid clotting, the needle is impregnated with a sterile sodium citrate solution by drawing the solution into the syringe three times.
7. When the effective concentration required corresponds to a dilution of 1:2000, the titer is decreasing. Serum has a high titer if one can employ a dilution over 1:2000 and if there is a strong specific signal without any background.
8. We have also immunized mice with nitrocellulose-bound protein by intraperitoneal injection of the powder. This is convenient when time and material are very limited, since very little protein is needed to induce a response (3–5 times less than for a rabbit) and since the time-lag for the response is shorter (the second immunization was two weeks after the first, and the third immunization, 10 days after the second). Mice have a high tolerance for the nitrocellulose powder. Unfortunately, the small amount of serum available is a limiting factor. This technique for immunizing mice can, of course, be used for the preparation of monoclonal antibodies.
9. Utilization of serum. The proper dilutions are determined. We routinely use 1:4000 for blots, 1:300–1:200 for immunofluorescence, and 1:50 for immunogold staining. Serum continues to recognize epitopes on tissue proteins after Lowicryl embedding. Labeling is highly specific, and gives a sharp image of *in situ* protein localization. *There is no need to purify the serum.* IgG purified from serum by whatever means usually gives poorer results than those obtained with diluted serum. Purification procedures

often give rise to aggregates and denaturation, always increase the background, and result in loss of specific staining.

10. Bacterial antigenic protein. When antibodies are used in screening cDNA libraries in which the host is *E. coli*, the antibodies produced against bacterial components of Freund's adjuvant may also recognize some *E. coli* components. An advantage of our technique is that it avoids the risk of producing antibodies against such extraneous components.
11. Is nitrocellulose antigenic? Some workers have been unable to achieve good results by immunization with nitrocellulose-bound protein. They reproducibly obtain antisera directed against nitrocellulose. We found that in every case, this resulted from injecting powdered nitrocellulose in Freund's adjuvant; using adjuvant actually increases the production of low affinity IgM that binds nonspecifically to nitrocellulose. We have never observed this effect in our experiments when the technique described here was followed strictly.
12. The purification step by 2-D electrophoresis implies the use of denaturing conditions (SDS), and thus, is not appropriate for obtaining antibodies directed against native structures. For that purpose, the protein should be transferred onto nitrocellulose after purification by classical nondenaturing methods and gel electrophoresis under nondenaturing conditions.

    However, it should be pointed out that, following the method of Dunn *(7)*, it is possible partially to renature proteins with modifications of the composition of the transfer buffer.
13. Second dimension electrophoresis can be carried out with a first electrophoresis under native conditions, followed by a second electrophoresis under denaturing conditions, i.e., with SDS.

    Because the resolution provided by a gradient is better, it should always be used in the 2-D electrophoresis. Agarose may also be used as an electrophoresis support.
14. If only a limited amount of protein is available, and/or if the antigen is weakly immunogenic, another procedure may be used. The first immunization is given as a single injection, of up to 0.8 mL, into the popliteal lymphatic ganglion *(8)*, using a 22-g needle, i.e., the finest that can be used to inject nitrocellulose powder. In this case, the antigen is delivered immediately into the immune system. If necessary, both ganglions can receive an injection. The small size of the ganglions limits the injected volume. The eventual excess of antigen solution is then injected

into the back of the rabbit, as described above. The boosters are given in the classic manner, i.e., in several subcutaneous injections into the rabbit's back.

If the amount of protein available is even more limited, a guinea pig may be immunized (first immunization in the lymphatic ganglion, and boosters, as usual).

15. The advantage of getting a high titer for the antibody of interest is that the amount of irrelevant antibodies is, by comparison, very low and, consequently, does not generate any background. Another advantage of using a crude serum with a high antibody titer, is that this serum may be used without further purification to screen a genomic bank *(9)*.
16. The time required for transfer and the voltage used is dependent on the molecular weight and the nature of the protein to be transferred. During transfer, the electrophoresis tank may be cooled with tap water.

References

1. Knudsen, K. A. (1985) Proteins transferred to nitrocellulose for use as immunogens. *Anal. Biochem.* **147,** 285–288.
2. O'Farrel, P. H. (1975) High resolution two-dimensional electrophoresis of proteins. *J. Biol. Chem.* **250,** 4007–4021.
3. O'Farrel, P. Z., Goodman, H. M., and O'Farrel, P. H. (1977) High resolution two-dimensional electrophoresis of basic as well as acidic proteins. *Cell* **12,** 1133–1142.
4. Laemmli, U. K. (1970) Cleavage of structural proteins during the assembly of the head of bacteriophage T4. *Nature (London)* **227,** 680–685.
5. Burnette, W. N. (1981) Electrophoretic transfer of proteins from sodium dodecylsulfate polyacrylamide gels to unmodified nitrocellulose and radiographic detection with antibody and radioiodinated protein A. *Anal. Biochem.* **112,** 195–203.
6. Vaitukaitis, J., Robbins, J. B., Nieschlag, E., and Ross, G. T. (1971) A method for producing specific antisera with small doses of immunogen. *J. Clin. Endocrinol.* **33,** 980–988.
7. Dunn, S. D. (1986) Effects of the modification of transfer buffer composition and the renaturation of proteins in gels on the recognition of proteins on Western blots by monoclonal antibodies. *Anal. Biochem.* **157,** 144–153
8. Sigel, M. B., Sinha, Y. N., and Vanderlaan, W. P. (1983) Production of antibodies by inoculation into lymph nodes. *Methods Enzymol.* **93,** 3–12.
9. Preziosi, L., Michel, G. P. F., and Baratti, J. (1990) Characterisation of sucrose hydrolising enzymes of *Zymomonas mobilis. Arch. Microbiol.* **153,** 181–186.

CHAPTER 4

Synthesis of Peptides for Use as Immunogens

David C. Hancock and Gerard I. Evan

1. Introduction

An increasing problem in cell and molecular biology is the preparation of antibodies specific to proteins that are present in minute quantities within cells or tissues. With the advent of recombinant DNA technology, it is now often possible to deduce the primary amino acid sequence of a polypeptide without its purification. Two strategies then exist to raise appropriate antibodies. Either the gene can be expressed in a heterologous species, usually bacteria, and the resultant protein used as an immunogen, or alternatively, small synthetic peptides can be made that contain amino acid sequences inferred from that of the gene. Such antipeptide antibodies crossreact with the intact native protein with surprisingly high frequency and have the additional advantage that the epitope recognized by the antibody is already well defined *(1)*. In this way, antibodies can be raised against novel gene products that are specifically directed against sites of interest, for example, unique regions, highly conserved regions, active sites, extracellular or intracellular domains. Moreover, the ready availability of the pure peptide immunogen against which the antibody was raised means that sera can be rapidly and easily screened, e.g., using an enzyme-linked immunosorbent assay (ELISA) for antipeptide activity. Free peptide can also be used to block antibody binding and so demonstrate immunological specificity, and it may be coupled to a solid support (e.g., agarose) to generate an affinity matrix for antibody purification. In this chapter, we describe the basic principles behind the design, synthesis, and

From: *Methods in Molecular Biology, Vol. 10: Immunochemical Protocols*
Ed.: M. Manson ©1992 The Humana Press, Inc., Totowa, NJ

use of synthetic peptides as immunogens, and in this and the following chapter, we give some of the basic methods used in our laboratory.

1.1. Choosing Peptide Sequences

Many peptide sequences are immunogenic, but not all are equally effective at eliciting antibodies that crossreact with the intact cognate protein (we term these cross-reactive peptides). Many factors can affect the success of using peptide immunogens to raise antiprotein antibodies. They include elements such as the number of peptides from one protein sequence to be used and the number of animals available for immunization (both of which may be determined by the existing facilities), the availability of sequence data, the predicted secondary structure of the intact protein, and finally, the ease of synthesis of specific sequences. However, there is no guarantee that antibodies raised against a particular synthetic peptide will crossreact with the intact protein from which the sequence is derived. In our experience, the probability of generating a successful antiprotein antibody by these methods may be somewhat less than 50%. Nonetheless, there are a number of ways of improving one's chances of success *(see below)*.

1.1.1. Predicted Structure of the Whole Protein

There are several predictive algorithms available that can provide data on antigenicity, hydrophilicity, flexibility, surface probability, and charge distribution over a given amino acid sequence. In addition, the algorithms of Chou and Fasman, and of Robson and Garnier *(2,3)* give a fair idea of where regions of particular secondary structure, such as α-helix, β-sheet, turns, and coils, are likely to form. Primary sequences can also indicate consensus sequences that may be involved in sites of posttranslational modification (e.g., *O*- and *N*-linked glycosylation sites and sites of phosphorylation) and that may therefore be immunologically unavailable in the mature protein. Clearly, accessibility on the external surface of the intact protein is, overall, the most important requirement for a cross-reactive peptide. Very frequently, the *C*-terminus of a protein is exposed, and this region makes a good first choice. The *N*-terminus can also prove to be a good candidate sequence, but in our experience, is less reliable and may anyway be modified or truncated. Regions with too high a charge or hydrophilicity are sometimes not as effective as might be expected, probably because almost all known antibody combining sites make contact with their epitope by polar and Van der Waal's bonds and not ionic interactions. Hydrophilic α-helical regions can be good peptide epitopes because, provided the synthetic peptide is itself long enough to form a helix, it often assumes an identical conformation to that in the intact protein.

1.1.2. Specific Requirements

By their nature, peptide-specific antibodies are *site-directed* probes for proteins. Both the sequence and position of the antibody epitope is predefined. It is therefore possible to target antipeptide antibodies to specific regions of interest in the intact protein, such as areas of high conservation or hypervariability, sequences whose consensus suggests them as binding sites, transmembrane domains, signal sequences, or any other functional regions of a protein. However, such criteria must also be consistent with the factors mentioned above.

1.1.3. Immunological Requirements

Peptides of 10–20 amino acids are optimal as antigens. Our standard is around 15 residues. Short peptides (below about 7 residues) are probably of insufficient size to function as epitopes. Larger peptides may adopt their own specific conformation in solution (which is often immunodominant over any primary structural determinants), which may not be reflected in the conformation of the sequence within the intact protein. Given the above criteria, it is fair to say that most peptide sequences are immunogenic if presented to the immune system in the right way (*see* Section 1.3.), but that not all will generate crossreactive antibodies. Probably the single most important factor in optimizing one's chances of making useful antibodies is to use several peptides from different regions of the protein sequence and immunize several animals with each. Often a given antipeptide antibody may work well in one assay, e.g., Western blotting, but not in another, e.g., immunoprecipitation.

1.1.4. Synthesis Requirements

The chemical difficulties of synthesizing certain amino acid sequences are complex, and briefly dealt with in the next section (Section 1.2.). In general, hydrophilic sequences are easier to synthesize, more soluble, and more likely to be exposed on the surface of the intact molecule. There appears to be little requirement for a high degree of purity for peptide immunogens. Our experience is that peptides of 75% purity (sometimes even less) generate effective polyclonal antisera, although criteria may need to be more stringent when making monoclonal antibodies.

1.2. Peptide Synthesis

Should synthetic peptides be needed, the major decision to be made is how to obtain them. The options are either to have them made commercially or to attempt to synthesize them in-house. Whichever option is chosen depends largely on current and future needs, local expertize, and funding. Peptides may be purchased from several companies specializing in contract synthesis, and if only a few are needed, this is clearly the easiest way to obtain the desired reagents. However, custom synthesis of peptides is very expensive, with many specific sequences or modifications costing even more. On the other hand, in-house synthesis is very labor-intensive, requires significant knowledge of peptide chemistry and, if carried out using an automated machine, involves large capital expenditure. In general, acquisition of an automated peptide synthesizer is probably best suited to laboratories or institutes with substantial and ongoing requirements for synthetic peptides and, preferably, with their own dedicated personnel. An in-house peptide synthesis facility is a particularly attractive alternative to custom synthesis, because it allows much greater flexibility in the design and production of peptides. This can be particularly important if specially derivatized peptides are needed, or if, for example, chemically defined immunogens, such as multiple antigen peptides (MAPS) *(4)*, or mimotopes *(5)*, or epitope mapping *(6)* are envisaged.

1.2.1. Principles of Peptide Synthesis

A wide choice of peptide-synthesizers is currently available, ranging from manual to fully automated. They are all based on solid-phase peptide synthesis methodologies in which either *t*-boc (*t*-butoxy carbonyl) *(7)* or Fmoc (9-fluorenylmethoxycarbonyl) *(8)* is the major protecting group during synthesis. A detailed description of peptide synthesis is clearly beyond the scope of this chapter and further information on practical and theoretical approaches to this chemistry may be found elsewhere *(9–11)*. However, a brief outline of solid-phase synthesis may prove useful.

The peptide is synthesized on a derivatized solid support, often a polyamide or polystyrene resin, and synthesis proceeds from C-terminus to *N*-terminus. The linker group, which provides the link between the resin and the peptide chain, varies and depends on the type of C-terminus desired, e.g., C-terminal acid or amide. In addition, resins are available that already have the first protected amino acid attached. This is often desirable, since the bond between the first amino acid and the resin linker is not a peptide bond and therefore, requires a different reaction to the subsequent peptide chain-assembly steps.

Each amino acid has its primary amine group protected (with either a *t*-boc or an Fmoc group). In addition, the amino acid side-chain may need protection. Commercially available amino acids are supplied either as free acids or with derivatized carboxyl groups, as required by the particular synthetic chemistry being used. After cleavage of the amine protecting group on the resin, the incoming amino acid is solubilized, activated in a polar solvent, and then allowed to react with the deprotected amine group of the growing peptide chain. After washing, usually with the same solvent, the peptide-resin is ready for another round of *N*-terminal deprotection and amino acid addition. Automated synthesis is generally performed in the 0.1–0.5 mmol range. The time taken to perform addition of each amino acid is of the order of 1 h although some smaller scale procedures can use shorter coupling times. The stepwise percentage yield is typically >95%.

When chain-assembly has been completed and the final *N*-terminal protecting group removed, the peptide is cleaved from the resin with strong acid (typically hydrogen fluoride (HF) or trifluoromethanesulphonic acid (TFMSA) with *t*-boc chemistry and trifluoroacetic acid (TFA) with Fmoc chemistry) to give a free acid C-terminus. This acid treatment also cleaves most side-chain protecting groups. Frequently, various scavengers have to be added during acid cleavage in order to minimize side-reactions and protect certain moieties: for example, thiols are used to preserve free sulfhydryl groups on cysteine, and ethyl methyl sulfide serves to protect tryptophan and methionine side-chains during Fmoc acid-cleavages. It may be necessary to perform pilot cleavages to determine the optimum conditions for a particularly difficult sequence. Once liberated from the resin, the peptide may be desalted, precipitated, and lyophilized. Clearly, the nature of the reagents used in these organic syntheses requires access to a good fume cupboard (in particular, hydrogen fluoride demands great care), and the postsynthesis workup typically utilizes a rotary evaporator and lyophilizer. The final product is usually evaluated by reverse-phase HPLC (C_8 or C_{18} columns with water/acetonitrile gradients) and amino acid analysis and/or protein sequencing.

The synthesis of most peptide sequences in the region of 20 residues in length may nowadays be considered quite routine. There are, however, always exceptions. Certain sequences can be extremely difficult to synthesize, and often dictate the choice of chemistry used. For example, tryptophan residues are often problematical with *t*-boc synthesis because of their instability during the harsher acid treatment during synthesis and deprotection. Peptides containing multiple arginine residues often present difficulties with Fmoc chemistry owing to poor substitution and incomplete deprotection. Very hydrophobic peptides may require unusual solvents to maintain their solubility during synthesis and longer peptides (above, say, 25 residues) often prove progressively more difficult to adequately solvate and therefore give low yields. Other problems encountered are frequently peculiar to a specific sequence made with a specific chemistry. Sometimes changing chemistries (i.e., *t*-boc to Fmoc or vice versa) solves the problem. Often it proves impossible to synthesize directly a pure peptide of the appropriate sequence, and the correct species has to be identified from among the reaction products and subsequently purified with concomitant loss of yield.

1.3. Conjugation of Peptides to Carrier Proteins

In general, short oligopeptides are poor immunogens, so it is necessary to conjugate them covalently to immunogenic carrier proteins in order to raise effective antipeptide antibodies. These carrier proteins provide necessary major histocompatability complex (MHC) Class II/T-cell receptor epitopes while the peptides can then serve as B-cell determinants. Keyhole limpet hemocyanin (KLH) and thyroglobulin are examples of carriers that we have used successfully to generate polyclonal antipeptide antibodies. We generally avoid using bovine serum albumin (BSA) because the high levels of anti-BSA antibody generated can interfere with subsequent studies on cells grown in media containing bovine sera.

The peptides are covalently conjugated to the carrier molecule using an appropriate bifunctional reagent—the most straightforward coupling methodologies involve the amine or sulfhydryl groups of the peptide. Not surprisingly, substantial antibody titers are also frequently generated against determinants present on these carrier molecules. In general, such anticarrier antibodies present few problems in polyclonal antipeptide antibodies and may, anyway, be adsorbed out on a matrix of carrier bound to agarose. When making monoclonal antibodies, however, the substantial anticarrier response may mask the frequently weaker antipeptide response, resulting in few, if any, peptide-specific hybridomas being isolated. For the preparation of monoclonal antibodies, therefore, we advise the use of the carrier tuberculin-purified protein derivative (PPD) *(12)*. The advantage of PPD is that it is an extremely poor B-cell antigen, yet elicits a strong helper T-cell response. Using PPD as the carrier thus gives an antibody response that is predominantly directed against the peptide of interest. The only drawback with routinely using PPD as a carrier is that animals first have to be primed with live attenuated tubercle bacilli at least 3 wk before immunization (*see* Chapter 5).

The most straightforward conjugation procedure uses glutaraldehyde as the bifunctional reagent that crosslinks amino groups on both carrier and peptide. In our experience, glutaraldehyde conjugation is reliable, easy, and effective, and generates good antipeptide antibodies even with short peptides containing internal lysine residues, doubtless reflecting the haphazard manner in which complex antigens are processed by the immune system. *m*-Maleimidobenzoyl-*N*-hydroxysuccinimide ester (MBS) can be used to crosslink the thiol group of cysteine on the peptide to an amino group on the carrier. The MBS method generates a somewhat better-defined conjugate, but it involves a slightly more complicated procedure and requires the presence of a *reduced* cysteine residue within the peptide (this is frequently added to the sequence during synthesis for conjugation purposes).

2. Materials

2.1. Conjugation of Peptides

1. Keyhole limpet hemocyanin (KLH): Purchased as a solution in 50% glycerol (Calbiochem) and stored at 4°C. If obtained as an ammonium sulfate suspension, the KLH will require *extensive* dialysis against borate buffered saline (20 m*M* Na Borate/144 m*M* NaCl) containing 50% glycerol.
2. Porcine thyroglobulin: Obtained as a partially purified powder.
3. Bovine serum albumin (BSA): Purified and essentially globulin-free.
4. Purified protein derivative (PPD of tuberculin): Derived from the tubercle bacillus and available from Statens Serum Institut, Copenhagen as a 1 mg/mL solution in PBS. Store at 4°C.
5. Glutaraldehyde (Sigma, Grade 1) stock: A 25% solution divided into 1-mL aliquots. It is stored at –20°C and never refrozen.
6. *m*-Maleimidobenzoyl-*N*-hydroxysuccinimide ester (MBS) stock (Pierce): Dissolve in dry dimethylformamide (DMF) at 3 mg/mL and store at –20°C.
7. Glycine ethyl ester: Make up as a 1*M* stock and adjust to pH 8.0 with NaOH.
8. Sodium bicarbonate stock (10X): A 1*M* solution adjusted to pH 9.6 with HCl.
9. Potassium phosphate buffers: Make up 1*M* stock solutions at pH 7.2 and pH 6.0 and dilute to 10 m*M* or 50 m*M* as required.
10. Saline: 0.9% NaCl.
11. Acetone.
12. Biogel P10 column.
13. Dithiothreitol (DTT), 100 m*M*.
14. 20 mL P30 (Bio-Rad) column.

3. Methods

3.1. Glutaraldehyde Conjugation Procedure (see *Note 1*)

1. Weigh out the peptide and an equal weight of KLH or thyroglobulin carrier. This gives an approximate ratio of 40–150 molecules of peptide to each molecule of carrier (2 mg of peptide per animal is ample). If

using PPD as carrier, weigh out at a molar peptide:PPD ratio of about 2:1 assuming the mol wt of PPD to be about 10 kDa (i.e., for a 15 amino acid peptide, use 3 mg of peptide for each 10 mg of PPD).

2. Dissolve the peptide and carrier protein in 0.1*M* (1X) sodium bicarbonate, using 1 mL for every 2 mg of carrier protein.
3. Thaw out a fresh vial of glutaraldehyde and add to the peptide-carrier solution to a final concentration of 0.05%. Mix in a glass tube, stirring with a magnetic stirring bar at room temperature overnight in the dark. The solution will usually go a pale yellow color.
4. *EITHER*: Dialyze against double-distilled water for 12 h and lyophilize the coupled carrier. Please Note—For PPD conjugates, use low-mol wt cut-off dialysis tubing. To assess the yield, weigh the lyophilized material and determine the percentage of peptide coupled.
   OR: Because coupling efficiency is usually reasonable, and not too critical anyway, it is easier to do the following. Add 1*M* glycine ethyl ester to a final concentration of 0.1*M* and leave for 30 min at room temperature. Then, precipitate the coupled carrier with 4–5 vol of ice-cold acetone at –70°C for 30 min. Pellet the protein at 10,000*g* for 10 min at room temperature, pour off the acetone, air-dry the pellet, and redisperse it in saline at 1 mg of carrier/mL. As the pelleted protein is rather sticky, this is best done using a Dounce™ homogenizer. If necessary, store the immunogen at –20°C and rehomogenize before use.

3.2. MBS Coupling Method (see *Note 2*)

1. Dialyze the carrier protein against 10 m*M* potassium phosphate buffer, pH 7.2. Adjust the carrier concentration to about 20 mg/mL. Weigh out the solid peptide and dissolve it in the same buffer at 5 mg/mL. Thiol-based scavengers, used to preserve free -SH groups in the peptide, are often present in peptide preparations and will need to be removed from the peptide solution by desalting on a Bio-Gel P10 column in 10 m*M* potassium phosphate buffer, pH 7.2. A brief nasal inspection is often sufficient to determine their presence. On the other hand, stock solutions of peptide (e.g., in PBS) may become oxidized if stored for extended periods, and so are best pretreated with 100 m*M* DTT (30 min at room temperature is adequate) to reduce the dissolved peptide before use. As the DTT will interfere with the conjugation, the reduced peptide should then be desalted on a P10 column as above.
2. About 4 mg of carrier is required for every 5 mg of peptide. Add 55 µL of 10 m*M* phosphate buffer, pH 7.2 to each 200 µL of carrier solution. Slowly add 85 µL of MBS solution to each 4 mg of carrier (*see* Note 3).

3. Stir for 30 min at room temperature in a glass tube.
4. Desalt the activated carrier on a 20 mL P30 (Bio-Rad) column. Pre-equilibrate and run the column in 50 m*M* phosphate buffer, pH 6.0. Collect 1-mL fractions (typically about 15). The protein elutes as a peak (visibly grey if the carrier is KLH) at the exclusion volume (i.e., in fractions 6–8). About 95% of carrier is recovered.
5. Add the activated carrier to the peptide solution while stirring at room temperature. Adjust the pH to 7.2 and stir at room temperature for 3 h.
6. Store the coupled carrier frozen. The conjugate can be used directly (i.e., there is no need to dialyze or precipitate the protein).

4. Notes

1. During glutaraldehyde conjugation, it is vital to exclude any buffers containing amino, imino (e.g., Tris), ammonium, or azide moieties as these will inhibit the crosslinking reaction. If the peptide or carrier are insoluble in coupling buffer, sodium dodecyl sulfate (SDS) may be added to 0.1% without affecting the conjugation.
2. It is often worth adding a cysteine residue at the C-terminus of the peptide to give the option of MBS coupling. This method couples the $-NH_2$ groups on KLH to -SH groups of cysteine in the peptide molecules.
3. During MBS coupling, the DMF concentrations should never exceed 30% or KLH will come out of solution. KLH concentrations in excess of 20 mg/mL will also cause it to come out of solution.

References

1. Lerner, R. A., Green, N., Alexander, H., Liu, F. T., Sutcliffe, J. G., and Schinnick, T. M. (1981) Chemically synthesized peptides predicted from the nucleotide sequence of the hepatitis B virus genome elicit antibodies reactive with the native envelope protein of Dane particles. *Proc. Natl. Acad. Sci. USA* **78,** 3403–3407 .
2. Chou, P. Y. and Fasman, G. D. (1978) Prediction of the secondary structure of proteins from their amino acid sequence. *Adv. Enzymol. Relat. Areas Mol. Biol.* **47,** 45–158.
3. Garnier, J., Osguthorpe, D. J., and Robson, B. (1978) Analysis of the accuracy and implications of simple methods of predicting the secondary structure of globular proteins. *J. Mol. Biol.* **120,** 97–120.
4. Tam, J. P. (1988) Synthetic peptide vaccine design: Synthesis and properties of a high-density multiple antigenic peptide system. *Proc. Natl. Acad. Sci. USA* **85,** 5409–5413.
5. Ghysen, H. M., Rodda, S. J., Mason, T. J., Tribbick, G., and Schoofs, P. G. (1987) Strategies for epitope analysis using peptide synthesis. *J. Immunol. Methods* **102,** 259–274.
6. Ghysen, H. M., Meloen, R. H., and Barteling, S. J. (1984) Use of peptide synthesis to

probe viral antigens for epitopes to a resolution of a single amino acid. *Proc. Natl. Acad. Sci. USA* **81,** 3998–4002.

7. Merrifield, R. B. (1963) Solid phase peptide synthesis. 1. The synthesis of a tetrapeptide. *J. Am. Chem. Soc.* **85,** 2149–2154.
8. Atherton, E., Gait, M. J., Sheppard, R. C., and Williams, B. J. (1979) The polyamide method of solid phase peptide and oligonucleotide synthesis. *Bioorg. Chem.* **8,** 351–370.
9. Bodanszky, M. (1984) *The Principles of Peptide Synthesis.* Springer-Verlag, Berlin.
10. Bodanszky, M. (1988) *Peptide Chemistry: A Practical Approach.* Springer-Verlag, Berlin.
11. Atherton, E. and Sheppard, R. C. (1984) *Solid Phase Peptide Synthesis: A Practical Approach.* IRL Press, Oxford.
12. Lachmann, P. J., Strangeways, L., Vyakarnam, A. and Evan, G. I. (1984) Raising antibodies by coupling peptides to PPD and immunising BCG-sensitised animals. CIBA Foundation Symposium. John Wiley, Chichester.

Chapter 5

Production and Characterization of Antibodies Against Synthetic Peptides

David C. Hancock and Gerard I. Evan

1. Introduction

1.1. Immunizations

Immunization protocols vary greatly between laboratories. In general, there are no hard and fast rules and most protocols give satisfactory results. The methods described below are designed to give optimal results with minimal injury to the test animal, and we have used them extensively and successfully for several years *(1–5)*. Peptide immunizations differ from those in which the immunogen is a larger macromolecule in that maximal antipeptide titers (which arise rapidly after 2–3 immunizations) do not always coincide with maximal titers against the intact protein (which tend to peak rather later at 4–6 immunizations). Thus, although antipeptide enzyme-linked immunosorbent assays (ELISAs) are useful gages of immune response, there is no substitute for eventual screening on the intact protein (e.g., by immunoprecipitation, Western blotting, and so on). Individual variation in antipeptide response is very marked, so it is advisable to use several animals (three to six) per immunogen. Rabbit responses are generally poorer in specific pathogen-free (SPF) animals, probably reflecting their greater immune naivity. Mouse responses are often best in F_1 crosses (e.g., BALB/c × C57Bl/6) rather than pure strains. Alternatively, SJL mice generally respond well.

Adjuvants stabilize immunogens so that they induce the immune system persistently over long periods. Oil–water adjuvants, such as Freund's, are extremely effective but must be prepared properly as *stable* emulsions. Such

From: *Methods in Molecular Biology, Vol. 10: Immunochemical Protocols*
Ed.: M. Manson

emulsions are thick, do not separate even after standing for long periods, and do not disperse if pipetted onto the surface of water. Immunogens administered in Freund's adjuvant can persist for weeks, and there is thus no point in repeating immunizations too frequently. Freund's complete adjuvant (FCA), which contains killed pertussis bacteria to induce massive nonspecific inflammation, causes ulceration (resulting in loss of immunogen and considerable discomfort to the animal) if administered in too large a bolus in one place or if given more than once. For this reason, subsequent immunizations are given using Freund's incomplete adjuvant (FIA), which contains no pertussis. An effective alternative to oil–water adjuvants is to administer the immunogen on alum as a fine adsorbed suspension. Precipitation of peptide-carrier conjugates with acetone also renders them partially insoluble, and thus more persistent immunogens. Acetone precipitates are particularly useful for priming prior to spleen fusions for the development of monoclonal antibodies.

Immunizations prior to such fusions should not be done with persistent adjuvants such as Freund's, because in this case, the aim is to induce a rapid and transient immune response whose early (lymphoproliferative) stage coincides with fusion to myeloma cells.

1.2. ELISA for Antipeptide Antibodies

Because the peptide immunogen used to generate antibodies is available in comparatively large amounts, it is seldom necessary to use highly sensitive radioimmunoassays to check antibody titers. For almost all purposes, a peroxidase- or β-galactosidase-based enzyme-linked colorimetric assay is adequate. The target antigen is passively adsorbed to the walls of microtiter wells, either as free peptide or as peptide conjugated to an *irrelevant* carrier protein (e.g., BSA). Usually the free peptide makes a perfectly effective antigenic target, but occasionally, important determinants on some peptides are masked by adsorption to the plate, in which case peptide-carrier conjugates should be used. Thus, if antibody titer is low, it is a good idea to try conjugated peptide as the target. Antibodies bound to the adsorbed peptide are detected with an appropriate enzyme-linked second-layer reagent, typically anti-immunoglobulin or protein A, and the assay developed with a colorimetric substrate.

1.3. Affinity Purification of Antipeptide Antibody

With a ready source of immunogen in the form of peptide, it is straightforward to purify antipeptide antibodies by affinity chromatography. The peptide is covalently coupled to agarose and the crude antibody passed down the column. Unbound material is washed away and the bound antibody eluted under denaturing conditions, for example, low pH (pH 2.5), high pH (pH

11.5), or 4*M* $MgCl_2$. Immunoglobulins are unusually resistant to permanent denaturation by pH extremes or chaotropic reagents, although there are always exceptions (especially with monoclonal antibodies). Elution by 4*M* $MgCl_2$ is the mildest, and therefore the first method to try; low pH followed by rapid neutralization is used most frequently in our laboratory.

The affinities of antibodies for their cognate peptides can be very high, so that it is sometimes difficult to quantitatively recover the higher-affinity antibodies in a polyclonal serum from the peptide-resin. For this reason, it is essential to use only low concentrations of peptide on the resin (typically 100–200 µg of peptide per mL of agarose gel) and to elute bound antibody in the *reverse* direction to which it was run into the column, to avoid driving eluting antibodies into a further excess of antigen. If both of these criteria are adhered to, yields are usually of the order of 60–80% recovery.

Suitable affinity resins are CNBr-activated Sepharose (Pharmacia) and similarly activated *N*-hydroxysuccinimide ester-based gels with spacer arms. These react with free amino groups on the peptide. If the peptide contains many lysine residues, alternative coupling systems may be used, for example, carbodiimide-activated agaroses. In our experience, however, even peptides with internal lysines make good immunoadsorbents, and we routinely use CNBr-activated Sepharose.

2. Materials

2.1. Immunizations

1. Freund's complete and incomplete adjuvants: FCA is a suspension of killed pertussis in oil. This should be shaken well before use. FIA is just oil. Freund's adjuvants should be mixed with the aqueous immunogen solution/suspension at a ratio between 1:2 and 2:1 and mixed until set. This is most easily achieved by passage back and forth between two *glass* syringes connected by a three-way luer fitting (*see* Chapter 1). After a while (1–5 min), the mixture should become noticeably "stiffer" and may then be used (*see* Section 3.).
2. Potassium aluminium sulfate (Alum) $AlK(SO_4)_2.12H_2O$: The aqueous immunogen solution/suspension is mixed with 0.3 vol 10% alum. The pH is then adjusted to about 8.0 with sodium hydroxide solution and the resultant precipitate washed in 0.9% NaCl solution and administered.
3. Acetone precipitates: The aqueous immunogen is precipitated with 4.5 vol of acetone at –20°C for 1 h. The precipitate is collected by centrifugation at 10,000*g* at *room temperature*, washed in 80% acetone, and air-dried. The pellet is resuspended in saline using a Dounce™ homogenizer and then administered directly or in association with

alum or Freund's adjuvants.

4. Bacillus Calmette-Guerin (BCG): An attenuated strain of bovine tubercle bacillus, is available from Glaxo or from outdated hospital supplies as a lyophilized powder. Before use, it is suspended in sterile distilled water.
5. SalineL 0.9% NaCl.
6. Glass syringes (2 and 10 mL).
7. Disposable three-way luer fitting taps.
8. Peptide-conjugate.

2.2. ELISA

8. 10X Adsorption buffer: 1*M* Sodium bicarbonate, pH 9.6 (adjust with HCl).
9. Peptide solution at 1 mg/mL in phosphate buffered saline (PBS) stored at –20°C.
10. Antipeptide antibody.
11. "Immulon 2" *hard* plastic 96-well microtiter plates (Dynatech Ltd.).
12. Rabbit antimouse Ig-horseradish peroxidase conjugate (RaMIg-HRP), swine antirabbit Ig-horseradish peroxidase conjugate (SwaRIg-HRP), protein A-peroxidase, or appropriate β-galactosidase-conjugated second-layer reagent.
13. Tris buffered saline (TBS): 25 m*M* Tris-HCl, 144 m*M* NaCl, pH 8.2.
14. Blocking buffer (TM): TBS containing 2% dried milk powder (e.g., Marvel).
15. Assay buffer (TMT): TBS containing 2% dried milk powder and 0.5% Tween-20.
16. Substrate solution for HRP: 1 m*M* ABTS (2,2' azinobis [3-ethylbenzthiazoline-6 sulfonic acid]) in 0.1*M* sodium acetate, pH 5.0. Add 1 μL of 30% hydrogen peroxide per 1 mL of ABTS solution *just before* use. Discard any old ABTS stocks that have a noticeable green color when dissolved in the absence of hydrogen peroxide.
17. Stop solution for peroxidase assays: 5% Sodium dodecyl sulfate (SDS).
18. Substrate solution for β-galactosidase: 4 mg/mL O-nitrophenyl-β-D-galactopyranoside (ONPG) dissolved in TBS containing 0.7% 2-mercaptoethanol and 1 m*M* $MgCl_2$.
19. Stop solution for β-galactosidase: 1*M* Na_2CO_3.

2.3. Affinity Purification

20. CNBr-activated Sepharose.
21. Affigel–10.
22. Sintered glass funnels.
23. Chromatography columns (Pharmacia C series).
24. Peristaltic pump.
25. Concentrated HCl.

26. PBS.
27. 100 m*M* sodium acetate, pH 4.0.
28. 2*M* NaCl in PBS.
29. TBS.
30. TBS containing 0.1% sodium azide (CARE—TOXIC).
31. TBS containing 0.1% NP40.
32. 250 m*M* EDTA, pH 8.0.
33. Saturated ammonium sulfate solution.
34. NP40 stock solution: 10% in water.
35. 100 m*M* sodium citrate, pH 2.5.
36. 2*M* Tris base.
37. Saline: 0.9% NaCl.
38. 4*M* $MgCl_2$.

3. Methods

3.1. KLH/Thyroglobulin–Peptide Conjugate Immunizations

1. Immunize rabbits as follows:
   a. Intradermally with 50–200 µg of peptide (as a conjugate) in FCA at multiple sites.
   b. After 2–3 wk, immunize again intradermally with 50–100 µg of conjugate in FIA.
   c. Repeat Step 2 at 3-wk intervals. Bleed 10 d after immunization. We usually test-bleed (from the ear) after the third and fifth immunizations, then completely exsanguinate after a further 1–4 immunizations as appropriate, depending on the efficacy of the antiserum.
2. Immunize mice as follows:
   a. In the tail-base with 20–100 µg of peptide (as a conjugate) in FCA.
   b. After 2 wk, immunize im (e.g., in the base of the tail) or sc (in the flanks) with about 50 µg of conjugate in FIA.
   c. Repeat Step 2 at 2–3-wk intervals until titers plateau (3–6 immunizations in total). Test-bleed (from the tail) 10 d after immunization.
   d. For fusions, 3–10 wk after the last immunization, immunize ip with about 50 µg of conjugate on alum or as an acetone precipitate in saline (day 6). Reimmunize 3 d later (day 3). Fuse 3 d later (day 0).

3. Immunization with PPD–peptide conjugates (*see* Note 1).
   a. Four to six weeks before start of immunization, inoculate with two adult doses of BCG (all animals) in saline sc *without* adjuvant.
   b. Administer the first PPD–peptide immunization in FIA.
   c. Continue immunizations in FIA at 3–4-wk intervals as above.
   d. For fusions, follow the same procedure used for KLH/thyroglobulin conjugates described in 2.d. above.

3.2. ELISA for Antipeptide Antibodies

3.2.1. Adsorption of Peptide to Microtiter Plates

1. If using free peptide as the antigenic target, dilute it to a final concentration of 50 mmol/mL in adsorption buffer (roughly a dilution of about 1 in 2000 for a 10-mer peptide). If using a peptide-carrier conjugate as the target, dilute the conjugate in adsorption buffer to a *peptide* concentration of about 10 mg/mL. The precise amount of peptide conjugate may need eventually to be titrated to give optimal signals.
2. Add 100 µL of diluted peptide solution to each well.
3. Leave at room temperature in a wet box overnight.
4. Shake out any unadsorbed peptide, and wash the plate three times in TBS by immersion of the plate in a TBS bath. Shake the plate dry.
5. Add 150 µL of T*M* buffer per well and leave at room temperature for at least 30 min. If required, store the plates at this stage at –20°C.

3.2.2. ELISA

1. Empty the wells and add 100 µL of antibody diluted in TMT per well. Suitable starting dilutions are 1 in 50 for antisera, 1 in 2 for hybridoma culture supernatants, and 1 in 500 for hybridoma ascites fluids. Serially dilute antibody in doubling dilutions down one row of the microtiter plate (i.e., 8 dilutions).
2. Leave for 30 min at room temperature.
3. Wash the wells three times in TBS as before.
4. Add 100 µL per well of appropriate second layer reagent diluted in TMT. Dilute the stock HRP-anti-Ig conjugates (from the manufacturer) 1 in 150, stock protein A-HRP solution (1 mg/mL) 1 in 150, and stock β-galactosidase conjugates 1 in 100 (*see* Note 2).
5. Leave for 30 min at room temperature.
6. Wash three times as before in TBS.
7. Add 100 µL of substrate solution per well.
8. Incubate at room temperature in the dark. Stop the reactions by adding 100 µL of appropriate stop solution to each well (5% SDS for peroxi-

dase, 1*M* Na_2CO_3 for β-galactosidase). The SDS also solubilizes any precipitated products formed in the HRP reaction. Peroxidase reactions take about 5–30 min to develop, β-galactosidase reactions can take longer. Judge the reaction time by eye (*see* Notes 3 and 4).

9. Read the optical density (OD) on an ELISA plate reader. Green ABTS reaction product and yellow ONPG reaction product may both be read at OD_{406}.

3.3. Affinity Purification

3.3.1. Preparation of Peptide-Agarose

1. Mix 1.5 g of CNBR-Sepharose and 200 mL of 1 m*M* HCl and leave for 15 min at room temperature.
2. Collect the slurry on a sinter funnel, drain until a moist cake is formed. Add the cake (typically ~ 5 mL vol) to 5 mL of PBS (pH 7.5–8.0) containing ~500 µg of peptide. Agitate *gently* for about 2 h at room temperature. Please Note—*Do not* use a magnetic stirrer as this fragments the agarose and generates fines that then slow or block the column flow.
3. Pour the slurry onto a sinter, wash sequentially with 20 mL of the following: PBS; 100 m*M* sodium acetate, pH 4.0; 2*M* NaCl in TBS. Store as a 50% slurry in TBS containing 0.1% sodium azide at 4°C.

3.3.2. Preparation of Serum or Ascites Fluid

1. Allow clots to form. Ring the tube to prevent the clot adhering and leave overnight at 4°C.
2. Pour off the supernatant and clarify the serum/ascites by centrifugation at 1000*g* for 5 min.
3. Adjust the sample to 5 m*M* EDTA and add 0.82 vol of saturated ammonium sulfate solution while stirring, and leave for 15 min at room temperature.
4. Collect the pellet by centrifugation (10 min, 10,000*g*, 4°C).
5. Redissolve the pellet in its original volume using TBS. Add 10% NP40 to a final concentration of 0.1% and spin in a microfuge to clarify.

3.3.3. Affinity Chromatography

1. Use a Pharmacia reversible column. Pack 2 mL of affinity matrix into the column (keep moist) in running buffer (TBS containing 0.1% NP40). Wash with 20 mL of running buffer over 20 min.
2. Run in the antibody solution as prepared in Section 3.3.2. The flow-rate should be about 1–2 mL/min (a peristaltic pump is useful to control the flow-rate). Run in the equivalent of about 1 mL of antiserum per mL of gel.
3. Wash with 10 column vol of running buffer.

4. Reverse the direction of flow. Wash with: 10 column vol of TBS containing 0.1% NP40 over 10 min; 5 column vol of TBS at the same flow-rate; and 5 column vol of 0.9% NaCl.
5. Elution of antibody may be achieved by either one of the following two procedures.
   a. *Low pH elution.* Elute the bound antibody with 4 column vol of 100 m*M* sodium citrate, pH 2.5 over 10 min, collecting the eluate and immediately neutralize to pH 5–8 with 2*M* Tris base. Adjust the pooled fraction to pH 6 and add 1 vol of saturated ammonium sulfate solution. Leave for 10 min at room temperature and pellet the antibody at 10,000*g* for 10 min at 4°C. Resuspend the antibody in water at about 1–5 mg/mL (OD_{280} of IgG is about 1.4 for a 1 mg/mL solution) and dialyze against TBS containing 0.1% sodium azide. Store in aliquots at –20°C.
   b. *Elution with 4M $MgCl_2$.* Elute the bound antibody with 4 column vol of 4*M* $MgCl_2$. Dilute the eluate 10× with distilled water. Add an equal volume of saturated ammonium sulfate and pellet the immunoglobulin at 10,000*g*. Resuspend the pellet in water and dialyze against TBS containing 0.1% sodium azide. Store in aliquots at –20°C.

4. Notes

1. If PPD is used as the carrier for peptide immunizations, the animals must first be primed with live attenuated tubercle bacteria (BCG strain) that express PPD on their surface. This priming step elicits a strong T-cell helper response against subsequent PPD-linked immunogens. Freund's complete adjuvant actually appears to interfere with this priming process and so *must* be avoided if using this method.
2. Protein A only binds certain Ig subclasses at neutral pH (for example, many mouse antibodies of the IgG_1 subclass fall into this category). Its binding range can, however, be greatly extended by using high pH (pH 8.5) buffers and high salt (1–2*M* NaCl), although these conditions can be inconsistent with antibody binding to peptide. For this reason, anti-Ig detection reagents are advised for primary testing of antipeptide antibodies.
3. β-Galactosidase reactions tend to have a lower spontaneous background than peroxidase reactions but take longer to develop. β-Galactosidase reactions can be speeded up by incubation at 37°C.
4. Avoid contamination of substrate solutions by skin contact since this can

sometimes increase background activity. Read HRP reactions immediately since atmospheric oxidation will gradually react with substrate in all wells.

References

1. Evan, G. I., Hancock, D. C., Littlewood, T. D., and Gee, N. S. (1986) Characterisation of human *myc* proteins. *Curr. Top. Micro. Immunol.* **132,** 362–374.
2. Evan, G. I., Lewis, G. K., Ramsay, G., and Bishop J. M. (1985) Isolation of monoclonal antibodies specific for the human c-*myc* protooncogene product. *Mol. Cell. Biol.* **5,** 3610–3616.
3. Hunt, S. P., Pini, A., and Evan, G. I. (1987) Induction of c-*fos*-like protein in spinal cord neurons following sensory stimulation. *Nature (London)* **328,** 632–634.
4. Moore, J. P., Hancock, D. C., Littlewood, T. D., and Evan, G. I. (1987) A sensitive and quantitative enzyme-linked immunosorbence assay for the c-*myc* and N-*myc* oncoproteins. *Oncogene Res.* **2,** 65–80.
5. Waters, C. M., Hancock, D. C., and Evan, G. I. (1990) Identification and characterisation of the *egr-1* gene product as an inducible, short-lived, nuclear phosphoprotein. *Oncogene* **5,** 669–674.

CHAPTER 6

Preparation and Testing of Monoclonal Antibodies to Recombinant Proteins

Christopher J. Dean

1. Introduction

Monoclonal antibodies (MAbs) are useful, often essential, reagents for the isolation, identification, and cellular localization of specific gene products and for the determination of macromolecular structure. The ability to clone and sequence specific genes has revolutionized our understanding of cellular structure and function, and the ability to prepare recombinant proteins or to synthesize peptides based on protein sequences derived from cDNA clones provides sufficient material for generating specific antibodies. The recombinant proteins may be derived from prokaryotic systems, such as *E. coli,* or from eukaryotic expression systems, such as Chinese hamster ovary cells (CHO) or insect cells expressing constructs in baculovirus. The eucaryotic systems are being used increasingly for expression of glycoproteins because the recombinant material is glycosylated. A number of protocols will be described here that we have used successfully with both rat (Y3 and IR983F) and mouse (SP2/0) myelomas to generate MAbs to recombinant material or peptides based on cDNA sequences.

Successful hybridoma production relies on the ability to (1) generate specific B-cells, (2) fuse them with a myeloma cell line, (3) identify the antibodies that are sought in culture supernatants, and (4) isolate and clone the

From: *Methods in Molecular Biology, Vol. 10: Immunochemical Protocols*
Ed.: M. Manson ©1992 The Humana Press, Inc., Totowa, NJ

specific hybridoma. Of particular importance is the elicitation of the specific B-cells required for fusion, and protocols to achieve this will be described. It should be remembered that the presence of specific antibody in serum is not necessarily a guarantee of success, nor is its absence a surety for failure.

The second important requirement is for a quick, reliable assay(s) for the specific antibody that can be applied to the large numbers of culture supernatants (≥96) that may be generated. Usually, the assays make use of a labeled second antibody (e.g., rabbit, sheep, or goat antibodies directed against the $F(ab')_2$ of mouse or rat immunoglobulins) to identify binding of monoclonal antibody to antigen. The second antibody can be detected because it is conjugated to a fluorescent marker, e.g., fluorescein, or a radiolabel, such as Iodine–125. Alternatively, conjugates of second antibody with enzymes such as alkaline phosphatase, peroxidase, or β-galactosidase may be employed.

Persistence is an important attribute of the hybridoma producer; fusions can fail for many reasons, and it is essential not to give up because of early failures. The methods described in this chapter include techniques for (1) preparation of antigen, (2) immunization, (3) hybridoma production, (4) assaying the MAb-producing hybridomas, and (5) isolation and purification of MAbs.

2. Materials

2.1. Generation of Immune Spleen or Lymph Node Cells

1. Rats of any strain aged 10–12 wk, BALB/c mice aged 6–8 wk (*see* Notes 1 and 2).
2. Phosphate buffered saline (PBS): Dissolve in water 1.15 g of Na_2HPO_4, 0.2 g of KH_2PO_4, 0.2 g of KCl, 8.0 g of NaCl, and make up to 1 L. The pH should be 7.4.
3. Antigen: One of the following sources of antigen can be used to raise MAbs.
   a. Cells, e.g., mouse 3T3-cells expressing recombinant human membrane protein (*see* Note 3).
   b. Soluble protein dissolved in PBS.
   c. Soluble recombinant protein extracted from cells or supernatants of eukaryotic cells (e.g., Chinese hamster ovary cells or insect cells expressing recombinant baculovirus) or bacteria, such as *E. coli,* harboring plasmids or recombinant viruses. In *E. coli,* recombinant material is often generated as a fusion protein with β-galactosidase (*see* Note 3).
   d. Protein separated electrophoretically in sodium dodecyl sulfate containing polyacrylamide gels (SDS-PAGE) and eluted

electrophoretically from gel slices. Often, β-galactosidase fusion proteins are prepared in this way because of their poor solubility.

e. Peptide conjugated to a protein carrier and dissolved in PBS (*see also* this vol., Chapter 4).

4. Freund's complete adjuvant (FCA).
5. Freund's incomplete adjuvant (FIA).

2.1.1. Conjugation to Carriers

6. Ovalbumin, bovine serum albumin (BSA), or Keyhole lympet hemocyanin (KLH), at 20 mg/mL in PBS.
7. PPD-kit (tuberculin-purified protein derivative, Cambridge Research Biochemicals, Ltd.)
8. Glutaraldehyde (specially purified grade 1): 25% Solution in distilled water.
9. 1*M* glycine-HCl, pH 6.6

2.2. Hybridoma Production

10. Dulbecco's Modified Eagle's Medium (DMEM): Containing glucose (1 g/L), bicarbonate (3.7 g/L), glutamine ($4 \times 10^{-3}M$), penicillin (50 U/mL), streptomycin (50 µg/mL), and neomycin (100 µg/mL), stored at 5–6°C and used within two weeks of preparation.
11. Fetal calf serum (FCS): Inactivated by heating for 30 min at 56°C and tested for ability to support growth of hybridomas (*see* Note 4).
12. HAT selection medium: Prepare 100× HT by dissolving 136 mg of hypoxanthine and 38.75 mg of thymidine in 100 mL of 0.02*M* NaOH prewarmed to 60°C. Cool, filter-sterilize, and store at –20°C in 1- and 2-mL aliquots. Prepare 100× A by dissolving 1.9 mg of aminopterin in 100 mL of 0.01*M* NaOH, then filter-sterilize and store at –20°C in 2-mL aliquots. HAT medium is prepared by adding 2 mL of HT and 2 mL of A to 200 mL of DMEM containing 20% FCS.
13. HT medium: Add 1 mL of HT to 100 mL of DMEM containing 10% FCS.
14. Feeder cells for fusion cultures (*see* Note 5. Essential for fusions employing rat myelomas): Three hours to one day before fusion, $2–4 \times 10^6$ rat fibroblasts in 10 mL of DMEM and contained in a 30 mL plastic universal are irradiated with about 30 Gy (3000 rad) of X- or γ-rays. Dilute with 90 mL of HT (3 h before fusion) or DMEM containing 10% FCS (1 d before fusion) and plate 1-mL aliquots into four 24-well plates or 0.2-mL aliquots into five 96-well plates. Alternatively, use thymocytes from spleen donors (mouse).

15. PEG solution: Weigh 50 g of polyethyleneglycol (1500 MW) into a capped 200-mL bottle, add 1 mL of water, and then autoclave for 30 min at 120°C. Cool to about 70°C, add 50 mL of DMEM, mix, and after cooling to ambient temperature, adjust the pH to about 7.2 with 1*M* NaOH (the mixture should be colored orange). Store as 1-mL aliquots at –20°C.
16. Freezer medium: Freshly prepared 5% dimethyl sulfoxide–95% FCS.
17. Myeloma cell line: Mouse SP2/0-Ag14 or rat IR 983F cells growing exponentially in 25- or 75-cm^2 flasks containing DMEM-10% FCS (dilute to 2–3 × 10^5 cells/mL the day before fusion). The rat myeloma Y3 Ag1.2.3. has to be grown in spinner culture to fuse well. Seven to ten days before cells are required, about 5 × 10^6 cells, stored frozen in liquid nitrogen as 1-mL aliquots in freezer medium, are thawed quickly at 37°C, diluted with 10 mL of DMEM–10% FCS, centrifuged (500*g* × 2 min), then resuspended in 100 mL of the same medium and placed in a 200-mL spinner flask. Stand for 2 d at 37°C to allow cells to attach to the base of the vessel, then place the spinner flask on a Bellco magnetic stirrer running at about 160 rpm. The Y3 myeloma has a generation time of about 10 h, and exponentially growing cultures require feeding daily by fourfold dilution with fresh medium.

2.3. Screening Culture Supernatants

18. PBSA: PBS containing 0.02% NaN_3 (NaN_3 should be handled with care. It is an inhibitor of cytochrome oxidase and is highly toxic and mutagenic. Aqueous solutions release HN_3 at 37°C).
19. PBST: PBSA containing 0.4% Tween-20.
20. PBS–BSA: PBSA containing 0.5% BSA.
21. PBS–Marvel: PBSA containing 3% Marvel (skimmed milk powder). Centrifuge or filter through Whatman No. 1 paper to remove undissolved solids. This is cheaper to use and just as effective as PBS–BSA.
22. PCB: Plate-coating buffer, pH 8.2, containing 0.01*M* Na_2HPO_4/KH_2PO_4 and 0.14*M* NaCl.
23. Alkaline sarkosyl: 1% Sodium dodecyl sarkosinate in 0.5*M* NaOH.
24. Plates containing cell-bound antigen: Monolayers of cells grown in 96-well polystyrene plates (*see* vol. 5, Chapter 54 for detailed instructions).
    a. Rodent cells expressing a recombinant protein, e.g., 3T3 cells transfected with a gene for a human transmembrane protein; or
    b. Tumor cell line overexpressing a transmembrane protein, e.g., the receptor for EGF or the product of the c-*erbB*-2 gene; or

c. Adherent cell line expressing high cytoplasmic levels of the specific antigen that can be accessed following fixation and permeabilization with methanol. Wash the cells with ice-cold DMEM, then add to each well 200 µL of methanol that has been precooled to –70°C by standing in cardice-ethanol. Leave at ambient temperature for 5 min, "flick" off the methanol, and wash twice with medium containing either 5% FCS (live cells) or PBS–Marvel (fixed cells); and
d. Control cell line that either does not express the specific antigen (e.g., normal 3T3 cells) or in which expression is at the normal one gene copy level. This will act as the negative control.

25. Plates coated with purified soluble antigen. As an alternative to 24 above, and providing purified soluble antigen is available, 96-well polystyrene (PS) or polyvinyl chloride (PVC) plates can be coated with protein, recombinant protein, peptide, or peptide conjugated to a carrier protein. Plates are coated as follows:
   a. Dissolve the antigen at 1 µg/mL in PCB.
   b. Coat by incubation with 50 µL of antigen per well for 2 h at 37°C or overnight at 4–6°C.
   c. Block the remaining reactive sites by incubation for 2 h at 37°C with 200 µL per well of PBS–Marvel.
   d. Wash the plates with PBST before use.

   In many cases, the coated plates can be stored at 4–6°C for several weeks (fill wells with PBST), but it is wise to check their antibody-binding capacity before use.
26. Plates for antibody capture assays: Where a specific antibody is available this may be a better assay to use than 25 above because the antigen is less likely to be subject to denaturation.
   a. Coat wells first by incubating for either 2 h at 37°C, or overnight at 4–6°C, with 50 µL (1–5 µg/mL in PBS, pH 8.0) of a polyclonal or monoclonal antibody to the specific antigen.
   b. Block the wells with PBSA–Marvel for 2 h at 37°C.
   c. Incubate with the specific antigen (50 µL per well of a 0.1–1 µg/mL solution in PBST) for 2 h at 37°C, or overnight at 4–6°C, then wash with PBST. For antigens that are not readily soluble, e.g., membrane proteins extracted in nonionic detergents, the extract in 0.5–1% Triton X-100 or Nonidet-P40 can be used after suitable dilution with PBSA containing 0.5% of the detergent. All subsequent procedures should be carried out

using buffers containing the detergent at 0.1–0.5%. These plates are best prepared within a day of use.

2.3.1. Immunoprecipitations

27. Protein A or specific anti-immunoglobulin covalently linked to Sepharose 4B or similar bead support for preparing immunoprecipitates.
28. CNBr-activated Sepharose 4B.
29. Radiolabeled (^{3}H, ^{14}C, ^{35}S, or ^{125}I) protein or cell extract prepared in PBSA containing 10^{-3} *M* phenylmethylsulfonyl fluoride (PMSF) as proteinase inhibitor and 0.5–1.0% nonionic detergent.
30. Purified specific monoclonal antibody.
31. Polyclonal antibody to mouse or rat immunoglobulins depending on the species in which the MAb was raised.

2.3.2. Western Blotting

32. Blotting membrane: Nitrocellulose or PVDF membrane.
33. Equipment for wet or semidry blotting.

2.3.3. Second Antibodies

34. Sheep, rabbit, or goat antibodies to rat or mouse $F(ab')_2$, IgG, IgA, and IgM for labeling with iodine-125 to carry out radioimmunoassay (RIA) or conjugated to alkaline phosphatase or biotin for an ELISA.
35. 1.5 mL Polypropylene microcentrifuge tubes coated with 10 µg of IODO-GEN (Pierce Chemical Co.) by evaporation, under a stream of nitrogen, from a 100 µg/mL solution in methylene chloride (*see* Note 7).
36. Carrier-free ^{125}I, radioactive concentration 100 mCi/mL (e.g., code IMS.30, Amersham International).
37. Gamma counter.
38. UB: 100 m*M* phosphate buffer (Na_2HPO_4/KH_2PO_4), pH 7.4, containing 0.5*M* NaCl and 0.02% NaN_3.
39. 30 cm × 0.7 cm Disposocolumn (Bio-Rad) containing Sephadex G-25, equilibrated before use with UB and pretreated with 100 µL of FCS to block sites that bind protein nonspecifically.
40. 3 mL Tubes for collection of samples.
41. Lead pots for storage of ^{125}I-labeled antibodies.
42. Streptavidin labeled with ^{125}I or conjugated to alkaline phosphatase or to fluorescein.
43. 96-well plate reader for ELISA.

2.3.4. Buffers and Substrates for Alkaline Phosphatase Used in ELISA

44. 10 m*M* Diethanolamine, pH 9.5, containing 0.5*M* NaCl.
45. 100 m*M* Diethanolamine, pH 9.5, containing 100 m*M* NaCl and 5 m*M* $MgCl_2$.

46. Substrate to give a soluble product (plate assays): 0.1% *p*-Nitrophenyl phosphate in 10 m*M* diethanolamine, pH 9.5, containing 0.5 m*M* $MgCl_2$.
47. Substrate to give an insoluble product (Western blots): NBT stock—5% nitroblue tetrazolium in 70% dimethyl formamide. BCIP stock—5% disodium bromochloroindolyl phosphate in dimethyl formamide. Alkaline phosphatase buffer—100 m*M* diethanolamine, pH 9.5, containing 100 m*M* NaCl and 5 m*M* $MgCl_2$. Just before use, add 66 µL of NBT stock solution to 10 mL of alkaline phosphatase buffer, mix well, and add 33 µL of BCIP stock solution.

2.4. Isolation and Purification of MAbs

48. PB: 0.175*M* phosphate buffer (Na_2HPO_4/KH_2PO_4), pH 6.6. Dilute 1:10 (PB/10) as appropriate.
49. 0.055*M* phosphate buffer (Na_2HPO_4/KH_2PO_4), pH 6.6, ± 1*M* NaCl.
50. Supernatant from cultures grown in DMEM containing as low a concentration of FCS as possible (e.g., rat hybridomas will grow at 2–3% FCS) and tested for maximum level of MAb. Roller cultures can yield milligram quantities (10–20 mg/L) of either rat or mouse antibodies.
51. Ascites from BALB/c mice or athymic rats. Grow up hybridomas in DMEM–10% FCS in roller culture (essential for rat hybridomas) and inject about 3×10^6 cells/mouse and 10^7 cells/rat into the peritoneal cavity. Harvest ascitic fluid 7–10 d later when tumor growth is visible by swelling of the abdomen. If hybridomas grow as solid tumors, then use animals that have been injected ip with 0.5 mL (mouse) or 1 mL (rat) of pristane 7–10 d previously (*see* Note 8).
52. Salt fractionation: Prepare a saturated solution of $(NH_4)_2SO_4$ by adding 777 g of the solid to 1 L of water. Place on a heated stirrer to dissolve, then allow to cool before use (free crystals of excess salt will precipitate). Store in a capped bottle at room temperature.
53. Ion-exchange chromatography: Place Whatman DE52 (preswollen, use 1 g/3 mL of ascites or 100 mL of culture supernatant) in a measuring cylinder and wash 2× with 5–6 vol of PB. Allow to settle between washes, but remove fines with supernatant. Resuspend in 5 vol of PB/10 and pour into a chromatography column using a reservoir. Attach the column to a peristaltic pump and pack at a flow rate of at least 45 mL/h/cm^2 of internal column cross-sectional area until the bed height is constant. Equilibrate the packed column with PB/10 until the effluent has the same pH and conductivity as the eluting buffer.
54. PA buffer: 10 m*M* Phosphate (Na_2HPO_4/NaH_2PO_4) buffer, pH 8.0, containing 150 m*M* NaCl and 0.02% NaN_3.
55. Protein A–Sepharose equilibrated to pH 8.0 with PA buffer.
56. CNBr activated Sepharose 4B.

57. Purified sheep, rabbit, or goat antibodies to rat or mouse $F(ab')_2$ for linking to Sepharose 4B.
58. 3*M* Ammonium thiocyanate (NH_4SCN) dissolved in water and stored in a brown glass bottle at 4°C.
59. 30 cm × 1 cm Column of Sephadex G–75 equilibrated with UB buffer.
60. Equipment for monitoring optical density at 280 nm and for collecting fractions of 0.5–20 mL.

3. Methods

3.1. Conjugation of Peptides to Carriers

Peptides that do not bear epitopes recognized by T-cells are poor immunogens and must be conjugated to carrier proteins or PPD to elicit good immune responses (*see also* this vol., Chapter 4).

1. Protein carrier, such as BSA or KLH: Mix the peptide and carrier in a 1:1 ratio, e.g., pipet 250 µL of each into a 5-mL glass beaker on a magnetic stirrer. Small fleas can be made from pieces of paper clip sealed in polythene tubing by heating. Add 5 µL of 25% glutaraldehyde and continue stirring for 15 min at room temperature. Block excess glutaraldehyde by adding 100 µL of 1*M* glycine and stirring for a further 15 min. Use directly or dialyze overnight against PBS and store at –20°C.
2. PPD kit: **Read the instructions supplied with the kit very carefully.** Inhalation of the ether-dried tuberculin PPD is dangerous for tuberculin-sensitive people to handle. Follow specific instructions to couple 2 mg of peptide to 10 mg of PPD and, after dialysis, store at –20°C.

3.2. Antigens for Immunization

1. Suspend whole cells in PBS or DMEM at 5×10^6–10^7 cells/mL; or
2. Mix proteins, peptide-conjugates, or eluates from polyacrylamide gels in PBS 1:1 with adjuvant (FCA for the first immunization, subsequently with FIA) in a capped plastic tube (LP3, Bijou or 30 mL universal) by vortexing until a stable emulsion is formed (*see also* this vol., Chapter 1). Check that phase separation does not occur on standing at 4°C for >2 h. Alternatively, allow a drop to fall from a Pasteur pipet onto a water surface; the drop should contract and remain as a droplet and not disperse.

3.3. Immunization Procedures

1. Anesthetize animals (*see* Note 1) and take a blood sample from the jugular or tail vein into a capped 0.5-mL or 1.5-mL microcentrifuge tube to act as a preimmune sample. Allow it to clot, centrifuge (1500*g*), remove the serum, and store at –20°C.

2. For fusions that will use spleen cells, immunize at five sites (4× sc and 1× ip) with a total of 50–500 µg of antigen in FCA per animal. Test-bleed 14 d later and reimmunize using the same protocol, but with antigen in FIA. Test-bleed and reimmunize at monthly intervals until sera are positive for antibodies to the antigen (*see* Section 3.5. *below*). Three days before the fusions are done, rechallenge the animals iv with the antigen in PBS alone.
3. For fusions that will use mesenteric lymph nodes of rats, the antigens are injected into the Peyer's patches that lie along the small intestine. The surgical procedures are described in vol. 5, p. 673. Again, test-bleed the rats, then immunize twice at 14-d intervals, and use the mesenteric nodes at day 17. This short protocol has resulted in good yields of specific IgG- and IgA-producing hybridomas *(1)*. If unsuccessful, give a third immunization using FIA via the Peyer's patches 1 mo after the second challenge and take the mesenteric nodes 3 d later.

3.4. Hybridoma Production

3.4.1. Preparation of Cells for Fusion

1. Centrifuge exponentially growing rat or mouse myeloma cells in 50-mL aliquots for 3 min at 400*g*, wash twice by resuspension in serum-free DMEM, count in a hemocytometer, and resuspend in this medium to 1–2 × 10^7 cells/mL.
2. Kill immune animals by cervical dislocation or CO_2 inhalation, test-bleed, and open the abdominal cavity. Remove spleens or mesenteric nodes by blunt dissection.
3. Disaggregate spleens or nodes by forcing through a fine stainless-steel mesh (e.g., a tea strainer) into 10 mL of serum-free DMEM, using a spoon-head spatula (dipped into ethanol and flamed to sterilize it).
4. Centrifuge cells for 5 min at 400*g*, wash twice in serum-free DMEM, and resuspend in 10 mL of the same medium.
5. Count viable lymphoid cells in a hemocytometer. Spleens from immune mice yield about 10^8 cells, from rats 3–5 × 10^8 cells, and the mesenteric nodes of rats, up to 2 × 10^8 cells.

3.4.2. Fusion Protocol

1. Mix 10^8 viable lymphocytes with 5 × 10^7 rat myeloma cells or 2 × 10^7 mouse myeloma cells in a 10-mL sterile capped tube and centrifuge for 3 min at 400*g*.
2. Pour off the supernatant, drain carefully with a Pasteur pipet, then release the cell pellet by gently tapping the tube on the bench.

3. Stir 1 mL of PEG solution, prewarmed to 37°C, into the pellet over a period of 1 min. Continue mixing for a further minute by gently rocking the tube.
4. Dilute the fusion mixture with DMEM (2 mL over a period of 2 min, then 5 mL over 1 min).
5. Centrifuge for 3 min at 400*g*, then resuspend the cells in 200 mL of HAT selection medium, and plate 2-mL aliquots into four 24-well plates seeded with irradiated fibroblasts, or if necessary, five 96-well plates (fusions with SP2/0 myeloma) (*see* Note 5, re: feeder cells).
6. Screen culture supernatants for specific antibody 6–14 d after commencement of incubation at 37°C in 5% CO_2–95% air (*see* Section 3.5. *below*).
7. With a Pasteur pipet, pick individual colonies into 1 mL of HT medium contained in 24-well plates. Feed with 1 mL of HT medium and split when good growth commences. Freeze samples in liquid nitrogen.
8. Rescreen the picked colonies and expand positive cultures. Freeze samples of these in liquid nitrogen and clone twice.

3.4.3. Cloning of Hybridomas

1. Prepare a suspension of mouse thymocytes (2.5×10^4/mL of DMEM–10% FCS) and use directly or seed irradiated rat fibroblasts in DMEM–10% FCS at 5×10^3 cells per well into 96-well plates for use the next day.
2. Centrifuge cells from at least two wells of a 24-well plate that contain confluent layers of hybridoma cells. Count the number of cells and then dilute to give about 50 cells in 20 mL of HT or DMEM–10% FCS.
3. Carefully "flick off" the supernatant medium from the rat feeder cells and plate 0.2-mL aliquots of hybridoma cells into each of the 96 wells.
4. Examine the plates 5–10 d later and screen those wells that contain only single colonies.
5. Pick cells from positive wells into 24-well plates, expand, and freeze in liquid nitrogen.
6. Reclone the best antibody-producing colonies (*see* Section 3.5. *below*).

3.5. Assays for Specific Monoclonal Antibodies

In most of the assays described the detection of rat or mouse MAb depends on the use of a second antibody reagent specific for $F(ab')_2$ or heavy chain isotype. These second antibodies are detected because they have either a radiolabel (^{125}I), fluorescent tag (fluorescein), or are conjugated to an enzyme (e.g., alkaline phosphatase, peroxidase, or β-galactosidase) either directly or indirectly via a biotin–streptavidin bridge. As examples of these procedures, two alternative types of methodology, i.e., radioimmunoassay

(RIA) using ^{125}I-labeled antibodies and enzyme-linked immunosorbent assay (ELISA) using alkaline phosphatase conjugates, will be described. It is assumed that in most cases, the second antibodies will be bought either as purified material for radiolabeling or already conjugated to fluorescein, biotin, or the enzyme of choice (*see* ref. 2 for additional methods).

3.5.1. Radiolabeling with Iodine-125

All manipulations should be carried out in a Class I fume hood according to local safety regulations.

1. Add 50 µg of purified antibody in 0.1 mL of PBS to an IODO-GEN-coated tube followed by 500 µCi of ^{125}I (e.g., 5 µL of IMS–30, Amersham International). Cap and mix immediately by "flicking" and keep on ice with occasional shaking (*see* Note 10).
2. After 5 min, transfer the contents of the tube with a polythene, capillary-ended, Pasteur pipet, to a prepared Sephadex G–25 column, then wash in, and elute with, UB.
3. Collect 0.5-mL fractions by hand, count 10-µL aliquots, and pool the fractions containing the peak of radioactivity.
4. Store at 4°C in a lidded lead pot (e.g., iodine–125 container).

3.5.2. RIA Using Antigens Bound Directly to PVC Multiwell Plates

1. Add 50 µL of antibody containing culture supernatant, purified antibody, or ascites, diluted to 1–10 µg/mL in PBS–Marvel, to each of the antigen-coated wells.
2. Incubate for 1 h at ambient temperature, then wash three times with 200 µL of PBST/well.
3. Add 50 µL of ^{125}I-labeled second antibody (10^5cpm/50 µL in PBS–Marvel) to each well and incubate for a further 1 h at ambient temperature.
4. Carefully discard the radioactive supernatant by inverting the plate over a sink, designated for aqueous radioactive waste, then wash the wells three times with PBST.
5. Cut the plates into individual wells with scissors and determine the ^{125}I-bound in a γ-counter.

3.5.3. RIA Using Antigens Bound via Antibody to PVC Multiwell Plates

The assays are done as described in 3.5.2. above with the modification that, if the antigens require the presence of a nonionic detergent to retain their solubility or native conformation, the detergent should be added to the PBS–Marvel diluent/wash solution.

3.5.4. Assay Using Live Adherent Cells Grown in Multiwell PS-Plates

All solutions should be prepared in DMEM or other suitable growth medium containing 5% FCS—newborn calf serum (NBS), tested for nontoxicity, is a cheaper alternative. If the effect of antibody binding on the behavior of the membrane protein is unknown, then all incubations should be carried out at 4°C (float plates on an ice-bath, precool diluents). Monolayers of cells vary widely in adhesion to plastic, and also, they may round-up after prolonged incubation at 4°C owing to depolymerization of microtubules. *See* Note 6 regarding the use of fixed cells.

1. Wash cell monolayers with DMEM–5% FCS or NBS to remove nonadherent/dead cells, then proceed as described for antigens bound to PVC plates (Section 3.5.2. above), but using DMEM–5% FCS/NBS for all diluents and washings.
2. After the final wash, lyse the cells by incubating for 15 min at room temperature with 200 µL/well of alkaline sarkosyl.
3. Transfer lysates to LP2 tubes and determine the amount of ^{125}I present. Depending on the cells used, the background binding will vary from 50–200 cpm/well, whereas positive wells will be at least five times this value.

3.5.5. Competitive Assays

These are useful for mapping epitopes, determining whether or not two antibodies crossreact, and for comparing antibody affinities. They are the basis of quantitative assays (RIA) and use multiwell plates coated either with antigen or antibody, as described in Section 2.3. above. The principle is to either:

a. Compete the test antibody with a ^{125}I-labeled specific monoclonal (*see* Section 3.5.1. for preparation) or polyclonal antibody for binding to antigen bound to a plastic surface. Antibody capture can be used to secure the antigen to the plastic well, and this form of assay is useful where it is necessary to retain the antigen in its native conformation; or
b. Compete test antibody with ^{125}I-labeled antigen for binding to a specific antibody bound to a plastic surface.

1. Make doubling dilutions in PBS–Marvel of culture supernatant or of purified antibody starting at 20 µg/mL to give 50 µL final vol.
2. Add to each well 50 µL of ^{125}I-labeled specific antibody or antigen ($2–4 \times 10^4$ cpm/mL in PBS–Marvel).
3. Transfer 50-µL aliquots of the mixtures to the antibody or antigen-coated PVC multiwell plate so that each well contains $1–2 \times 10^4$ cpm of radiolabel.

4. After 1–4 h at ambient temperature, wash the plates three times with PBST and determine the ^{125}I bound.

Controls to determine maximum binding should be included, together with a standard curve prepared from dilutions of unlabeled specific antibody or antigen. Comparison of the inhibition curve produced with the test antibody with that of the control yields information on the crossreactivity of the two antibodies and their relative affinities for antigen. Expect to get a maximum binding of between 1–5 × 10^3 cpm/well, depending on the purity and quality of the labeled antigen or antibody. Hybridoma supernatants containing good competing antibodies (1–10 µg/mL) will reduce binding to the background (50–100 cpm/well).

3.5.6. Immunoprecipitation

This is an essential procedure for the isolation of antigens from complex sources (e.g., cells), and for their subsequent separation by electrophoresis in SDS-containing polyacrylamide gels and analysis by Western blotting.

1. Label cellular proteins metabolically by incubating cultures for 4–24 h with ^{35}S-methionine, ^{3}H-lysine, or ^{14}C-amino acids in medium deficient in the relevant unlabeled amino acid.
2. Wash the cells 3× with complete medium, then incubate for a further hour in the same medium.
3. Wash the cells in ice-cold PBSA containing $10^{-3}M$ PMSF, then lyse with the minimum vol (1–2 mL/25-cm^2 flask) of PBSA containing $10^{-3}M$ PMSF and 0.5% Triton-X 100 by incubating for 30 min on ice.
4. Transfer the lysate to a centrifuge tube and spin at 30,000*g* for 30 min to remove cell debris.
5. Prepare immunoabsorbent beads by linking 5–15 mg of purified MAb or polyclonal antibody to mouse or rat $F(ab')_2$ to Sepharose 4B (about 3 mL of swollen gel, *see* Section 3.6.3.). Alternatively, use Protein A-beads for mouse antibodies.
6. Incubate 1 mL of cell lysate (ca 3 × 10^6 cells) with either 10 µg of MAb or 100 µL (packed vol) of Ab-beads overnight at 4°C.
7. When soluble MAb is used, add 50 µL (packed vol) of Protein A-beads or anti-Ig beads and incubate for a further 1 h at 4°C.
8. Wash the beads three times with PBSA containing 0.5% Triton-X 100 and PMSF, pelleting the beads by centrifugation.
9. Elute the antigen (± MAb) by heating the beads (5 min, 95°C) with an equal vol of SDS-sample buffer and apply to an SDS-containing polyacrylamide gel (*see* Chapter 24 for details of this procedure). Run prestained markers on these gels because they will transfer to blots, and assist in determining the molecular size of the proteins.

3.5.7. Western Blotting

This procedure is particularly useful where an antibody recognizes an amino acid sequence or carbohydrate moiety, but may be unsuitable for antibodies that recognize a conformational epitope on the protein.

1. Follow the instructions of the suppliers of the blotting equipment for the electrophoretic transfer of the proteins from SDS-containing polyacrylamide gels to either nitrocellulose or PVDF membranes (*see* ref. 3 and this vol., Chapter 24 for detailed instructions).
2. Carefully separate the membrane from the polyacrylamide gel. Block the blot by placing it in a polythene bag, add 20 mL of PBS–Marvel ± 3% BSA/180 cm × 150 cm blot, then after sealing the bag, incubate for 2 h at 37°C on a rocking platform.
3. Wash the blot twice with PBST, then cut into strips when necessary (after labeling the individual strips in pencil), using the gel comb as a guide. Alternatively, use a proprietory manifold that allows staining of individual tracks without the need for cutting the blot into individual strips.
4. Place each blot strip in a suitably-sized polythene bag and add 1 mL of neat culture supernatant or purified antibody (10 µg/mL in PBS–Marvel) for every 20 cm^2 of membrane.
5. Seal the bags and incubate for 1 h at room temperature on a rocking platform.
6. Cut off one end of each bag, discard the contents, and wash the blot strips three times with PBST. Transfer the strips to "communal" bags and add (1 mL/20cm^2 blot) of ^{125}I-labeled antibodies to rat or mouse $F(ab')_2$ (10^6cpm/mL in PBS–Marvel containing 3% normal rabbit, sheep, or goat serum).
7. After rocking for 1 h at room temperature, open the bags, dispose of the radioactive supernatant safely, then wash the blots four times with PBST and dry at 37°C.
8. Secure the blot strips to a sheet of Whatman 3MM paper using Photomount™ or other spray-on adhesive, then autoradiograph at –70°C using prefogged X-ray film and an intensifying screen *(4)*.

3.5.8. Enzyme-Linked Immunosorbent Assays (ELISA)

As an alternative to the use of ^{125}I, most, if not all, of the procedures described in the preceding section can be done using antibodies conjugated to an enzyme that converts a substrate into a colored or fluorescent product (*see also* this vol., Chapter 10) that is either soluble (plate assays) or insoluble (Western blots) (*see* Note 11). The protocols for the use of either radiolabeled or enzyme-conjugated antibodies are the same until the completion of washing following treatment with the second antibody or, in the case of competitive assays, the specific antibody. At this time, the substrate for the enzyme is added, the samples are incubated at room temperature or 37°C for a suitable time, then the reaction is terminated. Plates coated with proteins, peptides, or conjugated peptides and those used in antibody capture assays can usually be read directly in a multiwell plate reader (e.g., Titertek multiscan). For this reason, it is better to use flat-bottom PS multiwell plates because of their better optical qualities. Alternatively, as with tests using cell monolayers, the supernatants can be transferred to a new plate for reading. The final steps for use with alkaline phosphatase conjugates are given below (*see* Note 9).

Plate assays:

1. Wash the plate once with PBS, then twice with 10 m*M* diethanolamine, pH 9.5, containing 500 m*M* NaCl.
2. Add 50 µL of *p*-nitrophenyl phosphate substrate solution to each well and incubate at room temperature for 10–30 min.
3. Stop the reaction by addition of 50 µL of 100 m*M* EDTA/well.
4. Read at 405 nm (positives are bright yellow).

Western blots:

1. Wash the blots twice with 100 m*M* diethanolamine, pH 9.5, containing 100 m*M* NaCl and 5 m*M* $MgCl_2$.
2. Add 1 mL of BCIP-NBT substrate per 20 cm^2 of membrane and incubate at room temperature on a rocking platform until the purplish-black bands/spots are suitably developed.
3. Rinse the membrane in PBS containing 20 m*M* EDTA to stop the reaction.

3.6. Isolation and Purification of Monoclonal Antibodies

Antibodies sufficiently pure for radiolabeling, conjugation to fluorescein, or preparation of immunoaffinity columns can be obtained from ascitic fluid or culture supernatant by simple procedures using salt fractionation followed by either ion exchange-chromatography or immunoaffinity chromatography on columns consisting of immobilized antigen, Protein A, or antibodies to mouse or rat immunoglobulins.

3.6.1. Salt Fractionation

This is useful as a first step in all procedures, since it gives a partial purification and reduces bulk.

1. Measure the volume of antibody containing ascitic fluid or culture supernatant into a beaker on a magnetic stirrer.
2. While stirring, slowly add the required volume of saturated $(NH_4)_2SO_4$ to give a final concentration of 45% saturation.
3. Stir for a further 30 min, then centrifuge at 12000*g* for 30 min.
4. Pour off the supernatant, drain, add deionized water (1/3 the volume of ascitic fluid or 1/100 the volume of culture supernatant), then recap the centrifuge tube and rotate samples at 4–6°C until the precipitate has dissolved (4–16 h).
5. Measure the volume and reprecipitate with enough saturated $(NH_4)_2SO_4$ to give 45% saturation.
6. Redissolve the precipitate in water and dialyze against PBS for immediate use or PB/10 (five changes of 10 vol) for fractionation by ion-exchange chromatography.

3.6.2. Ion-Exchange Chromatography

1. Run the dialyzed 45% $(NH_4)_2SO_4$ fraction onto a DE52 column equilibrated with PB/10 and elute with the same buffer. Most rat and mouse antibodies of the IgG_{2a} and IgG_{2b} subclasses do not bind and appear in the eluate.
2. Change the buffer to 0.055*M* phosphate, pH 6.6 to elute IgGs with lower isoelectric points.
3. Apply a gradient of 100% A to 100% B, where A is 0.055*M* phosphate, pH 6.6 and B is the same buffer containing 1*M* NaCl. Antibodies of the IgG_{2c} and IgG_1 subclasses usually elute under these conditions, but they will be contaminated with other serum components.
4. Locate the antibody fractions by absorbance at 280 nm and by determining binding to antigen.
5. Dialyze the purified fractions against PBSA or UB.
6. Test for purity by SDS-PAGE.

When analyzed by SDS-PAGE under nonreducing conditions, IgG antibodies should give a single protein band of about 160–170 kDa. On reduction with DTT, two or more bands will be seen corresponding to the individual heavy chains (45–50 kDa) or light chains (25–30 kDa). Where the myeloma parent produces light chains (e.g., the rat Y3), these can be produced also by the hybridoma, and may be distinguished from those of the lymphoid parent owing to differences in size. Other protein bands that are visible only on reduction may point to discrete attack on the individual chains by proteases. It may help to prevent such degradation by treating the culture supernatants with PMSF (make $10^{-3}M$) before storage or processing. If the preparations are contaminated with proteins of different molecular size to the intact IgG, try fractionation by size-exclusion chromatography using one of the proprietory gel permeation media.

IgM and IgA antibodies are not readily purified by ion-exchange chromatography. Gel filtration or affinity chromatography are best for these isotypes.

3.6.3. Affinity Chromatography

Where sufficient antigen is available, e.g., a synthetic peptide, the monoclonal antibodies can be purified easily and quickly by affinity chromatography. Alternatively, purified preparations of monoclonal (e.g., MARK-1, ref. 5) or polyclonal antibodies to rat or mouse $F(ab')_2$ can be used following their immobilization to crosslinked agarose or polyacrylamide bead supports.

1. Using CNBr-activated Sepharose 4B, link the antigen by following the manufacturer's instructions to give 5–10 mg of protein or peptide/mL of swollen gel.
2. Before use, wash the gel with 3*M* NH_4SCN (1 mL/mL of gel bed), then elute with UB until the A_{280} is down to background.
3. Run on the 45% $(NH_4)_2SO_4$ cut of ascites or culture supernatant that has been dialyzed against UB.
4. Wash with UB until the A_{280} is down to background.
5. Connect the outlet of the affinity column to the base inlet of a Sephadex G–75 column that has been preequilibrated with UB. Connect the G-75-column to the monitor and fraction collector.
6. Elute first with 3*M* NH_4SCN (1 mL/mL bed vol) then change to UB. The first absorbance peak will contain antibody, the second, NH_4SCN. By fractionating the eluates directly on the Sephadex G-75 column, the antibody is exposed for a minimum time to the chaotropic ion and the antibody will suffer less denaturation than might occur if dialysis alone were used.
7. Pool the tubes containing the antibody protein peak and dialyze against at least four changes of UB (*see* Note 10).
8. The column is ready for reuse if necessary (store in UB).

3.6.4. Protein A Chromatography

Mouse antibodies of the IgG_{2a} and IgC_{2b} subclasses bind well to Protein A, but antibodies of the IgG_1 subclass bind less well. Only rat antibodies of the IgG_{2c} subclass bind with any affinity to Protein A, and this is a useful procedure for their purification.

1. Equilibrate the column containing Protein A beads with PA buffer.
2. Run on the 45% $(NH_4)_2SO_4$ cut of ascites or culture supernatant that has been dialyzed against PA buffer.

3. Wash the beads with ten column volumes of PA buffer or until the A_{280} is down to background.
4. Connect the Protein A column to a Sephadex G-75 column and fraction collector. Monitor the absorbance at 280 nm.
5. Elute the Protein A beads with one column volume of either UB containing 1.5*M* NH_4SCN or with 100 m*M* glycine, pH 3.0, then continue elution with UB. The first absorbance peak will contain antibody. Alternatively, elute with 100 m*M* glycine, pH 3.0, stepwise and collect 0.5-mL samples into tubes containing 50 µL of 1*M* Tris, pH 8.0.
6. Dialyze the first A_{280} peak containing antibody against UB.

3.6.5. Isotyping of Monoclonal Antibodies

Antibodies can be separated into different classes (IgM, IgG, IgA, and IgE) and subclasses (IgG_1, IgG_{2a}, IgG_{2b}, and so on) on the basis of the structure of their heavy chains. This separation into different isotypes can be achieved using antisera directed against antigens on the heavy chains that are specific for the particular isotype. Members of an isotype share important properties, such as the ability to bind components of the complement system, or to Fc-receptors present on certain effector cells of the immune system.

1. Coat PVC plates with monoclonal or polyclonal antibodies (5 µg/mL of PBS, pH 8.0) specific for rat or mouse heavy-chain isotypes using the protocol described in Section 2.3.25.
2. Block with PBS–Marvel.
3. Add 50 µL of test MAb to each well at 1 µg/mL in PBS–Marvel or as neat supernatant and incubate 1–4 h at room temperature or overnight at 4°C.
4. Wash three times with PBS–Marvel.
5. Add 50 µL per well (10^5 cpm in PBS–Marvel) of ^{125}I-labeled antibodies to rat or mouse $F(ab')_2$ and incubate 1 h at room temperature.
6. Wash three times with PBST and determine the amount of radioactivity-bound. Include MAbs of known isotypes as controls (*see* Note 12).

4. Notes

1. The use of animals for experimental purposes is under strict control in many countries, and licences are necessary before surgical procedures can be performed.
2. For growth of rat hybridomas as ascites, we recommend the use of nude rats since these have low levels of endogenous immunoglobulins (<1 mg/mL) and can yield up to 30 mL of ascitic fluid/rat (1–5 mg/mL

specific antibody).

3. Problems may be encountered when using β-galactosidase fusion proteins (noncleaved) as immunogens, because the response is directed predominantly against the bacterial enzyme. Also, with transfected cell lines, it may be desirable to eliminate responses against cell components other than the recombinant antigen. A protocol has been described *(6)* in which animals were stimulated first with unwanted antigen (e.g., normal 3T3-cells), then the responding lymphocytes were killed by treatment of the animals with cyclophosphamide. Subsequent immunization of these animals with the desired antigen led to successful MAb production.
4. Some batches of FCS are toxic to hybridomas, and it is important to test all new batches for their ability to support hybridomal growth. Alternatively, obtain samples of FCS that have been tested by the supplier.
5. Feeders are essential when the Y3 myeloma is the fusion partner. We use cell lines derived by trypsinization of the xyphoid cartilage that terminates the xiphisternum of adult rats. Chop the cartilages from 6–8 adult rats into 2–3 mm pieces with a scalpel, then transfer into 15 mL of DMEM containing 0.5% trypsin (bovine pancreas type III) and 1% collagenase (type II) and stir for 45 min at room temperature. Add FCS to 10% and filter through sterile gauze to remove debris. Wash the cells in DMEM–10% FCS and plate into the same medium. Passage the cells in DMEM–10% FCS after removing cells by incubation for 2–3 min in PBS–0.05% Na_2EDTA containing 0.2% trypsin. Store cells in liquid nitrogen as aliquots of 10^6 cells in 5% DMSO–95% FCS.
6. Cell monolayers fixed with glutaraldehyde or paraformaldehyde can overcome the problem of loss of cells from the wells, but the effect of fixative on the binding of antigen should be determined. Wash cells in PBS, then add to each well 50 μL of freshly prepared 0.5% glutaraldehyde in PBS, containing 2 m*M* $CaCl_2$, 2 m*M* $MgCl_2$, and 300 m*M* sucrose. After incubating on ice for 30 min, "flick off" the fixative and replace with 200 μL of PBS containing 0.1*M* glycine to block excess glutaraldehyde, leave for 30 min, then wash twice with PBS–Marvel. Fixed cells can be handled in the same way as antigens bound to PVC plates.
7. IODO-GEN (1,3,4,6-Tetrachloro-3-6-diphenylglycouril) is a better reagent for the iodination of proteins than chloramine-T and is less damaging and has fewer side reactions than the latter. The insolubility of IODO-GEN in water means that tubes can be precoated with the reagent dissolved in methylene chloride or chloroform, then the tubes are stored in the dark until required. The reaction is started by adding the protein

and radioiodide, and terminated by removing the sample from the reaction vessel.

8. Pristane (2,6,10,14-tetramethylenedecanoic acid) acts as an irritant, and as a result, macrophages and monocytes are recruited into the peritoneal cavity. The nutrients secreted by these cells provide a good environment for the growth of hybridomas in suspension.
9. When unfixed cells or frozen sections are used for ELISAs, it may be necessary to block endogenous alkaline phosphatase activity, and in this case, include 0.1 m*M* levamisole in the substrate solution.
10. Iodination reactions employing either chloramine-T or Iodogen are particularly sensitive to the presence of thiocyanate ions, and if antibodies are required for this purpose, it is essential that they are dialyzed thoroughly after elution from affinity columns with NH_4SCN.
11. The use of biotinylated antibodies provides perhaps the greatest versatility and sensitivity of all methods. The affinity of biotin for avidin or the more usually used streptavidin is very high, and the latter can be conjugated to radioisotope, fluorescent moiety, or enzyme. Again, the basic procedures are the same as outlined for ^{125}I-labeled antibodies with the additional steps required for streptavidin binding and subsequent incubation with enzyme substrate (*see also* this vol., Chapters 13–15).
12. Very occasionally, a MAb will show reactivity with more than one anti-isotypic antibody in this capture assay. In this case, try immunoprecipitation in agarose gels (Ouchterlony procedure, *see* ref. 7) using polyclonal reagents. Because a crosslinked lattice must be formed between several different epitopes on the antigen and several antibodies to give an immunoprecipitate, the Ouchterlony procedure gives few false positives.

References

1. Dean, C. J., Gyure, L. A., Hall, J. G., and Styles, J. M. (1986) Production of IgA-secreting rat × rat hybridomas. *Methods Enzymol.* **121,** 52–59.
2. Harlow, E. and Lane, D. P. (1988) *Antibodies, A Laboratory Manual.* Cold Spring Harbor Laboratory, Cold Spring Harbor, NY.
3. Towbin, H., Staehelin, T., and Gordon, J. (1979) Electrophoretic transfer of proteins from polyacrylamide gels to nitrocellulose sheets: Procedure and some applications. *Proc. Natl. Acad. Sci.* **76,** 4350–4354.
4. Laskey, R. A. (1980) The use of intensifying screens or organic scintillators for visualizing radioactive molecules resolved by gel electrophoresis. *Methods Enzymol.* **65,** 363–371.
5. Bazin, H., Xhurdebise, L. M., Burtonboy, G., Lebacq, A. M., De Clercq, L., and

Cormont, F. (1984) Rat monoclonal antibodies. I. Rapid purification from *in vitro* culture supernatants. *J. Immunol. Methods* **66,** 261–269.
6. McKenzie, S. J., Marks, P. J., Lam, T., Morgan, J., Panicali, D. L., Trimpe, K. L., and Carney, W. P. (1989) Production and characterization of monoclonal antibodies specific for the human *neu* oncogene product, p185. *Oncogene* **4,** 543–548.
7. Ouchterlony, O. and Nilsson, L. A. (1986) Immunodiffusion and immunoelectrophoresis, in *Handbook of Experimental Immunology*, 4th Ed. (Weir, D. M., ed.), Blackwell Scientific, Oxford, pp. 32.1–32.50.

CHAPTER 7

Screening of Monoclonal Antibodies Using Antigens Labeled with Acetylcholinesterase

Yveline Frobert and Jacques Grassi

1. Introduction

The production of large quantities of monoclonal antibodies (MAbs) of predetermined specificity has been rendered possible by the pioneering work of Köhler and Milstein *(1)*. These workers have shown that lymphocytes can be immortalized and subsequently cultured after somatic fusion with genetically selected myeloma cells. Usually, once fusion between spleen cells and myeloma cells has been performed, cells are suspended in a large volume of selective medium and distributed in culture wells, so that hybridomas are brought to clonal dilution. If fusion is successful, the first hybridoma colonies will be detectable within a few days (5–15 d). As fusion is a random process, most clones code for MAbs of unknown specificity, characterizing the immunological past of the host. It is then necessary to select the different colonies that secrete MAbs of the desired specificity. Owing to the great number of wells to be tested (often a few hundred), and to the small quantities of MAbs available (at best, 300 µL at a few µg/mL), it is not easy at this stage to characterize the fine specificity of the antibodies (i.e., recognition of a precise epitope, inhibitory effect on a biological system, properties suitable for purifying antigen, or for histochemical characterization, and so on). Initially, it is generally preferable to use a simple method to select all the hybridomas producing MAbs directed against the immunizing antigen. Further characterization of these MAbs is performed later, after expansion of the clones. Of the different

From: *Methods in Molecular Biology, Vol. 10: Immunochemical Protocols*
Ed.: M. Manson

steps involved in the production of MAbs, this "screening" step is certainly one of the most critical, since this is the moment when the good clones are kept or lost. Screening is preferably performed on all the culture supernatants regardless of whether hybridoma colonies are visible or not. This obviates the need for time-consuming microscopic examination of each well and limits the risk of losing a slowly growing clone.

An ideal screening method would include the following characteristics: (i) sufficient sensitivity to allow detection of antibodies below the µg/mL range; (ii) high specificity, resulting in strict and reliable characterization of clones secreting the expected antibodies, avoiding both underestimation and false positives; and (iii) simple methodology, since primary screening and cloning involve many assays (up to 4000 for the entire procedure, including recloning and final expansion of the clones).

Most of the methods described in the literature *(2–4)* are based on the use of solid-phase immobilized antigens. Specific antibodies in culture supernatants are bound to the solid phase coated with the specific antigen, and then detected by a reaction with a labeled reagent specific to mouse immunoglobulins (antimouse second-antibody or protein A from *Staphylococcus aureus*). One advantage of this approach is the use of a single-labeled reagent, regardless of the specificity of the MAbs. On the other hand, an appropriate solid phase must be developed for each antigen. This requires large amounts of purified antigen, which might be problematic in some cases. In addition, this approach may lead to the selection of falsely positive clones if one particular MAb is nonspecifically adsorbed on the solid phase.

We have developed a contrasting strategy based on the use of a solid phase coated with a universal reagent of mouse immunoglobulins (i.e., a second antibody). MAbs bound to this solid phase are detected by their ability to bind the labeled antigen specifically (*see* Fig. 1). This approach appears more specific in that nonspecific adsorption of unrelated MAbs on the solid phase will not be detected. In this method, the solid phase represents a universal reagent, but the appropriate antigen must be labeled for each particular screening.

The aim of this paper is to describe simple methods allowing the labeling of small quantities of antigen (or hapten) with the enzyme acetylcholinesterase (AChE). We show that this can be achieved in every case by using either covalent labeling methods or noncovalent coupling through avidin–biotin interactions. Depending on the labeling method used, three different screening procedures are described (*see* Fig. 1).

Acetylcholinesterase was chosen as the labeling enzyme because it presents many advantageous properties. First, AChE can be assayed at extremely low levels because of its very high turnover (amol amounts of the enzyme can

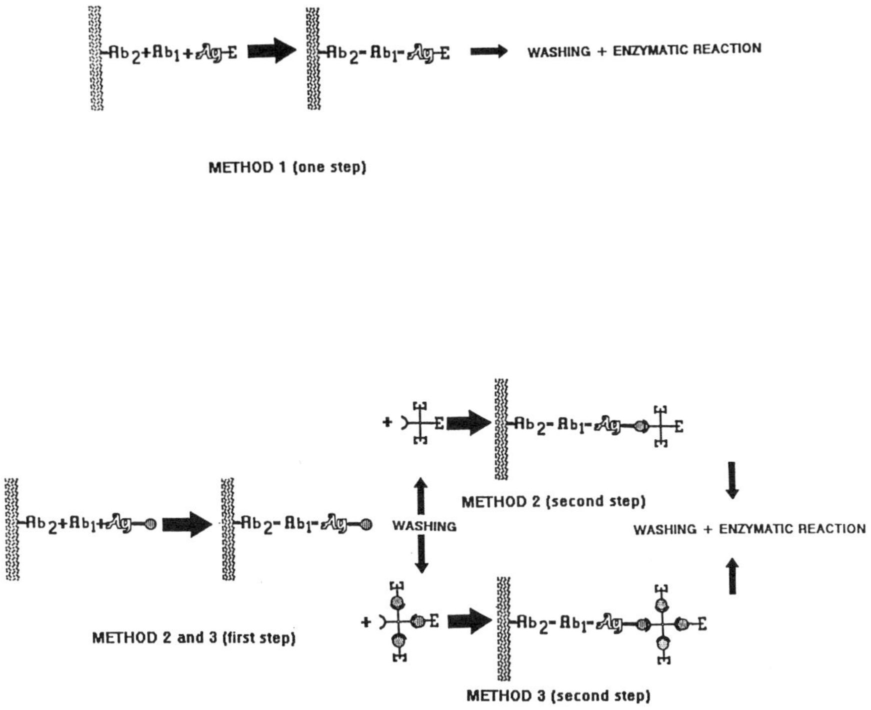

Fig. 1. Schematic representation of screening procedures. Monoclonal antibodies (MAbs) contained in culture supernatants are bound to a solid phase coated with a second antibody. The presence of MAbs on the solid phase is revealed by their ability to bind antigen labeled with the enzyme acetylcholinesterase (AChE). Three different methods for labeling antigens with AChE are described in this paper. Depending on the labeling method, three different screening procedures are used. METHOD 1: This is a one-step procedure using an antigen–AChE covalent conjugate as tracer. After completion of immunoreaction between MAb and tracer, the solid phase is extensively washed and the enzymatic activity immobilized on the solid phase is measured, using a colorimetric assay (Ellman's method). METHODS 2 and 3: These are two-step procedures using a biotinylated antigen (biot-antigen) as tracer. In the first step the biot-antigen is reacted with MAbs. After washing, biot-antigen immobilized on the solid phase is revealed by reaction with either an avidin–AChE conjugate (METHOD 2) or an avidin/biot-AChE complex (METHOD 3). Washing and enzymatic steps are performed as for METHOD 1.

be detected using a one-step, 1-h colorimetric assay *(5)*. This detection limit is lower than that observed for ^{125}iodine and other enzymes currently used in immunoassays (i.e., horseradish peroxidase, β-galactosidase, or alkaline phosphatase *(5)*. In addition, AChE provides a continuous signal for many hours (at least 24 h under the usual conditions), thus offering the possibility of increasing the sensitivity of the test by allowing the enzymatic reaction to proceed for a longer period. Finally, it has been shown that AChE can be used efficiently for labeling various molecules, including haptens *(6–9)*, antigens *(10,11)*, or antibodies *(11)*, coupling with the enzyme always being achieved without significant loss of enzyme activity. The corresponding conjugates are stable for years when kept frozen (–20°C or –80°C) or lyophilized (+4°C) under suitable conditions. Screening of MAbs using AChE-labeled antigens has already been described *(12)*.

2. Materials

1. Acetylcholinesterase (AChE): This is purified from the electric organs of the electric eel, *Electrophorus electricus*, using one-step affinity chromatography *(13)*. The resulting preparation has a specific activity of 10^5 Ellman U/mg of protein and shows the electrophoretic pattern characteristic of the pure enzyme. This preparation contains mainly asymmetric molecular forms of the enzyme (A_8 and A_{12} forms) and a small proportion of the G_4 globular form. Details of these molecular forms of AChE are reviewed in Massoulié and Toutant *(14)*. The major fraction of asymmetric forms is converted into the tetrameric form (G_4) by reaction with trypsin for 18 h at 25°C in 0.1*M* phosphate buffer, pH 7.0 (enzyme/protein ratio (w/w) 1/2000). This mixture of AChE forms is used for labeling without further purification (*see* Notes 1 and 2). For the measurement of AChE activity, *see* Note 3.
2. *N*-succinimidyl-4-(*N*-maleimidomethyl)-cyclohexane-1-carboxylate (SMCC) and *S*-succinimidyl-*S*-acetyl-thioacetate (SATA) are marketed by various companies.
3. *N*-hydroxysuccinimide ester of biotin (NHS-biotin).
4. Egg avidin.
5. Biogel A 0.5 m, A 1.5 m, and A 15 m are used for molecular sieve fractionation.
6. G25 gel.
7. 0.05*M* potassium phosphate buffer, pH 7.4
8. EIA Buffer: 0.1*M* Potassium phosphate buffer, pH 7.4, containing 0.1% BSA, 0.4*M* NaCl, 0.001*M* EDTA, and 0.01% sodium azide (toxic).
9. Washing buffer: 0.01*M* Potassium phosphate buffer, pH 7.4, containing 0.05% Tween-20.

10. 0.1 *M* Sodium phosphate buffer, pH 6, containing 0.005 *M* EDTA.
11. Polyspecific antimouse Ig antibodies are available from different commercial sources (use only affinity-purified quality).
12. 96-well microtiter plates (Immunonunc I 96F, NUNC, Denmark).
13. Anhydrous dimethylformamide (DMF).
14. 0.1 *M* Sodium borate buffer, pH 9.
15. 1 *M* Tris-HCl buffer, pH 7.4.
16. *N*-ethyl-maleimide 12.5 mg/mL (0.1 *M*) in 0.1 *M* potassium phosphate buffer, pH 7.4.
17. 1 *M* Hydroxylamine in 0.1 *M* potassium phosphate buffer, pH 7.
18. 0.001 *M* 5-5'-dithio-*bis*-nitrobenzoate (DTNB) in 0.1 *M* potassium phosphate buffer, pH 7.
19. Multidispenser (8 or 12 pipets)
20. Ellman medium: 7.5×10^{-4} *M* Acetylthiocholine and 5×10^{-4} *M* DTNB in 0.1 *M* phosphate buffer, pH 7.

3. Methods

3.1. Immobilization of Antimouse Immunoglobulins on the Solid Phase

The principle of the screening test implies that the antimouse immunoglobulin solid phase has the capacity to recognize all mouse immunoglobulin classes and subclasses (*see* Note 4).

1. Fill the wells of microtiter plates with 200 µL of a 10 µg/mL solution of affinity-purified second antibodies in 0.05 *M* phosphate buffer, pH 7.4.
2. Leave overnight at room temperature, then wash the plates extensively with washing buffer.
3. Saturate the solid phase by adding 300 µL of EIA buffer in each well and store the plates at +4°C. They can be used 8 h later or kept under these conditions at +4°C for at least 6 mo.

3.2. Labeling of Antigens with Acetylcholinesterase

Two methods for labeling antigens with AChE are presented.

3.2.1. Labeling of Antigens Through Avidin–Biotin Interactions

In this method, antigen and AChE (Fig. 1, METHOD 3) are covalently labeled with biotin molecules. AChE and antigen are coupled during the course of the screening test via high-affinity avidin interactions (*see* principle in Fig. 1, METHODS 2 and 3). Biotin is covalently linked to AChE or antigen by reaction of an activated *N*-hydrosuccinimide ester of biotin (NHS-biotin) with the primary amino groups of the proteins. NHS-biotin is quickly hydrolyzed in presence of water and is then dissolved in anhydrous

DMF a few minutes before use. Ideally, the reaction is performed in basic medium (borate buffer, pH 9), but can be realized at reduced efficiency in neutral medium (phosphate buffer, pH 7). (*See* further recommendations in Note 5.)

3.2.1.1. Labeling of Antigens with Biotin

1. To 100 µg of antigen dissolved in 200 µL of 0.1*M* borate buffer, pH 9 (10 nmol of a 10-kDa antigen), add 17 µL of a 1 mg/mL solution of NHS-biotin in anhydrous DMF (50 nmol). Allow the reaction to proceed for 30 min at room temperature.
2. Stop the reaction by adding 100 µL of 1*M* Tris-HCl buffer, pH 7.4.
3. After a further 15-min reaction period, dilute biotinylated antigen (biot-Ag) in EIA buffer up to a concentration of 100 to 10 µg/mL. Store concentrated biotinylated antigen at –20°C.

3.2.1.2. Labeling of AChE with Biotin

1. To 500 µL of a 250 µg/mL solution of AChE (31.000 Ell.U/mL, 0.4 nmol of G_4-AChE) in 0.1*M* borate buffer, pH 9, add 5 µL of a 8.5 mg/mL solution of NHS-biotin in anhydrous DMF (125 nmol). Allow the reaction to proceed for 30 min at room temperature.
2. Add 100 µL of 1*M* Tris-HCl buffer.
3. Add 1 mL of EIA buffer 15 min later.
4. Then purify biotinylated AChE (biot-AChE) on a Biogel A 0.5 m column (1.5 × 90 cm) equilibrated in EIA buffer. The enzymatic profile is established using Ellman's method, and fractions corresponding to the G_4 form (major peak) are pooled. Store concentrated biot-AChE at –20°C.

3.2.2. Covalent Labeling of Antigens with AChE

Antigens can be covalently coupled to AChE using the heterobifunctional reagent SMCC. This method involves the reaction of thiol groups (previously introduced into antigen) with maleimido groups incorporated into AChE after reaction with SMCC. This method has been applied successfully to the labeling of various haptens *(7,8)* and antigens *(10,11)*.

3.2.2.1. Introducing Maleimido Groups into AChE

Maleimido groups are introduced into AChE by reaction of their primary amino groups with the *N*-hydroxy-succinimide moiety of SMCC in neutral medium. This reaction is very similar to that used for incorporation of biotin molecules into proteins. Prior to this reaction, the enzyme is treated with *N*-ethyl maleimide in order to block any thiol groups. Finally, SMCC-AChE is purified by molecular sieve chromatography on a Biogel A 1.5 m column.

1. To 100 µL of a 2 mg/mL solution of G_4-AChE (0.62 nmol), add 10 µL of a 12.5 mg/mL solution of *N*-ethyl-maleimide in phosphate buffer, pH 7. Allow the reaction to proceed for 30 min at room temperature.
2. Add 10 µL of a 15 mg/mL solution of SMCC (449 nmol) in anhydrous DMF.
3. After a further 30-min reaction period at 30°C, chromatograph the mixture on a Biogel A 1.5 m column (1 × 30 cm) equilibrated in phosphate buffer, pH 6, containing EDTA.
4. Collect fractions of about 1 mL and determine their enzymatic activity using the Ellman method (*see* Note 3). Pool the fractions corresponding to the G_4 form (second major peak) and determine the AChE concentration of the pool enzymatically. G_4-SMCC can be used immediately. It can also be kept frozen at –80°C for at least six months.

3.2.2.2. Introducing Thiol Groups into Antigens

Thiol groups are introduced into antigens by reaction of their primary amino groups with SATA. This is the same kind of reaction as that used for incorporation of biotin molecules into antigens or of SMCC into AChE and requires the same precautions. The thioester function of SATA is then hydrolyzed in the presence of hydroxylamine in order to reveal the thiol groups. Excess reagents (SATA, hydroxylamine) are finally eliminated by molecular sieve chromatography. As an example, we shall describe the procedure used for the thiolation of egg-avidin (preparation of an avidin–AChE conjugate).

1. To 500 µg of avidin (8.7 nmol) dissolved in 250 µL of 0.1*M* borate buffer, pH 9, add 14 µL of a 2.5 mg/mL solution of SATA (150 nmol) in anhydrous DMF. Allow to react for 30 min at room temperature.
2. Add 250 µL of a 1*M* solution of hydroxylamine in 0.1*M* phosphate buffer, pH 7.
3. After a further 30-min reaction period, chromatograph the mixture on a G25 column (1 × 10 cm). Elute with phosphate buffer, pH 6, containing EDTA. Before and during chromatography, the eluant is kept under a continuous stream of nitrogen to eliminate dissolved oygen and avoid any oxidation of thiol groups. Collect fractions of about 1 mL and pool the fractions corresponding to the void volume (identified by measuring the absorbance at 280 nm or at 412 nm after reaction of thiol groups with DTNB (*see below*).
4. Measure the thiol content of the pool by mixing 100 µL of the solution with 100 µL of a 0.001*M* solution of DTNB in 0.1*M* phosphate buffer, pH 7 (reaction performed in wells of a microtiter plate).

5. After 5 min, read the absorbance at 412 nm and calculate the thiol concentration according the following equation:

$$[SH] = \frac{A \times 2}{\varepsilon \times 1}$$

where (A) is the measured absorbance, (ε) is the molar extinction coefficient of reduced DTNB (13600), and (l) the optical pathlength (0.439 cm if a microtiter plate is used).

Ideally, thiolated avidin will be coupled immediately to G_4-SMCC, although it can be kept for a few weeks at –80°C. If thawed, check its SH content as described above.

3.2.2.3. Coupling Thiolated Antigen with Maleimido-AChE

Coupling of AChE with the antigen is achieved by mixing G_4-SMCC (immediately after isolation by molecular sieve chromatography or from storage at –80°C) with an excess of thiolated antigen. We generally use a molar ratio (moles of SH/moles of G_4-SMCC) ranging from 10 to 50. Depending on the efficiency of the coupling reaction, the ideal ratio can be determined empirically.

In the case of avidin, we currently use a ratio of 50. In one typical experiment, this corresponded to a mixture of 395 µL of thiolated antigen (measured as containing $6.3 \times 10^{-5} M$ of SH groups) with 755 µL of a 26200 Ell.U/mL solution of G_4-SMCC.

1. Perform the coupling reaction at 30°C for 3 h.
2. Purify the enzymatic conjugate on a Biogel A 0.5 m column (Bio-Rad,USA, 1.5 × 90 cm) equilibrated in EIA buffer.
3. Collect fractions of about 2 mL, pool those containing AChE activity, and store the conjugate (at –20°C or –80°C) for months or years, depending on the stability of the antigen.

3.3. Performing the Screening

The screening procedure differs slightly according to whether the antigen is labeled with biotin or directly coupled to AChE (*see* Fig. 1). The use of biotinylated antigen requires one more step. In both cases, the first step consists of transferring an aliquot of the culture supernatant to second antibody-coated plates. This operation is straightforward since culture plates and coated plates are geometrically equivalent and the transfer can be performed using a multidispenser (8 or 12 pipets). Under these conditions, up to 2000 supernatants can be transferred in 1 h by a single experimenter. Usually, 50 µL of culture supernatants are used for the screening test.

3.3.1. Screening Using Covalent Antigen–AChE Conjugates (Method 1)

1. Transfer 50 µL of culture supernatants to second antibody-coated plates.
2. Add to each well 50 µL of the antigen–AChE conjugate previously diluted in EIA buffer. This tracer is generally used at a concentration of 2–10 Ell.U/mL. Allow the reaction to proceed overnight at +4°C.
3. Wash the plates extensively, using washing buffer. We recommend at least three washing cycles, including 3-min soaking for each step.
4. Add 200 µL of Ellman medium to each well. Allow the enzymatic reaction to proceed at room temperature.
5. Read the plates when a strong yellow color appears in any well. This usually occurs within 30 min or 1 h, but longer enzymatic reaction periods (up to 6 h) can be used.

3.3.2. Screening Using Biotin-Labeled Antigens (Methods 2 and 3)

1. After the transfer of culture supernatants, add 50 µL of biotin-antigen (0.1–1 µg/mL diluted in EIA buffer) to each well.
2. After overnight reaction at +4°C, wash the plates as described above (*see* Section 3.3.1.).
3. At this point, the presence of biotinylated antigen on the solid phase can be revealed in two ways:
   a. Method 2: Add 100 µL of avidin–AChE conjugate (2–5 Ell.U/mL diluted in EIA buffer), or
   b. Method 3: Add 100 µL of a prereacted mixture of avidin and biot-AChE. This mixture is obtained as follows (quantity for 100 tests): add 40 µL of a 15 µg/mL solution of avidin (in EIA buffer) to 80 µL of a 300 Ell.U/mL solution of biot-AChE (in EIA buffer) and allow to react for 30 min at room temperature.Then dilute the mixture with 10 mL of EIA buffer before adding 100 µL to each well.
4. After a further 2–4 h of reaction at room temperature with stirring, wash the plates, develop, and read as described above. For an example, *see* Note 6.

4. Notes

1. All the procedures described in this chapter have been especially optimized in terms of AChE labeling. We recommend the use of this enzyme that we feel is particularly well suited to this application.

However, the same basic principles can be applied (with appropriate modifications) to the labeling of antigens with other enzymes (peroxidase, alkaline phosphatase, β-galactosidase). This is particularly true for those labeling methods involving the use of biotinylated antigens, since biotinylated enzymes as well as avidin–enzyme conjugates are available from numerous commercial sources.

2. It is not advisable to use AChE from commercial sources (Sigma, USA or Boehringer, RFA), since these preparations are largely impure and possess a specific activity about tenfold lower than our preparations. Pure AChE can be obtained on request from our laboratory.
3. AChE activity is measured using the method of Ellman et al. (15); final concentrations of acetylthiocholine and DTNB are $7.5 \times 10^{-4} M$ and $5 \times 10^{-4} M$, respectively. One Ellman unit (Ell.U) is defined as the amount of enzyme producing an absorbance increase of 1 absorbance unit (AU) in 1 min, in 1 mL of medium and for an optical pathlength of 1 cm. It corresponds to about 8 ng of enzyme. AChE concentrations are determined enzymatically using a turnover number of 4.4×10^7 mol/h/site *(16)* and a molecular mass of 80 kDa for the catalytic subunit (G_4 = 320 kDa). According to these values, a detection limit of 1.8 amol of enzyme may be calculated for the G_4 form [i.e., the quantity of AChE producing an absorbance increase of 0.01 AU in 1 h, in 200 µL of Ellman medium, 0.44 cm pathlength *(5)*].
4. Polyspecific antimouse immunoglobulin antibodies may be obtained by immunizing rabbits with an appropriate mixture of MAbs. Ideally, this immunizing preparation should contain equivalent amounts of IgG_1, IgG_{2a}, IgG_{2b}, IgG_3, and IgM, some of these antibodies bearing lambda light chain. Primary immunization and booster injections are performed as described by Vaitukatis *(17)* in the presence of Freund's complete adjuvant. A dose of 50 µg is used for each injection. The first booster injection is given six weeks after immunization, and the rabbits are bled weekly from this date. Booster injections can be given every two months. The presence of antimouse Ig directed against the different classes and suclasses can be checked using conventional immunological techniques *(12)*. Specific antibodies from pooled sera will be further purified by affinity chromatography in order to prepare a solid phase with maximum binding capacity. Purification can be performed using an affinity column composed of mouse immunoglobulins (either polyclonal immunoglobulins or a mixture of MAbs) covalently linked to a solid matrix. In a previous paper, we described a method based on the use of polyclonal immunoglobulins immobilized on cyanogen bromide-activated Sepharose 4B *(12)*.

5. For the labeling of antigens with biotin, it is essential to avoid the presence of other compounds containing primary amino groups (Tris-HCl buffer for instance), since they interfere with the labeling reaction. Interfering substances may be eliminated by dialyzing the antigen preparation against borate buffer prior to labeling. In order to preserve the immunoreactive properties of the antigen, it is advisable to introduce only a few biotin molecules into the antigen. This can be achieved by using low NHS-biotin to antigen ratios. We usually recommend a molar ratio of 5 mol of NHS-biotin for each 10 kDa of antigen (e.g., 15 mol of NHS-biotin/1 mol of a 30-kDa antigen). This ratio can be modified within the range of 100 to 1 if, for any reason, the efficiency of the reaction is particularly low (poorly concentrated antigen, neutral pH) or high (highly concentrated antigen, primary amine enriched antigen).
6. As an example, we present the results of screening tests performed for detecting anti-interleukin 1β (IL1β) MAbs. The three methods described above have been tested, i.e., the use of an IL1β–AChE conjugate (Method 1), the use of biotinylated IL1β associated with either an avidin–AChE conjugate (Method 2), or a complex of avidin and biot-AChE (Method 3). Biot-IL1β was prepared using a NHS-biotin/IL1β ratio of 10 and was used at a concentration of 1 μg/mL. The preparation of IL1β–AChE conjugate has been described elsewhere *(11)*.

   IL1β–AChE and avidin–AChE conjugates were used at a concentration of 2 Ell.U/mL and 4.5 Ell.U/mL, respectively. The enzymatic reaction was allowed to proceed for 13 min in all three methods. The 96 wells of one entire plate were tested. Fifteen wells appeared clearly positive regardless of the method used. The signal observed with each method for these 15 wells is represented in Fig. 2. For each method, the "blank" was estimated by averaging the signal measured for the other 81 wells (*see* corresponding values in the legend of Fig. 2). These results show that Methods 1 and 2 provide an absolute signal significantly superior to that of Method 3. In terms of specific signals (signal/blank ratio), Method 1 is better than Method 2 because of lower nonspecific binding. In addition, from a practical point of view, the use of a covalent conjugate appears advantageous in that it requires one step less, thus simplifying the screening. The hierarchy observed in this test (Method 1 > Method 2 > Method 3) is representative of our general experience. It is worth noting, however, that the three methods allow correct selection of positive clones, and that methods using biotinylated antigens are more easily accessible to the experimenter because of a greatly simplified labeling procedure. The weaker signal observed with Method 3 cannot be regarded as a major drawback, since taking advantage of the enzy-

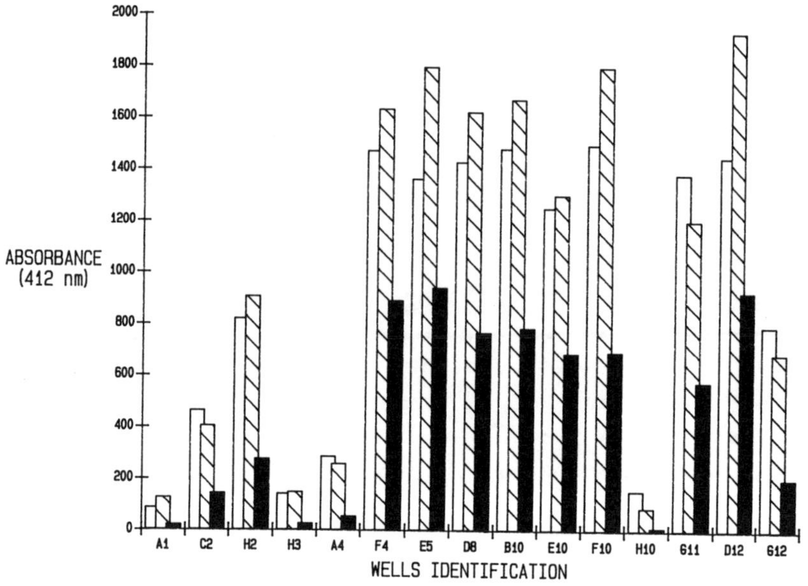

Fig. 2: Comparative results of screening tests performed with the three methods. The 96 culture supernatants of one entire plate were tested using the three screening procedures described in Fig. 1. Fifteen wells appeared clearly positive regardless of the method used. The signal observed with each method for these 15 wells is represented in this figure. For each method, the "blank" was estimated by averaging the signal measured for the other 81 wells and expressed in terms of mean ± 1σ *(see below)*. □, Method 1, blank = 12.5 ± 19 mUA; ▨ Method 2, blank = 37 ± 19 mUA; ■ Method 3, blank = 4 ± 4 mUA.

matic properties of AChE, a stronger signal, can be obtained using a longer reaction time (up to 6 h).

7. Maximum efficiency of the screening methods described in this paper will be obtained by optimizing the different steps of the labeling procedure and of the screening test itself. This includes:
   a. Optimization of the NHS-biot/antigen ratio for the biotin labeling of antigens.
   b. Optimization of the SATA/antigen ratio and of the thiolated-antigen/AChE-SMCC ratio for the preparation of covalent conjugates.
   c. Determination of the optimal concentration of the different reactants used in the screening test (biot-antigen, antigen–AChE conjugate, G4-avidin, and so on).

These prefusion operations can be performed using the sera of immunized mice. In addition, the availability of an optimized method will allow the experimenter to control the effectiveness of immunization and to select for fusion the mouse presenting the most marked immune response.

8. When optimized, the screening test can detect very low concentrations of antibodies (up to a few ng/mL *[12]*). This could lead to the detection of antibodies secreted by lymphocytes still present in the culture medium during the few days following fusion. We thus recommend that the screening test be performed at least 10 d after fusion and that the culture medium be changed at least twice during this period.

References

1. Köhler, G. and Milstein, C. (1975) Continuous cultures of fused cells secreting antibody of predefined specificity *Nature* **256,** 495–497.
2. Lane, D. P. and Lane, E. B. (1981) A rapid antibody assay system for screening hybridoma cultures. *J. Immunol. Methods* **47,** 303–307.
3. Huet, J., Phalente, L., Buttin, G., Sentenac, A., and Fromageot, P. (1982) Probing yeast RNA polymerase A subunits with monospecific antibodies. *EMBO J.* **1,** 1193–1198.
4. Hawkes, R., Niday, E., and Gordon, J. (1982) A dot immunoblotting assay for monoclonal and other antibodies. *Anal. Biochem.* **119,** 142–147.
5. Grassi, J., Maclouf, J., and Pradelles, P. (1987) Radioiodination and other labeling techniques, in *Handbook of Experimental Pharmacology,* vol. 82 (Patrono, C. and Peskar, B. A., eds.), Springer-Verlag, Berlin, pp. 91–141.
6. Pradelles, P., Grassi, J., and Maclouf, J. (1985) Enzyme immunoassay of eicosanoids using acetylcholinesterase as label. *Anal. Chem.* **57,** 1170–1173.
7. McLaughlin, L., Wei, Y., Stockmann, P. T., Leahy, K. M., Needleman, P., Grassi, J., and Pradelles, P. (1987) Development, validation and application of an enzyme immunoassay of atriopeptin. *Biochem. Biophys. Res. Commun.* **144 (1),** 469–476.
8. Renzi, D., Couraud, J. Y., Frobert, Y., Nevers, M. C., Gepetti, P., Pradelles, P., and Grassi, J. (1987) Enzyme immunoassay for substance P using acetylcholinesterase as label, in *Trends in Cluster Headache* (Sicuteri, F., Vecchiet, L., and Fanciullaci, M., eds.), Elsevier, Amsterdam, pp. 125–134.
9. Pradelles, P., Grassi, J., Chabardes, D., and Guiso, N. (1989) Enzyme immunoassay of cAMP and cGMP using acetylcholinesterase. *Anal. Chem.* **61,** 447–453.
10. Caruelle, D., Grassi, J., Courty, J., Groux-Muscatelli, B., Pradelles, P., Barritault, D., and Caruelle, J. P. (1988) Development and testing of radio and enzyme immunoassays for acidic fibroblast growth factor. *Anal. Biochem.* **173,** 328–339.
11. Grassi, J., Frobert, Y., Pradelles, P., Chercuite, F., Gruaz, D., Dayer, J. M., and Poubelle, P. (1989) Production of monoclonal antibodies against interleukin 1α and 1β. *J. Immunol. Methods* **123,** 193–210.
12. Grassi, J., Frobert, Y., Lamourette, P., and Lagoutte, B. (1988) Screening of monoclonal antibodies using antigens labeled with acetylcholinesterase: Application to the peripheral proteins of photosystem 1. *Anal. Biochem.* **168,** 436–450.

13. Massoulié, J. and Bon, S. (1976) Affinity chromatography of acetylcholinesterase. *Eur. J. Biochem.* **68,** 531–539.
14. Massoulié, J. and Toutant, J. P. (1988) Vertebrate cholinesterases: Structure and types of interactions, in *Handbook of Experimental Pharmacology* (Whitaker, V. P., ed.), Springer-Verlag, Berlin, pp. 167–224.
15. Ellman, G. L., Courtney, K., Andres, V., and Featherstone, R. (1961) A new and rapid colorimetric determination of acetylcholine esterase activity. *Biochem. Pharmacol.* **7,** 88–95.
16. Vigny, M., Bon, S., Massoulié, J., and Leterrier, F. (1978) Active site, catalytic efficiency of acetylcholinesterase molecular forms in electrophorus, torpedo, rat and chicken. *Eur. J. Biochem.* **85,** 317–323.
17. Vaitukaitis, L. (1981) Production of antisera with small doses of immunogen: Multiple intradermal injections, in *Methods in Enzymology,* vol. 73 (Langone, J. J. and Van Vunakis, H., eds.), Academic, New York, pp. 46–52.

CHAPTER 8

Purification of Immunoglobulin G (IgG)

Michael G. Baines and Robin Thorpe

1. Introduction

Several immunological procedures can be successfully carried out using nonpurified antibodies, such as unfractionated antisera, or ascitic fluid/culture supernatant containing monoclonal antibodies (MAbs). However, a much "cleaner" result can often be obtained if some form of enrichment or isolation of immunoglobulin is employed. Some procedures, such as conjugation with isotopes, fluorochromes, or enzymes, and preparation of immunoaffinity columns cannot usually be efficiently performed with nonpurified immunoglobulin, and some procedures may yield artifactual results if whole antiserum or ascitic fluid is used as a source of antibody. Purification of immunoglobulins is therefore essential or, at least, useful for a range of immunological methods. This process may consist of purification of total IgG or subpopulations (e.g., subclasses) of IgG from antisera/ascitic fluid/culture supernatant, or the isolation of a particular antigen binding fraction from such fluids. The former can be achieved by biochemical procedures, whereas the latter usually requires some form of affinity purification.

Many biochemical methods can be used for immunoglobulin purification. They range from simple precipitation techniques yielding an immunoglobulin enriched preparation, to more complex chromatographic techniques for the production of "pure" immunoglobulin. Most of these procedures can be applied to the purification of immunoglobulins from the more commonly used species; however, mouse and rat IgGs are relatively less stable than immunoglobulin from higher mammals, and are generally less easily purified. Each MAb has unique characteristics, and some procedures need to

From: *Methods in Molecular Biology, Vol. 10: Immunochemical Protocols*
Ed.: M. Manson ©1992 The Humana Press, Inc., Totowa, NJ

be "tailored" for individual monoclonals. Avian antibodies may require special conditions for efficient purification.

The procedures outlined in this chapter are mostly confined to purification of IgG; for isolation of other immunoglobulin classes, the reader is referred to a more detailed text *(see* ref. *1)*.

2. Materials

1. PBS: 0.14*M* NaCl, 2.7 m*M* KCl, 1.5 m*M* KH_2PO_4, 8.1 m*M* Na_2HPO_4. Store at 4°C.
2. PBSA: PBS containing 0.02% NaN_3 (CARE—TOXIC).
3. Phosphate buffer: 0.07*M* Na_2HPO_4, 0.07*M* $NaH_2PO_4 \cdot 2H_2O$, pH 6.3. Add the dihydrogen salt to the disodium salt until the required pH is obtained. Phosphate buffer is also used containing 1*M* NaCl for "purging" ion-exchange columns; and containing 0.1% NaN_3 for storage of ion-exchange columns. Store at 4°C.
4. HCl: 0.5 and 1*M*.
5. NaOH: 0.5, 1, 2, and 4*M*.
6. NaCl: 1 and 2*M*.
7. Tris-HCl buffer: 0.05*M*, pH 8.0, 1*M*, pH 8.8. Adjust pH with 1*M* HCl or concentrated HCl as appropriate. Store at 4°C.
8. Triethanolamine buffer: 0.02*M* triethanolamine, pH 7.7. For ionic gradient separations by fast-protein liquid chromatography (FPLC), also use triethanolamine buffer containing 1*M* NaCl. Store at 4°C.
9. Sodium carbonate buffer: 0.5*M* Na_2CO_3, pH 10.5. Adjust pH with 0.1*M* NaOH.
10. Sodium citrate buffer: 0.1*M* Trisodium citrate, pH 6.5. Adjust pH with 0.1*M* citric acid.
11. Ethanolamine buffer: 2*M* Ethanolamine.
12. Glycine-HCl buffer: 0.1*M* Glycine, pH 2.5. Adjust pH with 1*M* HCl.
13. Sodium acetate buffer: 0.06*M* CH_3COONa, pH 4.6. Adjust pH with acetic acid.
14. Glycine solution: 0.05% Glycine w/v.
15. Riboflavin solution: 4 mg/100 mL of distilled water. Freshly prepared.
16. Ampholytes.
17. Acrylamide solution: 28.5% w/v Acrylamide, 1.5% w/v *N,N'*methylene *bis*-acrylamide. Store at 4°C.
18. Anode solution: 1*M* H_3PO_4, for wide pH range focusing (*see* Note 28).
19. Cathode solution: 1*M* NaOH, for wide pH range focusing (*see* Note 28).

20. Fixing solution: 500 mL of distilled water containing 57.5 g of trichloroacetic acid and 17.25 g of sulfosalicylic acid.
21. Destaining solution: Add 500 mL of ethanol to 160 mL of acetic acid and make up to 2 L with distilled water.
22. Staining solution: Add 0.46 g of Coomassie blue R 250 to 400 mL of destaining solution. Filter before use.
23. Whatman No. 54 filter paper.
24. Saturated ammonium sulfate solution: Add excess $(NH_4)_2\ SO_4$ to distilled water (about 950 g to 1 L) and stir overnight at room temperature. Chill to 4°C and store this solution (in contact with solid salt) at 4°C.
25. "Mini-gel" solution: 1*M* Tris/1*M* Bicine (2.0 mL), 50% w/v acrylamide (2.5% w/v *bis*-acrylamide) (4.0 mL), 1.5% w/v $(NH_4)_2\ S_2O_8$ (0.4 mL), 10% w/v SDS (0.2 mL). Make up to 20 mL with distilled water.
26. "Mini-gel" running buffer: 1*M* Tris/1*M* Bicine (2.8 mL), 10% w/v SDS (1.4 mL). Make up to 140 mL with distilled water. Store at 4°C.
27. "Mini-gel" sample buffer: Sucrose (1 *g*), 1*M* Tris/1*M* Bicine (0.2 mL), 10% w/v SDS (1 mL), 2-mercaptoethanol (0.25 mL). Make up to 3 mL with distilled water and add 0.001% w/v bromophenol blue. Store at –20°C.
28. Sodium sulfate.
29. Caprylic acid.
30. Polyethylene glycol (PEG) solution: 20% w/v PEG 6000 in PBS.
31. Diethylaminoethyl (DEAE) Sepharose CL–6B.
32. Ultrogel AcA 44.
33. Cyanogen bromide.
34. Sepharose 4B.
35. *N,N,N',N'*-Tetramethylethylene diamine (TEMED).
36. High and low mol-wt markers.

3. Methods

3.1. Prepurification Techniques

3.1.1. Separation of Serum from Whole Blood

It is obviously necessary to separate the IgG containing serum from cells and other insoluble components of whole blood. This can be simply achieved by allowing the blood to clot and then centrifuging to yield serum as a supernatant. This should be carried out as soon as possible after collection to avoid hemolysis and degradation of IgG (*see* Note 1).

1. Allow blood to clot at room temperature (this takes approx 1 h). Leave overnight at 4°C to allow the clot to contract (*see* Note 2).
2. Detach the clot from the walls of the container using a wooden or plastic rod, and pour off all liquid into a centrifuge tube or vessel (leave the clot and adhering substances behind).
3. Pour the clot into a separate centrifuge tube and centrifuge for 30 min at 2500g_{av} at 4°C. Remove any expressed liquid and add this to the previously aspirated clot-free liquid.
4. Centrifuge the pooled liquid for 15–20 min at 1500g_{av} at 4°C. Aspirate the clear serum and store at –20°C (or –70°C). Alternatively, serum can be stored at 4°C if an antibacterial agent (e.g., 0.02% NaN_3) is added. Do not freeze chicken serum or IgG because the immunoglobulins are particularly subject to damage.

3.1.2. Clarification of Ascitic Fluid

Ascitic fluid derived from the peritoneal cavity of mice or rats that have been injected with hybridomas contains high concentrations of MAb, but this is usually mixed with variable amounts of blood cells, serum proteins, and fatty materials. It is therefore necessary to separate these components from the ascitic fluid before attempting purification of the MAb.

1. Allow the ascitic fluid to clot at room temperature (some samples will not clot; in this case proceed as in (2) below). Detach the clot from the sides of the container using a wooden or plastic spatula.
2. Pour the ascitic fluid into a centrifuge tube and centrifuge for 10–15 min at 2500g_{av} at 4°C. Aspirate the clear supernatant and store at –20°C or –40°C.

3.2. Preliminary Purification (Precipitation) Techniques

Addition of appropriate amounts of salts, such as ammonium or sodium sulfate, or other chemicals, such as rivanol, polyethylene glycol, or caprylic acid, causes precipitation of IgG from serum, hybridoma culture supernatants, or ascitic fluid. Although such IgG is usually contaminated with other proteins, the ease of these precipitation procedures coupled with the high yield of IgG produced has led to their very wide usage in producing enriched IgG preparations suitable for some immunochemical procedures, e.g., production of immunoaffinity columns, and as a starting point for further purification. The precipitated IgG is usually very stable, and such preparations are ideally suited for long-term storage or distribution and exchange between laboratories.

Ammonium sulfate precipitation is the most widely used and adaptable procedure, but other precipitation techniques can be the method of choice for some antibodies or purposes.

3.2.1. Ammonium Sulfate Precipitation

1. Prepare saturated ammonium sulfate at least 24 h before the solution is required for fractionation.
2. Centrifuge serum/ascitic fluid/culture supernatant for 20–30 min at 10,000g_{av} at 4°C. Discard the pellet (*see* Note 3).
3. Cool the serum/ascitic fluid/culture preparation to 4°C and stir slowly. Add saturated ammonium sulfate solution dropwise to produce 35–45% final saturation (*see* Note 4). Alternatively, add solid ammonium sulfate to give the desired saturation (2.7g of ammonium sulfate/10 mL of fluid = 45% saturation). Stir at 4°C for 1–4 h or overnight.
4. Centrifuge at 2000–4000g_{av} for 15–20 min at 4°C (alternatively, for small volumes of 1–5 mL, microfuge for 1–2 min). Discard the supernatant and drain the pellet (carefully invert the tube over a paper tissue).
5. Dissolve the precipitate in 10–20% of the original volume of PBS or other buffer by careful mixing with a spatula or drawing repeatedly into a wide-gage Pasteur pipet. When fully dispersed, add more buffer to give 25–50% of the original volume and dialyze against the required buffer (e.g., PBS) at 4°C overnight with 2–3 buffer changes. Alternatively, the precipitate can be stored at 4°C or –20°C if not required immediately.

3.2.2. Sodium Sulfate Precipitation

Sodium sulfate may be used for precipitation of IgG instead of ammonium sulfate. The advantage of the former salt is that a purer preparation of IgG can be obtained in some cases. The disadvantages are that yield may be reduced, and that fractionation must be carried out at a precise temperature that is warm (usually 25°C), as the solubility of Na_2SO_4 is very temperature-dependent. Sodium sulfate is usually employed only for the purification of rabbit or human IgG, but it can be used for other species. Sodium sulfate is not recommended for precipitation of most murine MAbs.

1. Centrifuge the serum at 10,000g_{av} for 20–30 min. Discard the pellet and warm the serum to 25°C. Stir.
2. Add solid Na_2SO_4 to produce an 18% w/v solution (i.e., add 1.8 g/10 mL) and stir at 25°C for 30 min to 1 h.
3. Centrifuge at 2000–4000g_{av} for 30 min at 25°C.
4. Discard the supernatant and drain the pellet. Redissolve in the appropriate buffer as described for ammonium sulfate precipitation, Step 5 in Section 3.2.1.

3.2.3. Precipitation with Polyethylene Glycol (PEG)

This procedure is applicable to both polyclonal antisera and most MAb-containing fluids.

1. Prepare 20% w/v PEG 6000 in PBS. Cool to 4°C.
2. Prepare serum/ascitic fluid, and so on for fractionation by centrifugation at 10,000g_{av} for 20–30 min at 4°C. Discard the pellet. Cool to 4°C.
3. Slowly stir the antibody containing fluid and add an equal vol of 20% PEG dropwise (*see* Note 5). Continue stirring for 20–30 min.
4. Centrifuge at 2000–4000g_{av} for 30 min at 4°C. Discard the supernatant and drain the pellet. Resuspend in PBS or other buffer, as described for ammonium sulfate precipitation, Step 5 in Section 3.2.1. (*see* Note 5).

3.2.4. Caprylic Acid Precipitation

Caprylic acid can be used to purify IgG, but the concentration required varies according to species. For monoclonal antibodies it is usually necessary to determine experimentally the quantity required to produce the desired purity/yield.

1. Centrifuge the serum at 10,000g_{av} for 20–30 min. Discard the pellet and add twice the vol of 0.06*M* sodium acetate buffer, pH 4.6.
2. Add caprylic acid dropwise while stirring at room temperature. For each 25 mL of serum, use the following amounts of caprylic acid: Human and horse, 1.52 mL; goat, 2.0 mL; rabbit, 2.05 mL; cow, 1.7 mL. Stir for 30 min at room temperature.
3. Centrifuge at 4000 g_{av} for 20–30 min. Discard the supernatant and drain the pellet. Resuspend in PBS or other buffer, as described for ammonium sulfate precipitation, Step 5 in Section 3.2.1.

3.3. Chromatography Techniques Based on Charge or Size Separation

3.3.1. Ion-Exchange Chromatography (IEC) (see Notes 6–8)

Ion-exchange chromatography (IEC) is a widely used method for the fractionation of IgG from both polyclonal antisera and preparations containing high concentrations of MAb. The separation of molecules in ion exchange is determined by the charges carried by solute molecules, and it is a technique of high resolving power. Ion-exchange groups are of two types, anion exchange groups, e.g., diethylaminoethyl (DEAE), and cation exchange groups, e.g., carboxymethyl (CM). The ion-exchange groups are covalently bound to an insoluble matrix, for example, Sephadex (crosslinked dextran), Sepharose (crosslinked agarose), or Sephacel (beaded cellulose), to form the ion-exchange resin (IER). After proteins have been adsorbed onto the IER, they can be selectively eluted by slowly increasing ionic strength (this disrupts ionic interactions competitively), or by altering the pH (the reactive groups on the proteins lose their charge). This is the basis of IEC, and the technique can be used to purify IgG from sera of most species.

3.3.1.1. Preparation and Equilibration of Ion Exchanger

1. Gently stir the IER into approx 5× its swollen vol of 0.5*M* HCl, for anion exchangers (use 0.5*M* NaOH for cation exchangers). Leave at room temperature with occasional swirling for 30 min.
2. Filter the resin by suction through a Buchner funnel, using a Whatman No. 54 filter paper. Then wash the resin cake with distilled water until the pH of the filtrate is greater than 4 (after acid) or less than 8 (after base).
3. Gently stir the IER into approx 5× its swollen vol of 0.5*M* NaOH, for anion exchangers (use 0.5*M* HCl for cation exchangers). Leave at room temperature with occasional swirling for 30 min.
4. Repeat Step 2.
5. Add the IER to an equal vol of 10X starting buffer (starting buffer is Tris-HCl, 0.05*M*, pH 8.0 or any other appropriate buffer). Mix thoroughly and leave at room temperature for 30 min.
6. The ion exchanger will adsorb some buffer ions in exchange for protons or hydroxyl ions and, hence, alter the pH. Restore the pH to its original value by gently stirring the slurry and adding the acid or basic forms of the buffer (1*M* HCl for an anion exchanger and 1*M* Tris for a cation exchanger).
7. Leave the slurry at room temperature for 30 min and then recheck the pH. Adjust if required. The ion exchanger is now at the correct pH with the counterion bound, but the ionic strength will be too high.
8. Wash the IER with 5× its vol of starting buffer through a Whatman No. 54 filter paper.
9. Degas the slurry using a Buchner side-arm flask under vacuum for 1 h with periodic swirling.
10. Resuspend the slurry in approx 5× its vol of starting buffer and leave to stand until most of the beads have settled. Remove the fines by aspirating the supernatant down to about 2× the settled slurry vol.
11. Carefully pack a clean column by first filling it with 10 mL of starting buffer, and then by pouring the resuspended slurry down a glass rod onto the side wall until the column is filled. Allow to settle under gravity.
12. Wash the column with starting buffer at the operating temperature until the pH and conductivity of the eluant are exactly the same as the starting buffer.

3.3.1.2. Sample Application and Elution

1. Dialyze the serum (ammonium sulfate fractionated) against phosphate buffer (pH 6.3, 0.07*M*) exhaustively (two changes of 1 L each over a 24-h period).

2. Apply the sample to the column (anion exchange columns, e.g., DEAE Sepharose CL–6B) and elute with phosphate buffer. Collect 2-mL fractions and monitor the absorbance at 280 nm (A_{280})
3. Collect the first protein peak and stop collecting fractions when the A_{280} falls to baseline.
4. Regenerate the column by passing through 3 column vol of phosphate buffer containing 1*M* NaCl.
5. Wash thoroughly in phosphate buffer (2–3 column vol) and store in buffer containing 0.1% NaN_3.
6. Pool the fractions from Step 3 and measure the A_{280}. The average extinction coefficient (E_{280}) of IgG is 13.6.

3.3.2. *Gel Filtration* (see *Notes 9–15*)

In gel filtration, a protein mixture (the mobile phase) is applied to a column of small beads with pores of carefully controlled size (the stationary phase). The movement of the solute is dependent on the flow of the mobile phase and the Brownian motion of the solute molecules, causing their diffusion into and out of the chromatographic bed. Large proteins, above the "exclusion limit" of the gel, cannot enter the pores and are hence eluted in the "void volume" of the column. Small proteins enter the pores and are therefore eluted in the "total volume" of the column and intermediate size proteins are eluted between the void and total volumes. Proteins are therefore eluted in order of decreasing molecular size.

Gel filtration is not especially effective for the purification of IgG, which tends to elute in a broad peak and is usually contaminated with albumin (derived from dimeric albumin, M_r 135,000). The technique is more useful for the purification of IgM. Some IgGs (monoclonal), however, possess pIs that make them unsuitable for fractionation using IEC. In such cases, it may be desirable to use gel filtration as a method of fractionation.

3.3.2.1. Preparation and Equilibration of Gel Filtration Column (*see* Note 12)

1. Gently stir the filtration medium (enough to fill the column plus 10%) into 2× the column vol of buffer.
2. Degas the slurry using a Buchner side-arm flask under vacuum for 1–2 h with periodic swirling. Do not use a magnetic stirrer since this may damage the filtration beads.
3. Resuspend the slurry in approx 5× its vol of buffer and leave to stand until most of the beads have settled. Remove the fines by aspirating the supernatant down to about 1.5× the settled slurry vol.

4. Carefully pack a clean column by first filling it to 10–20% of its height with buffer. Swirl the slurry to resuspend it evenly, and pour it down a glass rod onto the inside wall to fill the column. Allow to settle under gravity for 0.5–1 h and to let air bubbles escape.
5. Adjust the height of the outlet end of the column so that the vertical distance between it and the top of the column is less than the maximum operating pressure for the gel. Unclamp the bottom of the column and allow the gel to pack under this pressure.
6. Top up the column periodically by syphoning off excess supernatant, stirring the top of the gel (if it has settled completely) and filling the column up to the top with resuspended slurry.
7. Once the column is packed (gel bed just runs dry), connect the top of the column to a buffer reservoir, remove any air bubbles in the tube, and allow one column vol of buffer to run through the column.

3.3.2.2. Sample Application and Elution (*see* Note 13)

1. Disconnect the top of the column from the buffer reservoir and allow the gel to just run dry.
2. Apply the sample carefully by running it down the inside wall of the column so that the gel bed is not disturbed.
3. When the sample has entered the bed, gently overlay the gel with buffer and reconnect the column to the buffer reservoir.
4. Collect fractions (4–6 mL) and monitor the absorbance at 280 nm.

*3.3.3 HPLC and Related Techniques (*see *Note 16, refs. 2 and 3)*

Fractionation of immunoglobulins by high-performance liquid chromatography (HPLC) utilizes the same principles as chromatography in standard columns by gel filtration, ion-exchange, or affinity chromatography. In HPLC, protein separation occurs in a column of small cross-sectional area containing the chromatography matrix in the form of very fine particles (the stationary phase). The solvent (buffer) or mobile phase is pumped through the column using medium- to high-pressure pumps. This allows the sample molecules in the mobile phase to interact reversibly with the stationary phase in a continuous fashion. The advantages of HPLC over conventional chromatography techniques are speed, because of the small, high-capacity columns, improved reproducibility because of the sophisticated pumps and accurate timers, and in many cases, increased resolution because of the fine resins and control systems. Fast-protein liquid chromatography (FPLC) is a variant of HPLC, which has proved especially

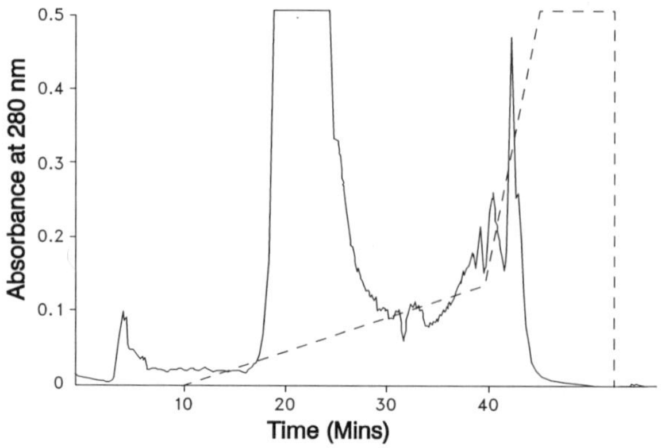

Fig. 1. Elution profile of FPLC-purified sheep antiserum. This profile represents the purification of an anti-p27 (core protein of simian immunodeficiency virus) polyclonal serum by the authors. The major peak is IgG, which elutes after about 18 min. The dotted line represents the profile of the ionic gradient used for elution of the IgG.

useful in the purification of murine MAbs, although the technique is applicable to IgG preparations from all species.

3.3.4. FPLC Purification of Sheep IgG (see *Fig. 1*)

1. Prepare serum by ammonium sulfate precipitation (45% final concentration). Redissolve the precipitate in triethanolamine buffer (pH 7.7, 0.02*M*) and dialyze overnight in this buffer at 4°C. Filter the sample before use (0.2 µm).
2. Assemble the FPLC system according to the manufacturer's (Pharmacia) instructions for use with the Mono-Q ion-exchange column.
3. Equilibrate the column (method programmed).
4. Assemble the "superloop" according to the manufacturer's instructions.
5. Load the sample manually.
6. Set the sensitivity in the UV monitor control unit (try 0.5) and zero the chart recorder.
7. Run the Mono-Q ion-exchange purification method (method programmed). Collect fractions (automatic). Elution buffers are triethanolamine (pH 7.7, 0.02*M*) and triethanolamine containing 1*M* NaCl.
8. Wash the Mono-Q ion exchange with 2*M* NaOH followed by 2*M* NaCl (method programmed).
9. Store the Mono-Q ion-exchange column in distilled water containing 0.02% NaN_3 (method programmed).

3.4. *Affinity Chromatography*

Affinity chromatography is a particularly powerful procedure that can be used to purify IgG, subpopulations of IgG, or the antigen binding fraction of IgG present in serum/ascitic fluid/hybridoma culture supernatant. This technique requires the production of a solid matrix to which a ligand, which either has affinity for the relevant IgG or vice versa, has been bound. Examples of ligands useful in this context are:

1. The antigen recognized by the IgG (for isolation of the antigen specific fraction of the serum/ascitic fluid, and so on).
2. IgG prepared from an anti-immunoglobulin serum, e.g., rabbit antihuman IgG serum or murine antihuman IgG monoclonal antibody for the purification of human IgG (*see* Note 17).
3. IgG binding proteins derived from bacteria, e.g., protein A (from *Staphylococcus aureus* Cowan 1 strain) or proteins G and C (from *Streptococcus; see* Note 18).

The methods for production of such immobilized ligands and for carrying out affinity purification of IgG are essentially similar regardless of which ligand is used. Sepharose 4B is probably the most widely used matrix for affinity chromatography, but there are other materials available. Activation of sepharose 4B is usually carried out by reaction with cyanogen bromide; this can be carried out in the laboratory prior to coupling or ready-activated lyophilized sepharose can be purchased. The commercial product is obviously more convenient than "home made" activated sepharose, but it is more expensive and may be slightly less active.

3.4.1. Activation of Sepharose with Cyanogen Bromide and Preparation of Immobilized Ligand

Activation of sepharose with CNBr requires the availability of a fume hood, and careful control of the pH of the reaction (failure to do this may lead to the production of dangerous quantities of HCN as well as compromising the quality of the activated sepharose). CNBr is toxic and volatile. All equipment that has been in contact with CNBr and residual reagents should be soaked in 1*M* NaOH overnight in a fume hood and washed prior to discarding/returning to the equipment pool. Manufacturers of ready activated sepharose provide instructions for coupling (*see* Note 19).

1. Wash 10 mL (settled vol) of Sepharose 4B with 1 L of water by vacuum filtration. Resuspend in 18 mL of water (do not allow the sepharose to dry out).
2. Add 2 mL of 0.5*M* sodium carbonate buffer, pH 10.5, and stir slowly. Place in a fume hood and immerse the glass pH electrode in the solution.

3. *Carefully* weigh 1.5 g of CNBr into an air-tight container (weigh in a fume hood; wear gloves)—remember to decontaminate equipment that has contacted CNBr in 1*M* NaOH overnight.
4. Add the CNBr to the stirred Sepharose. *Maintain* the pH between 10.5–11.0 by dropwise addition of 4*M* NaOH until the pH stabilizes and all the CNBr has dissolved. If the pH rises above 11.5, activation will be inefficient and the Sepharose should be discarded.
5. Filter the slurry using a sintered glass or Buchner funnel and wash the sepharose with 2 L of cold 0.1*M* sodium citrate buffer, pH 6.5—do not allow the sepharose to dry out. Carefully discard the filtrate (CARE, this contains CNBr).
6. Quickly add the filtered washed Sepharose to the ligand solution (2–10 mg/mL in 0.1*M* sodium citrate, pH 6.5) and gently mix on a rotator ("windmill") at 4°C overnight (*see* Note 20).
7. Add 1 mL of 2*M* ethanolamine solution and mix at 4°C for a further 1 h; this blocks unreacted active groups (*see* Note 21).
8. Pack the Sepharose into a suitable chromatography column (e.g., a syringe barrel fitted with a sintered disc) and wash with 50 mL of PBS. Store at 4°C in PBS containing 0.1% sodium azide.

*3.4.2. Purification of IgG Using Affinity Chromatography (*see *Fig. 2)*

Isolation of IgG by affinity chromatography involves application of serum, and so on, to a column of matrix-bound ligand, washing to remove non-IgG components and elution of IgG by changing the conditions such that the ligand–IgG interaction is disrupted. Affinity isolation of IgG can also be carried out using a batch procedure rather than on a column; this is particularly useful for large volumes containing a relatively small amount of IgG, e.g., cell culture supernatants, especially those produced by MAb secreting human cells.

1. Wash the affinity column with PBS. "Pre-elute" with dissociating buffer, e.g., 0.1*M* glycine-HCl, pH 2.5. Wash with PBS; check that the pH of the eluate is the same as the pH of the PBS (*see* Note 22).
2. Apply the sample to the column, close the column exit, and incubate at room temperature for 15–30 min (*see* Note 23).
3. Wash non-IgG material from the column with PBS; monitor the A_{280} as an indicator of protein content.
4. When the A_{280} reaches a low value (approx 0.02), disrupt the ligand–IgG interaction by eluting with dissociating buffer (*see* Note 21). Monitor the A_{280} and collect the protein peak into tubes containing 1*M* Tris, pH 8.8, 120 µL/1-mL fraction.

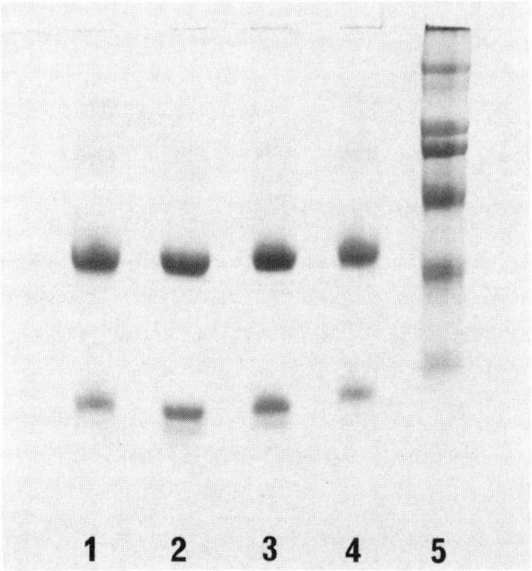

Fig. 2. SDS-PAGE of affinity-purified MAbs specific for human IL-3. Track 5 shows standard mol-wt markers (from top to bottom: myosin heavy chain, M_r 200 kD; β-galactosidase, M_r 116 kD; phosphorylase b, M_r 97.4 kD; bovine albumin, M_r 66 kD; egg albumin, M_r 45 kD; and carbonic anhydrase, M_r 29 kD). Track 1 shows Cohn fractionated human IgG. The remaining three tracks show three different IgG antibodies purified using a column of human recombinant DNA-derived IL-3 coupled to Sepharose 4B. The major bands are kappa and gamma chains; the faint higher M_r bands are caused by incomplete dissociation of heavy and light chains.

5. Wash the column with PBS until the eluate is at pH 7.4. Store the column in PBS containing 0.1% azide. Dialyze the IgG preparation against a suitable buffer (e.g., PBS) to remove glycine/Tris.

3.5. Monoclonal Antibodies (MAbs; see also *Vol. 5, Chapter 53 and this Vol., Chapter 6*)

Hybridoma technology has made possible the production of highly specific, homogeneous antibodies with predefined binding characteristics, which can be produced in large amounts, from immortal cell lines. MAbs can be exquisitely specific, but they are far from pure, being contaminated with tissue culture additives and nonimmunoglobulin secretion products when grown in vitro, and by host animal proteins when grown as an ascitic fluid.

MAbs have been routinely produced in mice and less easily in rats (*see* this vol., Chapter 6 and vol. 5, Chapter 54). There has also been some success in the production of human MAbs although this is still not a routine exercise. Rodent MAbs are not readily purified using conventional IEC and as discussed, gel filtration is not a particularly suitable method for the purification of IgG. FPLC is an ideal method for the purification of rodent MAbs; the technique has been described in Section 3.3.3., and with a few minor changes (*see* Note 23) remains, in principle, the same. The production of human MAbs and their subsequent purification presents some other difficulties (*see* Note 24).

3.6. Analysis of IgG Fractions

After the purification procedure, it is then necessary to obtain some index of sample purity. There are several methods available for this, and two of the most useful are isoelectric focusing and sodium dodecyl sulfate polyacrylamide gel electrophoresis (SDS-PAGE).

3.6.1. Analysis of IgG by Isoelectric Focusing (IEF) (see ref. 4 and Vols. 1 and 3)

The very high resolving power of isoelectric focusing (IEF) enables the technique to be used to demonstrate the charge-dependent heterogeneity of IgG. Focusing using polyacrylamide gel as the anticonvection medium/support is most usually employed for studies of IgG; this enables several samples to be analyzed on the same slab gel (*see* Fig. 3). The charge-dependent heterogeneity of IgG is caused by differences in charged amino acids and particularly, carbohydrate residues present on the individual IgG species; sialic acid residues are particularly important for this. Polyclonal antisera usually focus as a fairly broad "smear" with the major components superimposed as stronger bands, whereas oligoclonal IgGs focus as a series of intense bands. MAbs usually focus as 3–6 closely spaced bands, The pI of different murine MAbs varies very considerably between 5.5–8.5. The characteristic pI and banding pattern should be consistent for a given MAb and can be used as an identity test. This pattern is sometimes referred to as the "spectrotype" of the MAb. Gels for focusing can be purchased ready for use or can be prepared in the lab.

1. Dialyze samples against two changes of 100 vol of glycine solution overnight.
2. Assemble the gel mold and seal (*see* Note 25).
3. Prepare the gel solution. For each 10 mL, mix 1.8 mL of stock acrylamide solution with 0.25 mL of ampholyte solution. Make up to vol with water and degas for 2 min under vacuum. Add 0.1 mL of riboflavin, pour into the gel mold, and irradiate with the UV lamp about 50–70 cm from the gel. Polymerization takes about 1.5–2.5 h.
4. Turn off the lamp, remove the mold from the trough, and pour off residual fluid from the top of the gel. Cool the assembly in ice for 30 min and remove the plastic spacers and carefully prise the top glass plate

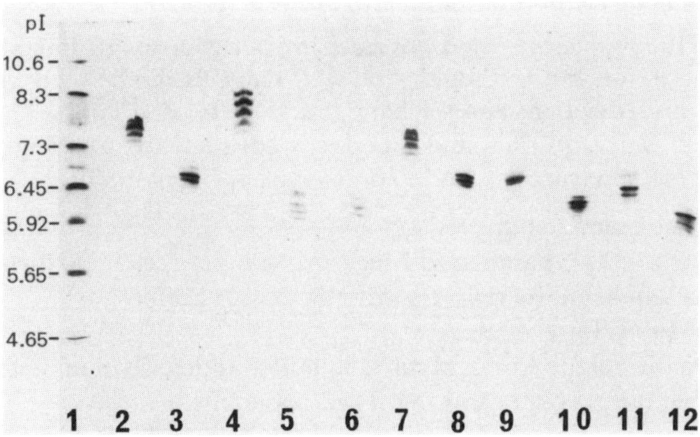

Fig. 3. Isoelectric focusing in polyacrylamide gel of different purified IgG mouse monoclonal antibodies. Track 1 shows marker proteins of known pI.

from the gel using a spatula (*see* Note 26). Place the gel on the cooling plate of the electrophoresis tank (or place the whole apparatus in a cold room) and allow to cool.

5. Prepare two electrode wicks (0.5 cm × width of the gel) from the filter paper and soak one in anode solution and the other in cathode solution. Place on the gel surface (*see* Note 27).
6. Apply samples to the gel. For this, soak samples onto rectangular pieces of filter paper (3 mm × 5 mm hold 10–15 µL). Blot off excess liquid and position them on the surface of the gel 0.5–1 cm from the cathodal wick and 2–5 mm apart. Samples should contain about 5–20 µg of protein.
7. Set up the chamber for IEF (platinum electrodes should press onto the wicks, *see* Note 28). Connect electrodes to the power pack and carry out focusing. If available, set 25 W constant power. For wide pH ranges (e.g., 3.5–9.5), focusing will be complete in 1.5–2 h, but for narrower pH ranges, focusing should be continued for 3–3.5 h. If constant power is not available, start at 200 V and increase up to 750–1500 V over 2 h. Alternatively, focus overnight at 150 V.
8. Turn off the power, remove the gel from the tank, and submerge it in the fixing solution for 0.5–1.0 h (agitation on an orbital shaker speeds up this process). Submerge the gel in stain at 60°C for 10 min. Destain in several changes of destaining solution until a clear background is obtained.

3.6.2. Sodium Dodecyl Sulfate Polyacrylamide Gel Electrophoresis (SDS-PAGE)

One of the simplest methods for assessing purity of an IgG fraction is by SDS-PAGE (*see* Note 29). Although "full size" slab gels can be used with discontinuous buffer systems and stacking gels, the use of a "mini-gel" procedure is much easier, quicker, and is perfectly adequate for assessing purity, monitoring column fractions, and so on (*see* vol. 5, Chapter 53).

1. Prepare the sample buffer.
2. Adjust the IgG preparation to 1 mg/ mL in 0.1*M* Tris/0.1*M* Bicine.
3. Mix the sample in the ratio 2:1 with the sample buffer.
4. Heat at 100°C for 2–4 min.
5. Prepare the gel solution and running buffer as described in Materials.
6. Assemble the gel plates and "mini-gel" apparatus.
7. Add 50 µL of TEMED to 10 mL of gel solution, and use this to set the trough in the "mini-gel" apparatus. Leave for 10 min.
8. Add 30 µL of TEMED to the remaining 10 mL of gel solution, pour the gel into the gel plates, and set the comb. Leave for 10 min.
9. Remove the comb and locate the gel plates in the electrophoresis tank. Add the running buffer.
10. Load the sample prepared as described in 1–4 (30–50 µL/track) and run standard mol wt markers (10 µL/track of manufacturer's recommended stock solution) in parallel.
11. Run at 150 V for 1.5 h.
12. Remove the gel and stain in excess Coomassie blue R stain for 2 h (gently rocking) or overnight (stationary).
13. Pour off the stain and rinse briefly in tap water.
14. Add excess destain and 2–3 pieces of either plastic foam packing or silk to the gel. Leave until destaining is complete (usually overnight, gently rocking).

4. Notes

1. Mouse and rat IgGs are less stable than IgG from higher mammals; it is best to separate mouse/rat serum from whole blood as soon as clotting has occurred.
2. If an anticoagulant has been added to the blood and the plasma isolated, it may be advisable to defibrinate the plasma to yield a serum analog. The method for this varies according to the anticoagulant; for citrate, add 1% of a solution of thrombin (100 IU/mL in 1*M* $CaCl_2$). For heparin, add 1% protamine sulfate solution (5 mg/mL) and thrombin as above.

3. If lipid contamination is a particular problem, add silicone dioxide powder (15 mg/mL) and centrifuge for 20 min at $2000g_{av}$.
4. The use of 35% saturation will produce a fairly pure IgG preparation but will *not* isolate all the IgG present in serum/ascitic fluid, and so on. Increasing saturation to 45% causes precipitation of nearly all IgG, but this will be contaminated with other proteins, including some albumin. Purification using $(NH_4)_2SO_4$ can be improved by repeating the precipitation, but this may cause some denaturation, especially of MAbs. Precipitation with 45% $(NH_4)_2SO_4$ is an ideal starting point for further purification steps, e.g., ion-exchange or affinity chromatography and FPLC purification.
5. Although the procedure works well for most antibodies, it may produce a fairly heavy contamination with non-IgG proteins with some samples; if this is the case, reduce the concentration of PEG. For this reason, it is best to carry out a pilot-scale experiment before fractionating all of the sample. PEG precipitation is a very mild procedure that results in little denaturation of antibody.
6. There are a number of criteria to consider when setting up an IEC system, e.g., choice of ion exchanger, matrix support, column and buffer. The most useful IERs for the purification of IgG are the anion exchanges DEAE cellulose, or sepharose, of which there are several forms. As a general rule of thumb, anion exchange resins are best used for proteins differing in the total number of aspartic and glutamic residues, whereas cationic exchange is preferable for proteins differing in their lysine, arginine, and histidine content. It should be remembered, however, that strength of binding is related also to the isoelectric point (pI) of the protein in question and to the total number of charges.

   Clearly, if two proteins of identical pI are present, they could be separated by using a pH at which there is a significant difference between them in terms of number of charged residues per molecule. The DEAE reactive moiety was originally coupled to fibrous cellulose (Whatman, Bio-Rad), which resulted in poor flow rates and clogging of the matrix. It appears that coupling the DEAE to the newer physical forms of cellulose, microgranular and beaded (DEAE-Sephacel, Pharmacia), results in resins with more reproducibility, higher capacity, and greater resolving potential than the older fibrous forms.

   The Sepharose-based DEAE CL–6B (Pharmacia) is also a very useful resin for ion exchange of high-mol wt proteins; it has a very low nonspecific adsorption of macromolecules, good flow properties, and is relatively stable to changes in ionic strength and pH. Sephadex-based IERs tend to alter volume drastically as a result of changes in ionic strength and pH. It should also be remembered that flow rate is not a particularly important criterion for an ion-exchange column to possess, unless very labile proteins are being separated. In most cases, the chromatography step is accomplished in a fairly short time-period, less than 24 h. A further consideration is the number of charged groups that are present in a unit volume of resin matrix and that are available to the protein for binding. This will govern the capacity of the column and will therefore determine the dilution of sample, loss of protein through nonspecific adsorption for a given fractionation, and not least, expense! The available protein capacity for microgranular/beaded cellulose and crosslinked agarose is very similar (0.11–0.15 g albumin/mL DEAE derivative).

In general, an ion-exchange matrix will require regeneration and swelling before use according to the manufacturer's instructions (*see* Method), although DEAE-Sephacel and DEAE-Sepharose CL–6B are supplied preswollen and ready to use.

When choosing a column for IEC, there are a few rules to follow. The column should be of controlled diameter glass with as low a dead space as possible. Column length should be 20–30 cm and the internal diameter 1.5–1.6 cm. Homemade columns are quite acceptable for most purposes, although commercial columns are available from several manufacturers.

When choosing a buffer, the pH should be within the operational range of the ion-exchange matrix (*see* manufacturer's guide) and suitable for use with the protein in question, in this case, IgG. For a protein to bind to the ion-exchange matrix, the pH should be approx 1 U above its pI for anion exchangers (1 U below for cation exchangers). The ionic strength of the application buffer should be between 0.01–0.05*M*, and for reproducible fractionations, the water used in the buffer should be as pure as possible. The addition of charged bacterial inhibitors, e.g., sodium azide or thimerosal, should be avoided.

7. An idiosyncratic and useful feature of IgG from many species is that they possess an "abnormal" pI (proteins are amphoteric and their net charge is zero at their pI). In such cases, the antiserum (fractionated) is simply passed through the anion (DEAE) exchange column under conditions that allow the IgG to pass straight through the column, whereas contaminating proteins, including albumin, bind to the column matrix. In these circumstances, there is no requirement for a pH or ionic gradient, although conventional adsorption/elution IEC can be used for the purification of IgG. The procedure works well with antisera from man, apes, monkeys, horses, goats, sheep, and rabbits. It does not, however,

work for rodent sera or preparations containing mouse or rat MAbs. One possible problem is that some immunoglobulins are unstable at low ionic strength, e.g., mouse IgG_3, and precipitation may occur during the ion-exchange procedure. Furthermore, in preparations of murine MAbs, it appears that DEAE cellulose chromatography after ammonium sulfate precipitation does not remove protease and nuclease activity, and does not adequately separate transferrin. Conventional IEC cannot therefore be considered to be a rapid and efficient general method for the purification of mouse or rat antibodies (especially MAbs).

8. If conventional adsorption/elution IEC is used for the purification of IgG, then there are a number of other technical points to consider. Once the sample has been applied to the column, the eluant should be monitored at A_{280}, ideally by passing the eluant through the flow cell of a UV absorbance monitor. If there is a high concentration of proteins in the eluant, then the ion exchanger or sample were not fully equilibrated or the absorbing capacity of the ion-exchanger matrix has been exceeded. After the sample has been loaded onto the column, the ion exchanger should be washed with two column-volumes of starting buffer to ensure complete elution of any unbound protein. The bound proteins are then eluted by increasing the ionic strength (preferable to variation of pH since it is easier to control). The ionic strength is best altered by increasing the concentration of other ions, e.g., NaCl, in which case the pH and buffering capacity are kept constant throughout the separation procedure. The entire process can be automated by using a commercial/homemade gradient marker and fraction collector. As a general rule of thumb, the total volume of the gradient should be between 5–10× the column volume, and the size of the collected fractions should be 10–20% of the total column volume. The elution profile can be plotted by recording the optical density at A_{280} of the eluted fraction and the first major peak contains the IgG.
9. The "void-volume" of the column is the volume of liquid between the beads of the gel matrix and usually amounts to about one-third of the total column volume.
10. The operating pressure of the gel is the vertical distance between the top of the buffer in the reservoir and the outlet end of the tube; this should never exceed the manufacturer's recommended maximum for the gel.
11. There are a number of criteria to consider when setting up a gel filtration system, e.g., choice of gel, column, and buffer. The gel of choice may be composed of beads of carbohydrate or polyacrylamide, and is available

in a wide variety of pore sizes and hence fractionation ranges. Useful gels for the separation of IgG include Ultrogel AcA 44 (mixtures of dextran and polyacrylamide) and Bio-Gel P 200 (polyacrylamide). For IgM fractionation, Sephacel S-300 (crosslinked dextran) is most useful.

When selecting a column for gel filtration, the column should be of controlled diameter glass with as low a dead space as possible at the outflow. Column tubing should be about 1 mm in diameter, which helps reduce dead space volume. A useful length for a gel-filtration column is about 100 cm, and the choice of cross-sectional area is governed by sample size, both in terms of volume and amount of protein. As a rough rule of thumb, the sample volume should not be greater than 5% of the total column volume (1–2% gives better resolution); and between 10–30 mg of protein/cm^2 cross-sectional area is a satisfactory loading. More protein will increase yield, but decrease resolution and hence purity, whereas less protein loaded improves resolution, but cuts back on yield. In general, homemade columns will suffice for gel filtration procedures, although commercial columns are available from a number of manufacturers. Ensure columns are clean before use by washing with a weak detergent solution and rinsing thoroughly with water. Gel filtration can be performed using a wide variety of buffers generally of physiological specification, i.e., pH 7.2, 0.1*M*.

12. Gel-filtration columns may also be packed by using an extension reservoir attached directly to the top of the column. In this case, the total volume of gel and buffer can be poured and allowed to settle without continual topping up. When using an extension reservoir, leave the column to pack until the gel bed just runs dry and then remove excess gel.
13. When loading commercial columns with flow adaptors touching the gel bed surface, the most satisfactory way to apply the sample is by transferring the inlet tube from the buffer reservoir to the sample. The sample enters the tube and gel under operating pressure and then the tube must be returned to the buffer reservoir. Ensure no air bubbles enter the tube. In the case of homemade columns, it is also possible to load the sample directly onto the gel bed by making it up in 5% w/v sucrose. Elution of samples is either by pressure from the reservoir (operating pressure) or by a peristaltic pump between the reservoir and the top of the column. In general, the flow rate should be the volume contained in 2–4 cm height of column/h. In order to prevent columns without pumps from running dry, the inlet tube should be arranged so that part of it is below the outlet point of the column. Eluted fractions should be collected by measured volume rather than time (this prevents

fluctuations in flow rate from altering fraction size). In general, for a 100-cm column, a column volume of eluant should be collected in about 100 fractions, the void volume is eluted at about fractions 30–35.

14. The protein yield in gel filtration should be greater than 80% and is often as high as 95%. Yield will improve after the first use of a gel, because new gel adsorbs protein nonspecifically in a saturable fashion. When very small amounts of protein are being fractionated, the column should be saturated with albumin prior to use.
15. Columns should be stored at room temperature or 4°C, usually in the presence of a bacteriostatic agent, e.g., 0.02% sodium azide.
16. The FPLC system may be used in a fully automated mode, a manual mode, or a combination of both. The details of the FPLC hardware and its use are comprehensively described in the manufacturer's manuals (Pharmacia). Briefly, the FPLC system consists of two high-precision pumps connected to a mixer for gradient formation, an automatic motor valve for injection of samples (samples are delivered manually to the valve via a loop with a syringe attachment), a single path UV monitor (and control unit) for monitoring samples, and a programmable liquid chromatography controller for the full or partial automation of the FPLC system. A fraction collector and chart recorder complete the hardware requirements together with the column of choice, e.g. ion exchanger, gel filtration, and so on.

    Rapid purification of IgG may be achieved using the above apparatus and the Mono Q HR5/5 (HR 10/10) anion exchanger column. The HR5/5 and HR10/10 columns are identical except in size, and thus capacity. If possible, it is advisable to use the HR10/10 column that can handle 10 mL of ammonium sulfate "cut" serum containing IgG at about 10–15 mg/mL. When using the HR10/10 column, it is necessary to use a "superloop," which must be manually injected with the sample. The superloops have a maximum capacity of 10-mL sample, although any smaller volume may be delivered. The loop sizes for the smaller columns have a maximum volume of 0.5 mL.

    Several criteria require further consideration. It is essential that the sample to be loaded and all buffers used are filtered using 0.2 μm millipore. All the methods for column equilibration, sample elution, column washing, and the like, are programmed into the liquid chromatography controller. The programs have to be put in by the user and the parameters of each program can be varied (e.g., flow rate, gradient times, and so on). In practice, for purification methods using the HR10/10 columns, we have found that a flow rate of 4.0–6.0 mL/min, with a salt gradient increasing from zero to 28% over about 30 min, is adequate for the single-step purification of IgG from serum. This is followed by a salt gradient of 100% for about 15 min to purge the column of remaining proteins. IgG elutes somewhere between 10–25% in the salt gradient, usually about 15%. When IgG elutes at 25% salt (depending on pI), then it will tend to coelute with albumin (which elutes at about 27% salt). When this occurs, an alternative method of purification should be employed, e.g., gel filtration. A typical elution profile of sheep IgG is shown in Fig. 1.

When setting the sensitivity in the UV monitor control unit, 0.5 is usually a sensible setting to start with. In cases where a large amount of IgG is expected (from a hyperimmune serum, for example), a setting of 1.0 may be more appropriate. It may be useful to run a diluted sample prior to the main purification step in order to ascertain the sensitivity of the system.

17. The use of subclass specific antibodies or MAbs allows the immunoaffinity isolation of individual subclasses of IgG.
18. Protein A does *not* bind all subclasses of IgG, e.g., human IgG_3, mouse IgG_3, sheep IgG_1, and some subclasses bind only weakly, e.g., mouse IgG_1. For some species, IgG does not bind to protein A well at all, e.g., rat, chicken, goat, and some MAbs show abnormal affinity for the protein. Protein A can be used to separate IgG subclasses from mouse serum (*see* ref. 1). Proteins C and G bind to IgG from most species, including rat and goat, and recognize most subclasses (including human IgG_3 and mouse IgG_1 and IgG_3). However, they also contain albumin binding sites, which makes the natural proteins unsuitable for affinity purification. Recombinant DNA technology has enabled mutant proteins to be produced in which the albumin binding sites have been spliced out. These are very good for affinity chromatography, but are expensive.
19. Coupling at pH 6.5 is less efficient than at higher pH, but is less likely to compromise the binding ability of immobilized ligands (especially antibodies).
20. Check the efficacy of coupling by measuring the A_{280} of the ligand before and after coupling. Usually, at least 95% of the ligand is bound to the matrix.
21. Elution of bound substances is usually achieved by the use of a reagent that disrupts noncovalent bonds. These vary from "mild" procedures, such as the use of high salt or high or low pH, to more drastic agents, such as 8*M* urea, 1% SDS, or 5*M* guanidine hydrochloride. Chaotropic

agents, such as thiocyanate or pyrophosphate, may also be used. Usually, an eluting agent is selected that is efficient, but that does not appreciably denature the purified molecule; this is often a compromise between the two ideals. In view of this, highly avid polyclonal antisera obtained from hyperimmune animals are often not the best reagents for immunoaffinity purification, as it may be impossible to elute the IgG in a useful form. The 0.1*M* glycine-HCl buffer, pH 2.5 will elute most IgG from immunoaffinity and protein A and G columns, but may denature some MAbs. Most IgGs can be eluted from protein A by using pH 3.5 buffer. "Pre-elution" of the column with dissociating reagent just prior to affinity chromatography ensures that the isolated immunoglobulin is minimally contaminated with ligand.

22. Incubation of the IgG containing sample with the ligand-matrix is not always necessary, but will allow maximal binding to occur. Alternatively, slowly pump the sample through the column. Flow rate depends on the IgG concentration in the sample and the binding capacity and size of the affinity column.
23. The technique described for the purification of sheep polyclonal IgG by FPLC is easily applied to the purification of rodent MAbs. Rodent MAbs are easily prepared as an ascitic fluid, which, like a polyclonal serum, should be ammonium sulfate "cut" at 45% final concentration prior to final application to the FPLC column. Some workers have used hydroxyapatite (hydroxylapatite) for the purification of murine MAbs, but the authors have not found this to be very effective in most cases.
24. The production of human MAbs *(5)* is a difficult procedure requiring a lot of hard work and some good fortune. Once a stable cloned antibody producing cell line has been produced, the problem of purification must be addressed. If the monoclonal producing cell line is an Epstein-Barr virus (EBV) transformed lymphoblastoid cell line, then the simplest way to prepare IgG for purification is to grow a large volume of culture supernatant (about 200 mL) and then reduce the volume to about 5–10 mL under nitrogen. The concentrated culture supernatant can be treated with saturated ammonium sulfate, and the resultant precipitate redissolved and used in conventional IEC or FPLC.

    In cases where the cell line is a heterohybrid, that is, human/mouse hybrid, it may be possible to grow the cell line as a tumor in sublethally irradiated mice and produce ascitic fluid. If this proves successful, then the ascitic fluid can be treated with saturated ammonium sulfate and the precipitate redissolved and used in conventional IEC or

FPLC. If for some reason these cell lines fail to produce ascitic fluid (this does happen), then the method described for the EBV-transformed lymphoblastoid cell line will suffice.

25. Apparatus for preparing IEF gels and carrying out focusing can be purchased from several suppliers, or gel molds can be made in the laboratory workshop. Volumes of gel solution and so on, vary according to the system used. Isoelectric focusing in polyacrylamide gel is usually carried out using a horizontal ("flat bed") system, although vertical apparatus can be used; the procedure described here is for a flat-bed system. Efficient cooling is essential, as high voltages are used. Ready prepared focusing gels can be purchased; these are easy to use but are fairly expensive.
26. If plates are difficult to separate and/or the polymerized gel sticks to plates and tears, siliconize the top plate before assembling the gel mold.
27. Platinum wire electrodes are expensive and carbon rods can be substituted. These should be soaked for 10 min in the electrode solutions prior to use, then placed directly on the gel surface and held in position by a glass plate placed on top of them. Samples should be applied at least 1 cm from the anode or cathode rod. Electrophoretic grade acrylamide and *bis*-acrylamide must be used. If poor results are obtained, use a new batch of acrylamide or deionize the stock solution using Amberlite MB-1 resin (BDH). Ampholytes are supplied as 40% (w/v) solutions and they should be used at 1–2% (w/v) final concentration in the gel. Try a 1% concentration first; if poor results are obtained, increase the concentration to 2% w/v.
28. For narrow range IEF, it may be inappropriate to use strong acids and bases as electrode solutions (*see* ref. 1 and the manufacturer's instructions).
29. The "mini-gel" is easily and quickly prepared, consisting only of a resolving gel, and takes approx 1.5 h to run once set up. The sample (IgG) is prepared for electrophoresis under reduced conditions and is run in parallel with standard mol-wt markers. The sample is loaded onto the "mini-gel" and run at 150 V for approx 1.5 h. The gel is then stained using Coomassie blue R to detect protein bands, followed by destaining, after which the gel can be photographed and/or dried onto filter paper. A typical gel showing three different loadings of an FPLC-purified mouse IgG MAb is shown in Fig. 4. Run under the reduced conditions described, the γ heavy chains have a characteristic mol wt of approx 50,000 and the light chains, a mol wt of approx 22,000.

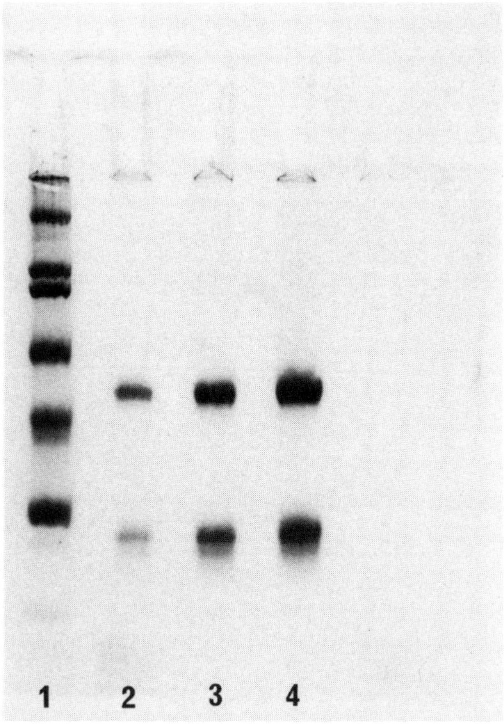

Fig. 4. SDS-PAGE of FPLC purified MAbs. SDS "mini-gel" electrophoresis analysis of an FPLC-purified mouse MAb stained with Coomassie blue R. Track 1 shows standard mol-wt markers (for details *see* Fig. 2). The remaining tracks show different loadings of the same MAb. Characteristic heavy chains (50,000) and light chains (22,000) are clearly seen to be free of contamination by other proteins.

Acknowledgments

We would like to thank Maryvonne Brasher for her part in the section on isoelectric focusing, Mark Page and Chris Ling for providing the sheep anti-p27, serum and Chris Bird for his part in the affinity purification of anti-IL3 MAbs. We are also grateful to Lisa Hudson for typing the manuscript.

References

1. Johnstone, A. and Thorpe, R. (1987) *Immunochemistry in Practice,* 2nd Ed. (Blackwell Scientific Publications, Oxford.
2. Burchiel, S. W., Billman, J. R., and Alber, T. R. (1984) Rapid and efficient purification of mouse monoclonal antibodies from ascites fluid using high performance liquid chromatography. *J. Immunol. Methods* **69,** 33–42.

3. Clezardin, P., McGregor, J. L., Manach, M., Boukerche, H., and Dechavanne, M. (1985) One-step procedure for the rapid isolation of mouse monoclonal antibodies and their antigen binding fragments by fast protein liquid chromatography on a Mono-Q anion exchange column. *J. Chromatography* **319,** 67–77.
4. Righetti, P. G. (1983) Isoelectric focusing: Theory, methodology and applications, in *Laboratory Techniques in Biochemistry and Molecular Biology* (Work, T. S. and Burdon, R. H., eds.), Elsevier Biomedical, Amsterdam.
5. Steinitz, M., Klein, G., Koskimies, S., and Makel, O. (1977) EB virus-induced B lymphocyte cell lines producing specific antibody. *Nature* **269,** 420.

CHAPTER 9

Epitope Mapping

Sara E. Mole

1. Introduction

Monoclonal antibodies (MAbs) are specific immunological tools because they bind to a precise determinant (the epitope) on the surface of a protein. The procedure of identifying the binding site of a MAb is often termed "epitope mapping."

Epitopes can be considered as a range of related forms that resemble one of three classes. Some epitopes consist of a linear or contiguous stretch of amino acids, which is not disrupted by denaturing agents (such as sodium dodecylsulfate, SDS), whereas others are disrupted because they consist of noncontiguous regions of the protein that come together to form the epitope. Yet others may not be recognized in the native protein, but are only exposed by denaturing agents.

However, for the purposes of epitope mapping, it is easiest to map epitopes present on the native protein that are resistant to denaturation, since these are easily identified and may consist of no more than a stretch of 4–10 amino acids that are in intimate contact with the antibody. I will be concentrating on this type of epitope.

The protein to which the MAbs bind and which is to be mapped may be large or small, and for the purposes of epitope mapping, is probably easily purifiable in the native form or as a bacterially expressed protein produced in large quantities. I will assume that the gene encoding the protein has been cloned and a cDNA clone is readily available. I will also assume that a panel of MAbs raised against the protein is to be mapped.

From: *Methods in Molecular Biology, Vol. 10: Immunochemical Protocols*
Ed.: M. Manson

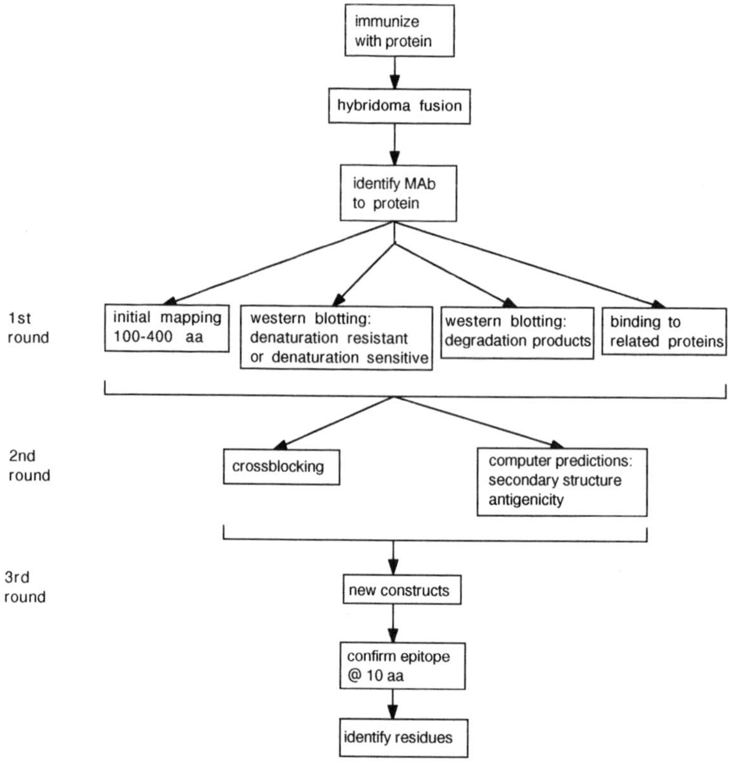

Fig. 1. Epitope mapping: an overall strategy.

Figure 1 summarizes the steps involved in epitope mapping. The first round of investigations serves to identify which MAbs are easy to map by using Western blotting of the native protein or of subcloned fragments to identify those MAbs that recognize denaturation sensitive or resistant epitopes, respectively. The mapping of denaturation-resistant epitopes can then be pursued.

This first round also includes the expression of the protein, if large, in smaller, more manageable fragments, such as 100–400 amino acids. Vectors suitable for this should possess strong and inducible promoters, allow expression of the insert in bacteria, and allow easy sequencing of the insert without the need for subcloning. Examples include vectors such as pUR *(1)*, pUC *(2)*, pT7 *(3)*, or pBluescriptII *(4)*.

Initial mapping of a MAb to the expressed protein fragments can be done by colony blotting and Western blotting. Colony blotting is a simple way to identify which expressed fragment of the protein binds the panel of MAbs. In this context, it is the primary epitope-mapping strategy. Initial mapping using Western blotting identifies degradation products of the expressed protein. This can differentiate between MAbs whose sites are only present in the full-length expressed protein and which are probably cleaved away from the main bulk of the protein, and those located in a more central region of the protein whose sites are also present in degradation products.

The different approaches used in the first round of epitope mapping generate information that should be analyzed in preparation for the next round.

Parallel approaches in this first round can also generate useful information. Checking the binding of MAbs to related proteins will identify those MAbs that bind to conserved regions and those that bind to dissimilar regions. Comparison of the sequences of related proteins indicates these regions. This information is of particular use in the second round of investigations in which computer predictions of secondary structure and antigenicity are used to indicate the possible location of epitopes, particularly as conserved regions of a protein may be of structural or functional significance. It may be best initially to concentrate on the epitope mapping of MAbs binding to these regions as potentially being of greatest interest.

The second round also includes crossblocking of MAbs to define classes of MAbs that bind to identical or overlapping regions. Epitope mapping is facilitated by choosing only one or two MAbs from each class for subsequent rounds of mapping.

Assimilating and analyzing the information gained from the first two rounds of mapping will suggest new constructs that can be used in the finer mapping of the binding sites. Obvious methods to express smaller fragments of the protein are: cloning of restriction fragments that encode part of the protein; deletion of regions from existing constructs by using deletion polymerase chain reaction (PCR); cloning designed oligonucleotides (20–70 nucleotides). These complementary methods should be used until the epitope has been localized.

The predicted epitope should be confirmed using cloned oligonucleotides or synthetic peptides. Important residues can be identified by mutagenesis or cloning of incorrect oligonucleotides or by using incorrect synthetic peptides.

Other, perhaps alternative, approaches that could be used for epitope mapping include homolog-scanning mutagenesis *(5),* which could also be a means of complementing the secondary structure predictions. This involves the synthesis of an array of small peptides that completely span the protein and so allows identification of many small continuous epitopes of a protein or protein fragment. Another method would be the random cloning of very small fragments of DNA encoding parts of the protein and screening each clone obtained with each available MAb. The fragments of DNA could be generated by DNase digestion of the purified open reading frame encoding the total protein.

2. Materials

All reagents should be of AR or ultrapure grade. All solutions should be made with double-distilled water and autoclaved or filter-sterilized where appropriate. Sterile pipetman tips and Eppendorfs, and so on, should be used.

2.1. Cloning of Oligonucleotides

1. Oligonucleotides: These should be complementary and encode the amino acid sequence suspected of containing an epitope.
2. Vector: This should contain a suitable restriction site for the cloning of the oligonucleotides, e.g., pUR *(1)*, pUC *(2)*, pT7 *(3)*, pBluescriptII *(4)*.
3. SOC medium: 2% Bactotryptone, 0.5% yeast extract, 10 m*M* NaCl, 2.5 m*M* KCl, 10 m*M* $MgCl_2$, 10 m*M* $MgSO_4$, 20 m*M* glucose, pH 7. The last three ingredients should not be autoclaved and can be added last after filter-sterilizing. Alternatively the complete medium can be filter-sterilized.
4. Competent *E.coli:* Can be purchased from Gibco/BRL (Paisley, Scotland) and stored at –70°C or they can be made in the laboratory. Competent cells should always be kept cold.
5. Ligation buffer (10X): 500 m*M* Tris-HCl, pH 8, 100 m*M* $MgCl_2$, 200 m*M* DTT (dithiothreitol; stock is 1*M* in 0.01*M* sodium acetate, pH 5.2), 10 m*M* ATP (adenosine 5'-triphosphate; stock is 100 m*M* in water at pH 7), 500 µg/mL BSA (bovine serum albumin). Store in aliquots at –20°C. Note that DTT and ATP are not stable and do not withstand repeated freezing and thawing.
6. Siliconized Eppendorfs: Coat Eppendorf tubes with a commercial siliconizing agent as recommended by the manufacturer. Rinse well with distilled water before autoclaving.
7. TE: 10 m*M* Tris-HCl, 1 m*M* EDTA, pH 8.
8. T4 ligase: Use the amount recommended by the manufacturer.
9. L-Agar plates: 1% bactotryptone, 0.5% yeast extract, 1% NaCl, pH 7.5 with NaOH, 1.5% agar, autoclave to sterilize. Pour 25 mL/90 mm plate.
10. Selective media: Include the appropriate antibiotic in L-agar plates.

2.2. Deletion PCR

11. TE: As in 7 above.
12. CAP/kinase buffer (10X): 500 m*M* Tris-HCl, pH 7.5, 100 m*M* $MgCl_2$, 50 m*M* DTT, 10 m*M* ATP.
13. T4 kinase: Use the amount recommended by the manufacturer.
14. T4 ligase: Use the amount recommended by the manufacturer.
15. Primers: Oligonucleotide primers are usually 21-mers. Their sequence is determined by the construct to be made. Here, they are identified as primer 1 and primer 2. Avoid the possible formation of hairpin structures within a primer, as this will inhibit the reaction.
16. Template DNA: Use less than 100 ng of the construct. If possible, linearize this with an appropriate restriction enzyme since this increases the efficiency of the reaction.
17. PCR buffer: As recommended by the manufacturer.
18. *Taq* polymerase: As recommended by the manufacturer.
19. Phenol/chloroform: Phenol equilibrated with TE should be mixed with an equal volume of chloroform.
20. Chloroform.
21. 3*M* Sodium acetate, pH 5.2.
22. Ethanol.
23. 80% Ethanol.

2.3. Cross Blocking

24. 96-Well microtiter PVC plates.
25. MAb: Purified MAb at a concentration of 5 µg/µL in a suitable buffer (such as 10 m*M* sodium phosphate, pH 7.2) is preferable. MAb supernatant may be sufficient. Store at 4°C.
26. PBS: 150 m*M* NaCl, 3 m*M* KCl, 8 m*M* Na_2HPO_4, 1.5 m*M* KH_2PO_4, pH 7.2.
27. Blocking buffer: 3% BSA in PBS containing 0.02% sodium azide (CARE—TOXIC).
28. Fetal calf serum (FCS).
29. NP-40.
30. Biotinylated MAb.
31. PBS containing 10% FCS and 0.1% NP-40.
32. ^{125}I streptavidin.

2.4. Peptide Binding

33. 96-Well microtiter PVC plates.
34. Synthetic peptide (nonconjugated) suspended in PBS at approx pH 7.

The concentration should be greater than 30 µg/mL.

35. Hydrogen peroxide (30% solution) stored at 4°C.
36. PBS: As in 26.
37. Blocking buffer: As in 27.
38. NP-40.
39. MAb as in 25.
40. Rabbit antimouse immunoglobulin conjugated to horseradish peroxidase (RbaMIgHRP). Store at 4°C.
41. Substrate stock: 10 mg of 3',3',5',5',-tetramethylbenzidine (TMB) in 10 mL of DMSO. Store in the dark at RT.
42. Substrate solution: 100 µL of substrate stock in 10 mL of 0.1*M* sodium citrate/phosphate buffer, pH 6, with 2 µL of hydrogen peroxide. This should be prepared fresh.
43. 1*M* H_2SO_4.

3. Methods

The nature of the topic necessarily spans a wide range of techniques. For simplicity, I have described only those methods that are not found in standard laboratory manuals or in previous chapters of this and earlier volumes of the same series, and which I felt merited a simple step by step method. This chapter, therefore, departs from the usual format, the alternative being a book within a book. For standard molecular biology techniques, I recommend ref. *6* and vols. 2 and 4 in this series, and for immunological techniques, ref. *7*. The methods I have included cover techniques for creating new constructs that are particularly relevant for epitope mapping and some immunological techniques that are best used as a basis for the development of an assay specific to the protein under investigation.

3.1. Cloning of Oligonucleotides

This method can be applied at any stage when a sequence suspected of containing an epitope is identified. This occurs in the third round in Fig. 1. The exact position in the overall procedure will depend on the size of protein under investigation. *See* Note 4 for additional information.

3.1.1. Annealing of Oligonucleotides

1. Mix 2 µg of each complementary oligonucleotide in 20 µL of 1X ligation buffer in siliconized Eppendorfs.
2. Heat at 65°C for 2 min and then incubate at 22°C for 30 min to allow reannealing.
3. Dilute 1 µL in 200 µL of TE to give a concentration of 10 ng/µL.
4. Store at –20°C.

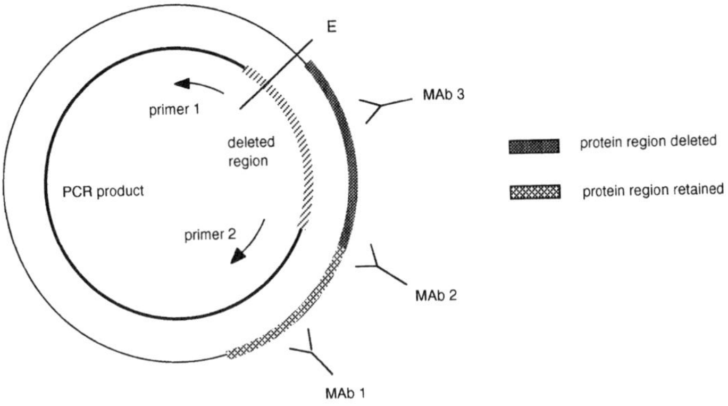

Fig. 2. Diagram illustrating the technique of deletion PCR. PCR is performed using primers flanking the region to be deleted. The template is an existing construct expressing the protein under investigation. Self-ligation of the PCR product results in the deletion of a region present in the original construct. This can be detected by the binding of MAb 1 and the absence of binding of MAb 3. The binding site of MAb 2 may be mapped in this new construct.

3.1.2. Ligation of Oligonucleotides with Vector

5. Mix 100 ng of digested vector with 10 ng or 100 ng of annealed oligonucleotides in 1X ligation buffer in a siliconized Eppendorf. Make up the volume to 10 µL. Include a tube with vector, but no oligonucleotides.
6. Add T4 ligase (use the manufacturer's recommended amount).
7. Incubate at 14°C overnight.

3.1.3. Transformation of E.coli

8. Transfer 2–4 µL of ligated mixture into 50 µL of competent cells.
9. Incubate on ice for 30 min.
10. Heatshock at 42°C for 90 s.
11. Incubate on ice for 2 min.
12. Add 4 vol of SOC medium and incubate at 37°C for 45 min with shaking.
13. Plate out on selective media. Colonies may then be screened with labeled oligonucleotide and suitable inserts sequenced.

3.2. Deletion PCR

This procedure is used to make new constructs as illustrated in the third round of investigations in Fig. 1. The procedure *(8)* is summarized in Fig. 2. Two primers flanking the region to be deleted are used in the PCR reaction. The orientation of these primers is such that amplification of the vector plus some of the cloned DNA occurs, resulting in the absence of a specific region of the cloned DNA.

3.2.1. PCR Reaction

1. Set up the following reaction (to 100 µL total volume): 10 µL of primer 1 (50 µg/mL stock); 10 µL of primer 2 (50 µg/mL stock); template DNA; PCR buffer as recommended by the manufacturer; *Taq* polymerase as recommended by the manufacturer.
2. Perform 20 cycles of the form: 92°C for 2 min; 50°C for 1 min; 72°C for 10 min. The exact conditions will depend on the make of the temperature cycling machine and should be varied accordingly.
3. Run 10 µL of the PCR product on a 1% agarose gel for verification of the reaction (include linearized template DNA as a control). The PCR product should be smaller by the size of the deletion compared to the control.
4. Extract once with phenol/chloroform; twice with chloroform.
5. Precipitate the DNA by adding sodium acetate, pH 5.2 to 0.3*M* and 2 vol of ethanol. Incubate on dry ice for 15 min. Spin, wash with cold 80% ethanol, and respin. Dry the pellet under vacuum and suspend the precipitated DNA in 20 µL of TE.

3.2.2. Processing and Identifying the PCR Product

The PCR product should be kinased before self-ligating, as follows.

6. To 10 µL of PCR product, add 1 µL of CAP/kinase buffer, 1 µL of 10 m*M* ATP, and 0.5 µL of T4 kinase. Incubate at 37°C for 1 h. Inactivate by heating to 70°C for 10 min.
7. Add 1 µL of 10 m*M* ATP, 1 µL of DTT, and 0.5 µL of T4 ligase. Incubate at 22°C for 1 h. Inactivate by heating to 70°C for 10 min.
8. If possible, digest this product using a restriction enzyme (E in Fig. 2) that has had its site deleted by the PCR reaction. This should decrease the number of background parental construct colonies obtained.
9. Transform *E. coli* and plate on a selective medium.
10. Screen the colonies using colony blotting.
11. Sequence those colonies that appear to be of the correct construction.

3.3. Crossblocking

This method is applied during the second round of investigations in Fig. 1. I will describe a typical assay. For each system, the optimum concentration range of reagents, such as the concentration of antigen for coating the plates and the concentration of primary Ab used must be identified. This is done by setting up duplicate dilution series in a pilot experiment (*see* Note 7). I have included the details of a typical assay here. The basic assay is summarized in Fig. 3.

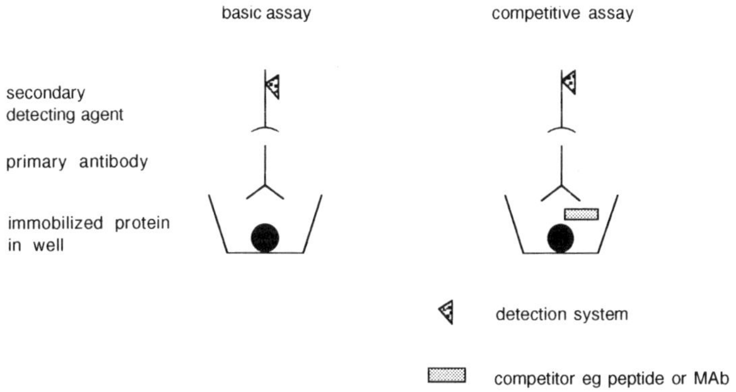

Fig. 3. Diagrammatic representation of a typical immunoassay used in epitope mapping. The antigen under investigation is immobilized in the well of a microtiter plate (*see* Note 6). To this is bound the primary antibody. This may have a ligand attached, such as biotin. The binding of the primary antibody to the antigen is detected using a secondary reagent coupled to a detection system, such as peroxidase, streptavidin, or a radiolabel. In the competitive assay, the competitor, such as another antibody or a peptide, is included. The inhibition of binding of the MAb under investigation is detected as before.

1. Coat the wells of a microtiter plate with antigen at a concentration of about 15–20 μg/mL in PBS using 50 μL per well. Leave overnight at RT in a humid environment.
2. Wash two to three times with PBS using a squirt bottle to fill up each well, and flick the contents out over a sink.
3. Block the remaining protein binding sites by filling each well with blocking buffer. Incubate for at least 3 h at RT in a humid environment.
4. Wash the wells two to three times with PBS and store plates at –20°C if not required the same day.
5. Add an appropriate dilution of MAb to the microtiter plate. Allow the MAb to bind for 1 h at RT (*see* Note 7).
6. Add 50 μL of biotinylated MAb diluted appropriately (such as 1 in 1000) in PBS containing 10% FCS and 0.1% NP-40. Incubate at 4°C overnight in a humid environment.
7. Wash three times in PBS containing 0.1% NP-40 and three times in PBS.
8. Add 50 μL of ^{125}I-streptavidin diluted 1 in 200 in PBS containing 10% FCS and 0.1% NP-40. Incubate for 1 h at RT in a humid environment. Wash three times in PBS containing 0.1% NP-40 and three times in PBS.
9. Cut off the wells using a hot wire cutter into a microtiter template and count the wells individually. The results can be plotted to show the MAb(s) that inhibit the binding of the biotinylated MAb.

3.4. Peptide Binding

This method is applied at the later stages of the epitope mapping procedure to precisely identify an epitope. I will describe a typical assay. As in the crossblocking assay, the optimum concentration range of reagents must be identified for each system. Figure 3 summarizes the assay.

Steps 1–4 are as in Section 3.3. above.

5. Mix the MAb (to give 5 µg/mL) with the synthetic peptide in doubling dilutions of peptide across a fresh microtiter plate. Use 10× excess of peptide initially and make the dilutions in blocking buffer. This is best done in duplicate. Include controls, such as an irrelevant peptide, irrelevant MAb, and so on. Allow the MAb and peptide to bind for at least 3 h.
6. Transfer 50 µL of the MAb-peptide solution to the corresponding wells of the antigen plate and incubate overnight at RT in a humid environment.
7. Wash three times in PBS containing 0.1% NP-40 and three times in PBS.
8. Add 50 µL of RbaMIgHRP diluted 1 in 500 in PBS containing 3% BSA and 10% FCS (no azide; *see* Note 7). Incubate for 1 h at RT in a humid environment.
9. Add 50 µL of substrate solution and incubate for 10–30 min at RT until a blue color develops.
10. Add 50 µL of 1*M* H_2SO_4 to stop the reaction and convert the color to bright yellow. Read the plate at 405 nm on an ELISA plate reader.

4. Notes

1. Positive and negative binding. It is important to use only positive binding data in epitope mapping. Although lack of binding may be caused by the absence of residues required for binding, this failure to bind, particularly in the case of discontinuous epitopes, may simply reflect a different conformation to the native form. The use of overlapping constructs to map epitopes can be very useful, since it avoids the temptation to use negative data in assigning a binding site, because the site is defined by the region of overlap between two constructs that bind the same MAb. When using small fragments, the flanking regions of the protein may also help stabilize the actual epitope region as compared with the expression of an oligonucleotide encoding, for example, only 10 amino acids.
2. Importance of controls. In colony blotting in particular, it is important to include both positive and negative controls, since some MAb supernatants are more crossreactive than others, although still able to be used

providing the level of background can be checked. The level of background binding can also be a problem in the immunoassays. This can usually be solved by altering the concentrations of the various reagents.

3. Mapping denaturation-sensitive epitopes. The mapping of denaturation sensitive epitopes is much more difficult, since the spatial relationship of different residues must be maintained in order for the MAb to bind. There is no reason why MAbs recognizing discontinuous epitopes should not be mapped, but the extent to which this is successful will depend on the ability of small fragments of the protein to fold up as if they are part of the native protein. One possibility is to separate fragments by stretches of residues that bear no resemblance to the native protein but which serve to keep the same conformational structure.
4. Cloning of oligonucleotides. I have successfully cloned oligonucleotides up to 70 bp in length. There should be no limit apart from those associated with synthesis. Any problems are likely to arise from inaccuracies in the sequence and length of long oligonucleotide products. Cloning is facilitated by incorporating overhanging ends that are part of restriction sites in the design of the oligonucleotides. The vector is then cleaved with the appropriate restriction enzyme. Dephosphorylation of the linearized vector increases the efficiency of cloning.
5. Binding to related proteins. A simple cell-staining assay can be used to investigate the binding of MAbs to virally produced proteins as a result of infection with related viruses. If the protein under investigation is a member of a cellular family of proteins, then immunoprecipitation using one MAb and Western blotting of the different proteins pulled out using other MAbs or a crossreacting serum would give information on the epitopes conserved within the family.
6. Immunological techniques. The source of antigen could be overproduction by a bacterial expression system or purified native antigen.

   Some epitopes, usually those sensitive to denaturation, are lost on binding to plates. The assay should be modified, if this is a problem, so that the protein is not used to coat the plates.
7. Crossblocking. The identification of the appropriate dilution of MAb to be used in the assay can be done by performing the basic assay using a fixed concentration for the antigen, and performing a dilution series covering a range of MAb concentrations. Plotting the results in log form should result in a sigmoid-shaped graph. The dilution of MAb suitable for the assay is one found on the steep part of the graph.

   Controls, as in any experiment, are essential and should include MAbs that do not bind to the antigen. Each well should be set up in duplicate.

Azide should not be included in the solution containing RbaMIgHRP, as this inhibits the action of the enzyme.

References

1. Rüther, U. and Müller-Hill, B. (1983) Easy identification of cDNA clones. *EMBO J.* **2**, 1791–1794.
2. Yanisch-Perron, C., Vieira, J., and Messing, J. (1985) Improved M13 phage cloning vectors and host strains: Nucleotide sequences of the M13mp18 and pUC19 vectors. *Gene* **33**, 103–119.
3. Tabor, S. and Richardson, C. C. (1985) A bacteriophage T7 RNA polymerase/promoter system for controlled exclusive expression of specific genes. *Proc. Natl. Acad. Sci.* **82**, 1074–1078.
4. Stratagene catalogue (1989), pp. 156–159.
5. Cunningham, B. C., Jhurani, P., Ng, P., and Wells, J. A. (1989) Receptor and antibody epitopes in human growth hormone identified by homolog-scanning mutagenesis. *Science* **243**, 1330–1336.
6. Sambrook, J., Fritsch, E. F., and Maniatis, T. (1989) *Molecular Cloning: A Laboratory Manual*, 2nd Ed. Cold Spring Harbor Laboratory, Cold Spring Harbor, New York.
7. Harlow, E. and Lane D. P. (1989) *Antibodies: A Laboratory Manual*, 2nd Ed. Cold Spring Harbor Laboratory, Cold Spring Harbor, New York.
8. Mole, S. E, Iggo, R. D, and Lane D. P. (1989) Using the polymerase chain reaction to modify expression plasmids for epitope mapping. *Nucleic Acids Res.* **17**, 3319.

CHAPTER 10

Enzyme–Antienzyme Method for Immunohistochemistry

Michael G. Ormerod and Susanne F. Imrie

1. Introduction

Immunohistochemical stains use antibodies to identify specific constituents in tissue sections. In order to detect the site of reaction, the antibody is labeled with an enzyme that can be reacted with a suitable substrate to give a colored product. The alternative is to use a fluorescent label. The advantage of an enzyme label is that the nuclei can be counterstained, thereby revealing the tissue architecture, and that the stain fades slowly, if at all, with time, allowing the slides to be stored.

In the original method, the section was incubated with the primary antibody, followed by a secondary antibody to which a suitable enzyme had been attached. (For example, if the primary antibody was a mouse monoclonal, the secondary antibody could be a goat antimouse immunoglobulin.) It was later found that the number of enzyme molecules per molecule of primary antibody could be increased by using an enzyme–antienzyme method, thereby increasing the sensitivity.

The basic enzyme–antienzyme method is outlined in Fig. 1. The primary antibody is bound to the antigen of interest followed by incubation with an appropriate anti-immunoglobulin antibody, and finally with a complex of enzyme–antienzyme antibodies. The primary antibody and the antienzyme antibodies are raised in the same species so that the anti-immunoglobulin will link the two together. The enzyme used is usually either horseradish per-

From: *Methods in Molecular Biology, Vol. 10: Immunochemical Protocols*
Ed.: M. Manson

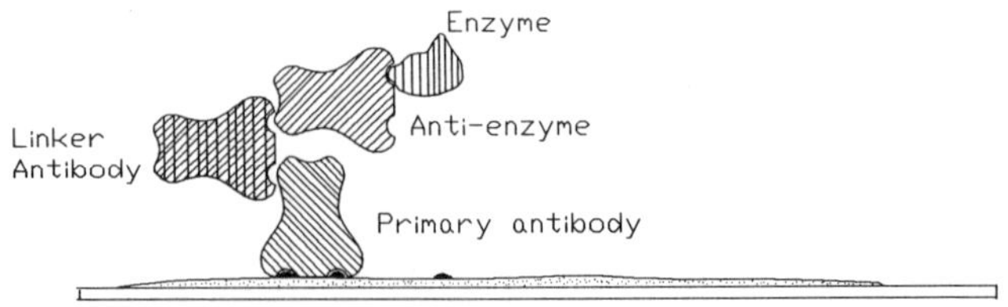

Fig. 1. The enzyme–antienzyme method. The primary antibody and the antienzyme antibody must have been raised in the same species. Because there are several epitopes on the enzyme and the immunoglobulins are divalent, a large complex of enzyme–antienzyme molecules builds up. This increases the number of enzyme molecules linked to each primary antibody molecule. This effect is not shown in the figure because it would make the diagram too complicated.

oxidase, in which case the method is called peroxidase–antiperoxidase (PAP) or alkaline phosphatase, where the method is referred to as APAAP. These enzymes are chosen because they have a high turnover number (giving a high yield of product) and have substrates that can give an insoluble, colored product.

For further general reading about immunohistochemistry, *see* refs. *1–3*. For a more detailed discussion of the methods involved in immunohistochemistry, *see* ref. *4*.

2. Materials

1. Xylene or Histoclear.
2. Ethanol.
3. Phosphate buffered saline (PBS): 8.5 g of sodium chloride, 1.07 g of disodium hydrogen orthophosphate (or 2.7 g of $Na_2HPO_4{\cdot}2H_2O$), 0.39 g of sodium dihydrogen orthophosphate (or 0.51 g of $NaH_2PO_4{\cdot}2H_2O$). Make up to 1 L with distilled water.
4. Bovine serum albumin (BSA).
5. Detergent—BRIJ or Tween.
6. Primary antibody (*see* Notes 3 and 4).
7. Anti-immunoglobulin (linking) antibody raised against the species used to produce the primary antibody. (*See* Introduction and Note 4.)
8. Enzyme–antienzyme complexes. The antienzyme antibody must have been raised in the species used to produce the primary antibody. (*See* Introduction and Note 4.)
9. Diaminobenzidine (DAB) solution. Dissolve *y* mg of DAB in *y* mL of 0.1 *M* Tris buffer, pH 7.2. Add *y* mL of distilled water containing 2/3 *y* µL of 30% H_2O_2 (*see* Note 1).

10. Mayer's hemalum: 1 g of hematoxylin, 50 g of aluminium potassium sulfate, 0.2 g of sodium iodate, 1 g of citric acid, 50 g of chloral hydrate, and 1 L distilled water. Dissolve the hematoxylin in distilled water, using gentle heat if necessary. Add the alum, heat if necessary. Add the sodium iodate, mix well and leave overnight. Then add citric acid, mix well and add the chloral hydrate.
11. Hanker Yates Reagent (HYR): Dissolve 7.5 mg of HYR in 5 mL of 0.1*M* Tris-HCl buffer, pH 7.6, and add 6 µL of 30% H_2O_2. (HYR contains paraphenylenediamine plus pyrocatechol.)
12. Carbazole: Dissolve 2 mg of 3-amino, 9-ethyl carbazole in 0.5 mL of dimethyl formamide (DMF) in a glass tube. Add 9.5 mL of 0.2*M* acetate buffer, pH 5. Just before use, add 5 µL of 30% H_2O_2.
13. DPX—mountant for cover slips: As supplied by the manufacturer.
14. Glycerin jelly (alternative mountant): 10 g of gelatin, 70 mL of glycerin, 0.25 g of phenol, and 60 mL of distilled water. Dissolve the gelatin in distilled water using gentle heat. Add glycerin and phenol and mix well. Aliquot into 10 mL of batches and store in the refrigerator. For use, melt in a water bath at 60°C—avoid shaking as this creates air bubbles.
15. Chloronaphthol: Dissolve 20 mg of 4 chloro-1-naphthol in 40 mL of 20% methanol in 0.05*M* Tris saline, pH 7.6 (0.6 g of Tris base plus 3 g of NaCl in 100 mL of distilled water, adjust the pH to 7.6 with HCl). Add 13.5 µL of 30% H_2O_2 and heat gently to 50°C before use.
16. Tetramethyl benzidine: Dissolve 5 mg of tetramethyl benzidine in 2 mL of DMSO. Add to 50 mL of 0.02*M* acetate buffer, pH 3.3, containing 20 µL of 30% H_2O_2 imediately before use.
17. Methyl green.
18. Veronal acetate buffer: 0.97 g of sodium acetate (trihydrate), 1.47 g of sodium barbitone, 250 mL of fresh distilled water (CO_2-free) and 2.5 mL of 0.1*M* hydrochloric acid, pH 9.2.
19. Fast Red salt: Dissolve 5 mg of naphthol AS BI phosphoric acid sodium salt in 1 drop of DMF in a glass tube. Dissolve 5 mg of Fast Red TR salt in 10 mL of veronal acetate buffer, pH 9.2. Mix the two solutions together and filter.
20. Fast Blue salt: Dissolve 5 mg of naphthol AS BI in DMF in a glass tube and add to 5 mg of Fast Blue BB salt in 10 mL of 0.1*M* Tris-HCl buffer, pH 9.
21. New Fuchsin: Mix 250 µL of a 4% solution of New Fuchsin in 2*M* HCl with 250 µL of 4% sodium nitrite. Leave the mixture to stand in the cold for 5 min, then add it to 40 mL of 0.2*M* Tris-HCl buffer, pH 9, and add 10 mg of naphthol AS TR phosphoric acid dissolved in 0.2 mL of DMF in a glass tube.

22. Periodic acid.
23. Potassium borohydride.
24. Acetic acid: A 20% solution in distilled water.
25. Pronase solution: 50 µg of pronase per mL of PBS.

3. Methods

3.1. The Basic Method

1. If the section has been cut from a paraffin block, take it through xylene (or Histoclear) and ethanol to water (*see* Note 2).
2. Block any endogenous enzyme activity as appropriate (*see* Section 3.3.). Wash the section in water.
3. Use proteolytic enzyme if appropriate (*see* Section 3.4.). Wash the section in water.
4. Rinse the section in PBS, then wipe any excess from the slide, so that the antiserum is not diluted too much on the slide.
5. Incubate for 1 h at room temperature in a moist chamber with 100 µL of primary antibody appropriately diluted in either PBS, 0.5% BSA, or (preferably) PBS containing 5% serum, the serum being obtained from the species in which the second antibody was raised (*see* Notes 3–6).
6. Wash the section with PBS, 0.5% BSA followed twice by PBS containing 0.01% detergent (BRIJ or Tween-80).
7. Repeat Step 4.
8. Incubate for 1 h at room temperature in a moist chamber with 100 µL of second, anti-immunoglobulin, antibody appropriately diluted (*see* Note 4).
9. Repeat Steps 6 and 7.
10. Incubate for 1 h at room temperature in a moist chamber with 100 µL of the enzyme–antienzyme complex appropriately diluted.
11. Repeat Steps 6 and 7. Wash in distilled water.
12. Develop the color and counterstain and mount as appropriate. (*See* Notes 7 and 8 and Section 3.2.)

3.2. Developing the Colored Product

These procedures relate to Step 12 in Section 3.1.

For all the methods listed below, the substrate solutions should be prepared fresh and the sections should be at room temperature.

*3.2.1. Peroxidase Substrates (*See *Note 9)*

1. Diaminobenzidine (DAB): Cover the section with this substrate and incubate the slide for 5 min. The product is brown and is stable in alcohols

and in xylene. Counterstain with Mayer's hemalum for 5 min. Wash in tap water. Dip the slide in saturated lithium carbonate for a few seconds; this makes the nuclear stain blue. Wash in tap water. Dehydrate through ethanol and xylene (or Histoclear) and mount in a permanent mountant, e.g., DPX.

2. Hanker Yates: Incubate the slides for 15 min. The product is blackish-brown and is stable in xylene and alcohols. Counterstain with hemalum and mount in DPX.
3. Carbazole: Incubate the slides for 15 min. The product is red, insoluble in water but dissolves in organic solvents. Counterstain with hemalum and mount in a water-based mountant, e.g., glycerin jelly.
4. Chloronaphthol: Incubate the slides for 8 min. The product is blue and soluble in xylene. We do not know a suitable counterstain; hemalum being blue is unsuitable and methyl green is water soluble. Mount in glycerin jelly.
5. Tetramethyl benzidine: Incubate the slides for 15 min. The product is blue and stable in xylene and alcohols. Counterstain with methyl green and mount in DPX.

3.2.2. Alkaline Phosphatase Substrates

1. Fast Red salt: Cover the section with this substrate and incubate for 45–60 min. The product is red and is soluble in organic solvents. Counterstain with hemalum and mount in glycerin jelly.
2. Fast Blue salt: Use two or three 5-min incubations, each in fresh substrate. The product is blue and soluble in organic solvents. There is no satisfactory counterstain. Mount in glycerin jelly.
3. New Fuchsin: Incubate the slides for 10 min. The product is red and stable in xylene and alcohols. Counterstain with hemalum and mount in DPX.

3.3. Blocking Endogenous Enzymes

The tissue under study may contain endogenous enzyme that is the same as that used in the enzyme–antienzyme stain. It is desirable that this should be destroyed otherwise interpretation of the stained slide is difficult.

1. For peroxidase, incubate the section in one of the following solutions for the time indicated and then wash in tap water before starting the method in Section 3.1.
   a. 2.3% Periodic acid in distilled water for 5 min or
   b. 0.03% Potassium borohydride in distilled water (freshly prepared) for 2 min or

c. 0.1% Phenylhydrazine in PBS for 5 min.

The last treatment is the most gentle and should be used for labile antigens.

2. For alkaline phosphatase, incubate the section in 20% acetic acid for 5 min, then wash it in tap water. This treatment may destroy the antigen. An alternative for tissues other than the intestine is to make the substrate solution (Section 3.2.2.) 1 m*M* in levamisole (increase to 2 m*M* for tissues rich in alkaline phosphatase, e.g., kidney or placenta).

3.4. Revealing "Hidden" Antigens

In fixed tissue, some antigens can be revealed by treatment with a proteolytic enzyme. This protocol uses pronase.

1. Dewax the section in xylene or Histoclear and take it through ethanol to water.
2. Incubate the section in PBS at 37°C for 5 min, then in pronase solution at 37°C for 20 min.
3. Wash the section in running tap water for 5 min, then wash twice in PBS.

   After enzyme treatment, the sections are very fragile and must be handled with care.

4. Notes

1. DAB is a suspect carcinogen and must be treated with care. Use in a fume hood.
2. For routine histopathology, tissues are usually fixed in formalin and embedded in paraffin wax. Sections cut from this material must be dewaxed in xylene and taken through ethanol to water. This gives sections of high quality but, unfortunately, many antigens are destroyed by this process. Some will survive fixation in ethanol-based fixatives, such as Methacarn (60% methanol, 30% chloroform, 10% glacial acetic acid). Others will not survive embedding in paraffin and must be studied in sections cut from unfixed, frozen tissue. These sections are often briefly fixed before use. The correct fixative has to be determined empirically for each antigen under study.
3. The primary antibody can be in the form of a polyclonal antiserum or a monoclonal antibody produced either in a culture supernatant or in an ascitic fluid. The concentration of the antibody in an ascitic fluid will be an order of magnitude greater than that in the culture supernatant but the latter will be free of other immunoglobulins.
4. The appropriate concentration of the primary antibody must be determined by using serial dilutions on a set of sections cut from a tissue

known to carry the antigen of interest. The linking antibody and the enzyme–antienzyme complex are best used at the concentration recommended by the manufacturer. If desired, this can be checked using serial dilutions.

5. For incubations in antibody, the slides are usually placed in a suitable box in which they can be kept horizontally in a humid atmosphere. This is important because the solutions must not be allowed to dry out during incubation.
6. In the methods described, incubation at room temperature is recommended. However, room temperature can vary considerably and, if the final product is to be quantified (for example, by image cytometry), the temperature of incubation should be carefully controlled.
7. The substrate chosen for color development depends on the enzyme used and the final color required by the investigator. For routine work, diaminobenzidine is normally used with peroxidase and Fast Red with alkaline phosphatase.
8. The choice of mountant depends on the solubility characteristics of the colored product. Permanent mountants, such as DPX, are based on organic solvents. If the colored product is soluble in such solvents, then a water-based mountant must be used. Guidance is given for each substrate described.
9. When using peroxidase and diaminobenzidine, it is sometimes necessary to bleach the brown pigment found in fixed tissue sections, e.g., in formol saline fixed red blood cells. Bleach with 7.5% hydrogen peroxide in distilled water for 5 min. Wash well in tap water.
10. Different tissues contain different isoenzymes of alkaline phosphatase and the intestinal isoenzyme is not inhibited by levamisole. The enzyme used in immunohistochemistry is extracted from calf intestine so that levamisole can be used as an inhibitor without effecting the desired reaction. For labile antigens in the intestine, it is better to switch to the peroxidase method.
11. Control sections should be included with each set of stained slides. A positive control from a tissue known to contain the antigen should be included for each primary antibody used. The negative control is normally a section stained with the omission of the primary antibody.
12. The negative control should be quite clean. If a reaction is observed, check for the presence of endogenous enzyme or pigment. In their absence, try to reduce the background by more careful washing and increasing the protein in the washing solution. Finally, try fresh secondary reagents purchased from a different source.

13. If no staining is observed in the positive control, check whether a reagent has been inadvertently omitted or whether the wrong reagent has been used (for example, antimouse Ig on a rat monoclonal). This is easy to do if several different antibodies from different species are being used.
14. If staining is weak, one of the reagents may have deteriorated. For example, stock solutions of H_2O_2 should be renewed regularly.

Acknowledgments

MGO was supported by a program grant from the Cancer Research Campaign.

References

1. Sternberger, S. S. and De Lellis, R. A., eds. (1982) *Diagnostic Immunohistochemistry.* Masson Publishing USA, New York.
2. Bullock, G. R. and Petrusz., P., eds. (1982 and 1983) *Techniques in Immunocytochemistry,* vols. 1 and 2. Academic, London.
3. Polak, J. M. and van Noorden, S., eds. (1983) *Immunocytochemistry.* John Wright and Sons, Bristol.
4. Ormerod M. G. and Imrie S. F. (1989) Immunohistochemistry, in *Light Microscopy in Biology. A Practical Approach* (Lacey, A. J., ed.), Oxford University Press.

CHAPTER 11

Double Label Immunohistochemistry on Tissue Sections Using Alkaline Phosphatase and Peroxidase Conjugates

Jonathan A. Green and Margaret M. Manson

1. Introduction

One of the most common methods in immunohistochemistry involves the use of an antibody to the antigen of interest detected indirectly with an enzyme-labeled antispecies secondary antibody. The enzyme catalyzes the formation of a colored insoluble reaction product at the antigen site. It is possible, with careful choice of reagents, to label two antigens simultaneously, resulting in two different colored reaction products *(1)*. Cells or tissue sections can also be double-labeled with two antispecies secondary antibodies carrying different fluorochromes (*see* this vol., Chapter 42), or by using suitable antibodies conjugated to different sizes of colloidal gold (*see* this vol., Chapter 19).

Described here is an indirect method for detecting two different cellular antigens in acetone-fixed tissue, using a rabbit polyclonal antibody, and a murine monoclonal antibody on the same section. One secondary antispecies antibody is conjugated with alkaline phosphatase, the other with peroxidase, thus resulting in two differently colored products showing the localization of the two antigens (Fig. 1).

2. Materials

1. Acetone-fixed, paraffin-embedded tissue sections (*see* Notes 1 and 2).
2. Xylene.
3. 70 and 100% Ethanol.

From: *Methods in Molecular Biology, Vol. 10: Immunochemical Protocols*
Ed.: M. Manson ©1992 The Humana Press, Inc., Totowa, NJ

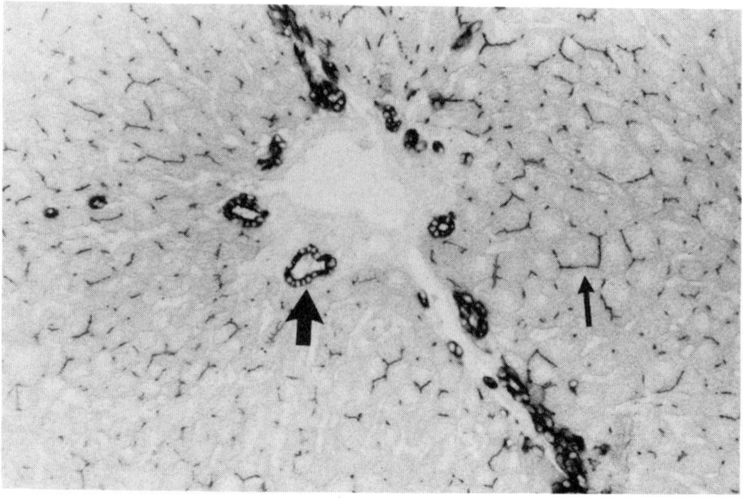

Fig. 1. Double label immunohistochemistry on rat liver. An acetone-fixed rat liver section was incubated with a polyclonal antiserum raised in rabbit to a hepatocyte cell surface protein (courtesy of Dr. S. Stamatoglou) and a mouse monoclonal antibody against a bile duct specific cytokeratin (courtesy of Dr. E. B. Lane). The hepatocyte protein was localized by use of a secondary peroxidase conjugated antibody resulting in a red/brown product (thin arrow). The bile duct cytokeratin was identified by using an alkaline phosphatase conjugated secondary antibody giving a blue color (thick arrow).

4. 15% Glacial acetic acid in water.
5. 0.3% Hydrogen peroxide in water.
6. Phosphate buffered saline (PBS), pH 7.4: 8 g of NaCl, 0.2 g of KH_2PO_4, 2.8 g of $Na_2HPO_4 \cdot 12H_2O$, and 0.2 g of KCl dissolved and made up to 1 L in distilled water.
7. PBS–BSA: PBS containing 1% bovine serum albumin.
8. PBS–Tween: PBS containing 0.05% (v/v) Tween-80.
9. Tris buffer: 0.1*M* Tris-HCl, pH 8.2.
10. Acetate buffer: 0.2*M* Sodium acetate, pH 5.0.
11. Primary antibody raised in rabbit against antigen 1 (*see* Note 3).
12. Peroxidase-conjugated antirabbit IgG raised in sheep (*see* Note 3).
13. Primary mouse monoclonal antibody against antigen 2.
14. Alkaline phosphatase conjugated antimouse IgG raised in goat.
15. BCIP/NBT substrate: Dissolve 5-bromo-4-chloro-3-indolyl phosphate at 50 mg/mL in dimethyl formamide. Dissolve nitroblue tetrazolium at 75 mg/mL in 70% (v/v) dimethyl formamide. Add 30 μL of BCIP and 45 μL of NBT to 10 mL of Tris buffer, pH 8.2 immediately before use.

16. AEC substrate: Dissolve 2 mg of 3-amino, 9-ethyl carbazole, in 500 µL of dimethyl formamide. Add 9.5 mL of acetate buffer and 10 µL of 30% H_2O_2. Use immediately.
17. Aqueous mountant: Apathy's mountant.

3. Method

1. Place the sections (*see* Notes 2 and 3) in a slide rack, dewax in xylene for 20 min, followed by transfer to a fresh xylene bath for a further few minutes. Then rehydrate through an ethanol series, 2 × 100%, 1 × 70%, 1 × distilled water (*see* Note 4).
2. Rinse the sections in PBS–Tween. This helps to prevent nonspecific binding, and also makes subsequent solutions easier to apply to the slide.
3. Place the sections in the acetic acid solution for 15 min to block endogenous alkaline phosphatase activity (*see* this vol., Chapter 10).
4. Rinse as in Step 2.
5. Place the sections in 0.3% hydrogen peroxide for 15 min to block endogenous peroxidase activity.
6. Rinse as in Step 2.
7. Place the slides horizontally in a moist chamber and incubate for 30 min with 100 µL of PBS–BSA to block nonspecific binding sites.
8. Remove the solution from the slide and wipe the excess from around the section so that the antiserum is not diluted.
9. Incubate with both primary antibodies, suitably diluted in 100 µL of PBS–BSA (*see* Note 5), for 2 h in the moist chamber.
10. Rinse the slides three times with PBS–Tween, 2 min for each rinse.
11. Wipe around the sections as before and apply 100 µL of liquid containing both secondary antibodies diluted 1:100 in PBS–BSA and incubate in the humid chamber for 1 h.
12. Rinse the sections as in Step 10.
13. Flood the sections with AEC substrate solution and incubate for 15 min (*see* Note 6).
14. Rinse the sections as in Step 2.
15. Flood the sections with BCIP/NBT substrate solution and incubate for 15 min.
16. Wash as in Step 2.
17. Dry around the sections and mount in Apathy's mountant (*see* Note 7).

Sites of positive reaction product are seen as a red/brown granular deposit where the peroxidase conjugated second antibody is bound, and a purple/blue deposit where the alkaline phosphatase conjugated antibody is bound.

4. Notes

1. To prepare tissue, 2–3 mm slices are removed as quickly as possible into ice-cold acetone and left on ice for 4–6 h. (Less than 4 h can result in incomplete penetration of fixative and subsequent formation of ice crystals). Samples are then stored in a –20°C freezer until they can be processed using routine paraffin embedding techniques.
2. Care should be taken not to allow the temperature of the paraffin bath to exceed 60°C, since most laboratories use embedding media containing plastics. Above 62°C, these will begin to form polymers, which are very difficult to remove, and may give rise to background staining. This will show up as a pale ring of staining extending beyond the tissue section.
3. One of the most likely reasons for failure with this technique is the wrong choice of antibodies. It is important that the two primary antibodies have been raised in different species and that the two secondary antibodies have been raised in species distinct from either of the primary antibody species.
4. Xylene should be changed frequently. Ideally, not more than 100 slides should be dewaxed per 400 mL. This helps to reduce background by removing the wax more efficiently. Xylene is toxic and should ideally be handled in a fume hood. Histoclear is a commercially available, safer alternative to xylene.
5. The dilution of the antibodies has to be determined empirically. Polyclonal antibodies can usually be diluted between 1:50 and 1:250. Monoclonal antibodies in tissue culture supernatant may need to be concentrated by centrifuging a frozen sample in a microfuge for 10 min and discarding the top third of the solution. This solution (100 µL) can then have the other primary antibody diluted into it in place of PBS–BSA. Monoclonal antibodies produced as ascites can usually be diluted between 1:200 and 1:1000.
6. It is important to carry out the reaction for peroxidase first since the H_2O_2 contained in this reaction mixture might bleach any color that was already deposited on the section.
7. No satisfactory counterstain is known. The blue of hematoxylin is too close to that of the alkaline phosphatase product to allow good photography.

Acknowledgments

The authors are grateful to Stamatis Stamatoglou, National Institute for Medical Research, Mill Hill, London and Birgitte Lane, University of Dundee, U.K. for supplying antibodies.

Reference

1. Mason, D. Y. and Sammons, R. E. (1978) Alkaline phosphatase and peroxidase for double immunoenzymatic labelling of cellular constituents. *J. Clin. Pathol.* **31**, 454–460.

CHAPTER 12

Immunohistochemical Detection of Bromodeoxyuridine-Labeled Nuclei for In Vivo Cell Kinetic Studies

Jonathan A. Green, Richard E. Edwards, and Margaret M. Manson

1. Introduction

The classical technique for identifying cells engaged in DNA synthesis is by their uptake of [^{3}H]-thymidine, detected using autoradiography. However, this method can be inconvenient, as specialized darkroom and radioisotope facilities are required, with the potential health hazard that handling isotopes entails. Bromodeoxyuridine (BrdU), the halogenated 5-substituted derivative of deoxyuridine, is a thymidine analog specifically incorporated into the DNA of proliferating cells during S phase. This is now a well-established alternative to ^{3}H thymidine, since it has been shown that labeling indices for the two molecules are the same *(1,2)*. The development of a monoclonal antibody *(3)* that recognizes BrdU incorporated into single-stranded DNA has resulted in several techniques using immunocytochemical staining to detect incorporated BrdU in frozen, paraffin- and plastic-embedded sections of tissue by light microscopy. It has also proved extremely valuable for studies in conjunction with flow cytometry and even, for in vivo studies of human tumor cell kinetics (*see* this vol., Chapter 43). We describe here a method to detect DNA synthesis by in vivo labeling of nuclei with BrdU, followed by indirect immunological detection in paraffin-embedded tissue *(4)*.

From: *Methods in Molecular Biology, Vol. 10: Immunochemical Protocols*
Ed.: M. Manson ©1992 The Humana Press, Inc., Totowa, NJ

2. Materials

1. Bromodeoxyuridine: 50 µg/g Body wt dissolved in 1 mL of sterile 0.9% saline. Prepare fresh.
2. Carnoys fixative: 60% Absolute alcohol, 30% chloroform, and 10% glacial acetic acid.
3. Glass microscope slides that have been coated with 3-aminopropyltriethoxysilane (APES) (*see* Note 1).
4. Xylene.
5. 100, 90, and 70% ethanol series.
6. Phosphate buffered saline (PBS), pH 7.4: 8 g of NaCl, 0.2 g of KH_2PO_4, 2.8 g of $Na_2HPO_4 \cdot 12H_2O$, 0.2 g of KCl dissolved and made up to 1 L in distilled water.
7. 0.3% H_2O_2 in distilled water.
8. 1*M* HCl.
9. Monoclonal antibody to BrdU: A rat monoclonal obtained from Sera-Lab, Crawley Down, Sussex, UK.
10. Rabbit antirat peroxidase conjugated secondary antibody.
11. Substrate solution: A stock solution of diaminoazobenzidine (DAB) is made up at 5 mg/mL in PBS, aliquoted into 0.5 mL amounts, and stored frozen. For use, add 4.5 mL of PBS to 0.5 mL of DAB stock and add 50 µL of freshly prepared 1% H_2O_2 immediately before use.
12. Hematoxylin.
13. Mountant: DPX.

3. Method

3.1. Preparation of Tissue Sections

1. Dissolve the required amount of BrdU in 1 mL of sterile saline immediately prior to use and protect the solution from light to prevent photodecomposition.
2. Inject the solution ip and sacrifice the animal 1 h later (*see* Notes 2 and 3).
3. Dissect out the liver and gently wipe excess blood from the outside of the organ (*see* Note 4).
4. Using a single edge razor blade, cut thin (2–3 mm) slices of liver and place them immediately into Carnoys fixative.
5. Leave the tissues in the fixative for a minimum of 2 h to a maximum of 18 h and then transfer them to 95% ethanol (*see* Note 5).
6. Embed the tissues in paraffin wax by routine histological procedures as soon as possible to avoid excessive hardening.

7. Cut sections 5-μm thick and mount on coated glass slides. Dewax the slides in xylene for 3 × 20 min and then rehydrate in 100% ethanol (2 × 2 min), then 95% and 70% ethanol and water for 2 min each.

3.2. Immunohistochemistry

1. Wash the sections in PBS for 2 min. It is convenient to place the slides on a rack over the sink to carry out all washing steps.
2. Inhibit endogenous peroxidase activity in the tissue sections by incubating in 0.3% H_2O_2 for 20 min.
3. Wash the sections in running tap water for 10 min, then rinse in distilled water at 60°C.
4. Incubate the sections in 1*M* HCl at 60°C for exactly 8 min (*see* Notes 6 and 7). This step denatures the DNA just sufficiently to allow access of the anti-BrdU antibody.
5. Wash briefly in tap water, then distilled water, and finally, PBS.
6. Dry around the section with a tissue (*see* Note 8) and incubate in monoclonal antibody against BrdU, diluted 1:100 in PBS, for 30 min at room temperature in a humid box to prevent the sections from drying out. One hundred microliters of solution should be enough to cover the tissue.
7. Wash 3 × 5 min in PBS.
8. Dry around the sections again and apply rabbit antirat IgG peroxidase conjugate diluted 1:100 in PBS and incubate as in Step 6.
9. Wash 3 × 5 min in PBS.
10. Incubate the sections in freshly prepared DAB substrate solution for 10 min (*see* Note 9). A brown product will be deposited at the site of BrdU incorporation.
11. Wash the sections in PBS.
12. Counterstain lightly with hematoxylin.
13. Dehydrate the sections in increasing concentrations of ethanol (70, 90, 100%), clear with xylene, and mount in DPX (*see also* this vol., Chapter 10).

4. Notes

1. Clean the slides in 5% DECON 90 or an equivalent detergent overnight. Rinse in hot water for 30 min, then 2× in deionized water. Dry at 60°C. Dip the slides in a 4% solution of APES in acetone for 2 min, then rinse 2× in acetone. Dry the slides in an oven at 60°C. They can then be stored at room temperature indefinitely.

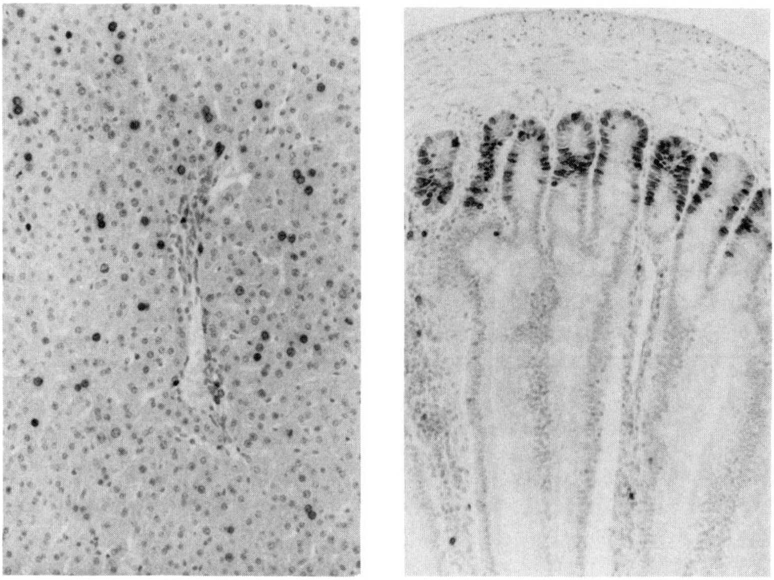

Fig. 1. Nuclei in liver (left), and small intestine (right) that have incorporated BrdU during DNA synthesis are stained more darkly.

2. It is vital that the BrdU is injected into the peritoneal cavity and not the gut. Injection of BrdU into the gut is the most common reason for failure to see any stained nuclei.
3. After injection of the BrdU, the time before sacrifice may be extended to give a higher degree of incorporation into the dividing nuclei. However, a single injection of BrdU will be undetectable after three cell divisions.
4. This method can be applied to other tissues in the animal where active DNA synthesis is occurring, for instance, the small intestine. In fact, it is good practice to prepare a composite block containing the tissue of interest and a piece of small intestine as a positive control (*see* Fig. 1).
5. We have also used this method on acetone-fixed paraffin-embedded material with equal success. Slices of fresh liver are fixed in ice-cold acetone for at least 4 h. Material can then be transferred to a –20°C freezer prior to routine embedding and sectioning. In this case, uncoated glass slides were used.
6. This is the most important step in the procedure. The exact time (usually between 2 and 10 min) must be determined empirically, if other

fixatives such as acetone are used.

7. Some workers neutralize the HCl with 0.1*M* borax buffer, pH 8.5 for 5 min.
8. A pen for drawing a hydrophobic ring around the tissue section is marketed by Dako. It obviates the need to wipe around the sections after each step and reduces the amounts of antibody solution required.
9. The rabbit antirat antibody may alternatively be conjugated with alkaline phosphatase and detected with naphthol AS-BI and fast red. In this case, incorporation will be indicated by a red color. *See also* this vol., Chapter 10 for alternative conjugates and substrates.

References

1. Schutte, B., Reynders, M. M. J., Bosman, F. T., and Blijham, G. H. (1987) Studies with antibromodeoxyuridine antibodies: II Simultaneous immunocytochemical detection of antigen expression and DNA synthesis by *in vivo* labelling of intestinal mucosa. *J. Histochem. Cytochem.* **35,** 371–374.
2. Lanier, T. L., Berger, E. K., and Eacho, P. I. (1989) Comparison of 5-bromo-2-deoxyuridine and [^{3}H]thymidine for studies of hepatocellular proliferation in rodents. *Carcinogenesis* **10,** 1341–1343.
3. Gratzner, H. G. (1982) Monoclonal antibody to 5-bromo and 5-iododeoxyuridine: A new reagent for detection of DNA replication. *Science* **218,** 474,475.
4. Wynford-Thomas, D. and Williams, E. D. (1986) Use of bromodeoxyuridine for cell kinetic studies in intact animals. *Cell Tissue Kinetics* **19,** 179–182.

CHAPTER 13

Avidin–Biotin Technology

Preparation of Biotinylated Probes

Edward A. Bayer and Meir Wilchek

1. Introduction

1.1 Avidin–Biotin Technology

During the past decade, the avidin–biotin complex has become a useful and versatile tool for application in a variety of biological disciplines *(1–3)*. One of the major applications is in immunology and related areas *(4)*.

In essence, the introduction of the avidin–biotin complex into a given system does not contribute a factor of specificity to an immunochemical reaction; rather, its involvement serves to mediate between the recognition system (i.e., antibody–antigen complex) and a given probe or reporter group. Such mediation usually results in an amplified or improved signal.

The widespread application of avidin–biotin technology is an outgrowth of the remarkable affinity of the egg-white glycoprotein avidin for the vitamin biotin *(5)*. More recently, the bacterial analog of avidin, called streptavidin, has been successfully used as a substitute for avidin and, in some cases, streptavidin may be the preferred reagent. Both proteins recognize biotin with similarly high affinity constants ($K_A \sim 10^{15}\ M^{-1}$); both are tetrameric, possessing a single biotin-binding site per subunit—four per tetramer.

In applying this system, a biotinylated binding molecule (e.g., antigen, primary or secondary antibody) is allowed to interact with a target system (cells, microtiter plates, and so on). By using an appropriate avidin-conjugated probe (e.g., fluorescent or electron-dense marker, solid support, enzyme), the system can be used for a variety of different applications.

From: *Methods in Molecular Biology, Vol. 10: Immunochemical Protocols*
Ed.: M. Manson

In this chapter, we describe some of the methodologies that have been developed for attaching biotin to antibodies, antigens, and other probes. The preparation of selected avidin-containing conjugates and complexes will also be described in the following chapter. Protocols for their use in immunoaffinity chromatography, immunocytochemistry, immunoassay, and other immunochemical techniques are presented in Chapter 15. In most cases, a single example of the application will be given, although the methods are universally applicable and subject to innumerable strategies depending on the imagination and savvy of the investigator.

1.2. Preparation of Biotinylated Probes

One of the two major steps in the application of avidin–biotin technology is to incorporate the biotin moiety into the experimental system *(6)*. This is usually accomplished by covalently attaching biotin to a biologically active binder. The binder can be an antigen, a primary monoclonal or polyclonal antibody, a complement, interleukin-2, or any other component of the immune system.

There are many ways of attaching biotin to such molecules via a variety of chemically reactive biotin-containing derivatives. Many of these are commercially available from Sigma Chemical Co. (St. Louis, MO), Pierce Chemical Co. (Rockford, IL), Calbiochem AG (Lucerne, Switzerland), Molecular Probes, Inc. (Eugene, OR), and others.

The most popular method of attaching biotin to proteins is via amino groups of lysine using biotinyl *N*-hydroxysuccinimide (BNHS) ester or longer-chained analogs (e.g., biotinyl ε-aminocaproyl *N*-hydroxysuccinimide ester). Other biotin-containing group-specific reagents have been used to attach the biotin moiety to other functional groups of protein. For example, maleimido derivatives of biotin result in the biotinylation of sulfhydryl groups (cysteines); diazobenzoyl derivatives are appropriate for histidines and tyrosines. Biotin can also be attached via saccharide constituents of polysaccharides, glycoproteins, and other glycoconjugates, using periodate oxidation and hydrazido derivatives of biotin.

We describe simple procedures for attaching biotin to antigens, antibodies, and the enzyme alkaline phosphatase.

2. Materials

For all procedures:

1. Phosphate-buffered saline (PBS), pH 7.4.
2. Dialysis tubing (for retention of M_r 12,000 proteins): boiled in distilled water immediately before use.

2.1. Preparation of Biotinylated Antibodies and Antigens via Lysine Residues

3. Antibody or antigen: (2 mg/mL) dissolved in 0.1*M* sodium bicarbonate, pH 8.5.
4. Biotinyl *N*-hydroxysuccinimide (BNHS) ester or biotinyl ε-aminocaproyl *N*-hydroxysuccinimide ester: 2 mg/mL freshly dissolved in dimethylformamide (heat if necessary) (*see* Notes 1–4).

2.2. Preparation of Biotinylated Antibodies and Antigens via Sugar Residues

5. Antibody or antigen: (2 mg/mL) dissolved in PBS.
6. Sodium periodate: 0.1*M*, freshly prepared aqueous solution.
7. Biotin hydrazide: 10 mg/mL dissolved by heating in PBS.

2.3. Preparation of Biotinylated Alkaline Phosphatase

8. Alkaline phosphatase (Sigma Type VII-NT, from bovine intestinal mucosa, 3*M* sodium chloride solution, 5000 U)
9. Distilled water.
10. Sodium bicarbonate solution (1*M*).
11. BNHS ester: 34 mg/mL of dimethylformamide in a glass test tube, dissolved by heating briefly in hot tap water (60°C).

3. Methods

3.1. Preparation of Biotinylated Antibodies and Antigens via Lysine Residues (see Notes 5 and 6)

1. For every milliliter of antibody or antigen solution, add 25 µL of BNHS ester or long-chain analog.
2. Leave the solution at room temperature for 1 h.
3. Dialyze (1 L, two buffer changes) at 4°C against PBS.
4. Store in aliquots at –20°C.

3.2. Preparation of Biotinylated Antibodies and Antigens via Sugar Residues

1. For every milliliter of antibody or antigen solution, add 110 µL of the periodate solution.
2. Let the reaction stand at room temperature for 30 min.
3. Dialyze for 4 h at room temperature against 2 L of PBS.
4. Add 250 µL of the biotin hydrazide solution per mL of protein.

5. Incubate at room temperature for 1 h.
6. Dialyze exhaustively (100–1000 vol, 3–5 buffer changes) against PBS at 4°C.
7. Store in aliquots at –20°C.

3.3. Preparation of Biotinylated Alkaline Phosphatase

1. Dilute the enzyme with 3.7 mL of distilled water, and add 0.4 mL of 1*M* sodium bicarbonate solution.
2. Add 9 µL of the BNHS ester solution.
3. Leave the solution at room temperature for 2 h.
4. Dialyze (1 L, two buffer changes) at 4°C against PBS.
5. Bring to a final concentration of 500 U of biotinylated enzyme per mL, and store in aliquots at –20°C.

4. Notes

1. BNHS and its homologs can be stored at –20°C for long periods of time in dry dimethylformamide, although in our laboratory we prefer to prepare fresh solutions of the reagent. For concentrated solutions, the preparation can be heated in order to dissolve the reagent.
2. Recently, there has been a trend to replace BNHS with long-chain homologs, such as biotinyl-ε-aminocaproyl-*N*-hydroxysuccinimide ester. The addition of a spacer in biotinylating reagents facilitates subsequent interaction with avidin probes.
3. Water-soluble analogs of BNHS and its derivatives, i.e., the sulfosuccinimide reagents, are available from Pierce Chemical Co., and may be substituted for BNHS in the biotinylation procedure. In certain cases, these may be advantageous, especially when working with proteins or cells that are sensitive to organic solvents, such as dimethylformamide or dimethylsulfoxide. The water-soluble biotin-containing derivative may thus be dissolved directly in aqueous solution before combining with the desired protein solution or cell suspension. Such reagents may, however, be disadvantageous in other ways. For example, if they are predissolved in aqueous solution, they may undergo substantial autohydrolysis before interacting with the protein.
4. *p*-Nitrophenyl esters of biotin and its long-chain derivatives may be used instead of the *N*-hydroxysuccinimide derivatives. The levels of biotinylation obtained are essentially identical in both cases.
5. Biotinylation via amino groups of lysine is generally the preferred approach, since lysines are usually prevalent and characteristically occupy an "exposed" position in most proteins. Lysines are often not directly

involved in the binding or catalytic activity of many proteins, and their modification usually has but a nominal effect on the interaction of the biotinylated protein with its target substrate. Nevertheless, there are cases in which modification of lysines is deleterious to either the physicochemical or biological properties of the protein. Their modification by BNHS and similar reagents results in a reduction in charge and consequent decrease in pI of the biotinylated protein. In other instances, biotinylation of an essential lysine may cause inactivation of the protein. In recent years, other biotinylating reagents for derivatization via other amino acid residues have been described. Proteins may now be selectively biotinylated via cysteines, cystines, tyrosines, histidines, glutamic, and aspartic acids. For a more extensive coverage of such procedures, *see* ref. 3.

6. The above-described procedures for biotinylation of antibodies and antigens are representative, and have been shown to provide optimal results for a variety of antibody types. Accordingly, approx 10 biotin moieties on average are covalently linked to an antibody using this procedure. This value can be altered by simply altering the relative concentration of reagent or antibody. Likewise, the concentration of antibody used in the reaction may be varied greatly. We have biotinylated protein solutions containing but fractions of a milligram of material per milliliter of solution, and others of up to 100 mg/mL of protein.
7. Although it is of academic interest to determine the exact number of biotin molecules per protein molecule, this value is sometimes difficult to assess and is not particularly relevant. In our experience, it is more important to determine two other characteristics of the biotinylated preparation: (1) the percentage of biotinylated protein molecules that interact with avidin, and (2) the relative activity of the biotinylated preparation. The first characteristic provides an indication of the availability of the biotin moieties for interaction with an avidin probe. The easiest way to determine this value is simply to pass a calibrated amount of the biotinylated preparation over an avidin-containing resin and to determine the percent of material that fails to bind to the column. The second value gives an assessment of whether the biotinylation has affected the activity of the protein and can be determined by comparative assay. *See* ref. 3 for a detailed discussion of the analysis of biotinylated proteins.
8. In some instances, purification of the antibody or antigen is unnecessary and whole antiserum or crude extracts containing the antigen of interest can be biotinylated. Although all of the proteins in the crude material would be expected to undergo extensive biotinylation, the required separation would ensue on interaction with the respective target mol-

ecule and the remainder of the irrelevant biotinylated species would be removed *in situ* during the washing procedure. In any event, the final procedure should include control (e.g., unbiotinylated) samples to guard against false positive results emanating from nonspecific interactions.

9. Using this procedure, most biotinylated antibodies and many antigens are incredibly stable, and can be stored at –20°C for many years. Preparations of biotinylated alkaline phosphatase have been stored under similar conditions for periods of over five years with no observed alteration in their properties and/or performance in assays.
10. Some enzymes are particularly stable to biotinylation procedures, and relatively harsh procedures can be employed. Other enzymes are exceptionally sensitive, such that biotinylation cannot be obtained under conditions that result in an active preparation. When biotinylating a particular enzyme for the first time, this information can be obtained only empirically.

References

1. Bayer, E. A. and Wilchek, M. (1978) The avidin–biotin complex as a tool in molecular biology. *Trends Biochem. Sci.* **3,** N237–N239 .
2. Bayer, E. A. and Wilchek, M. (1980) The use of the avidin–biotin complex in molecular biology. *Methods Biochem. Anal.* **26,** 1–45.
3. Wilchek, M. and Bayer, E. A., eds. (1990) Avidin–Biotin Technology. *Methods in Enzymology,* vol. 184, Academic, Orlando, FL, pp. 3–746.
4. Wilchek, M. and Bayer, E. A. (1984) The avidin–biotin complex in immunology. *Immunol. Today* **5,** 39-43.
5. Wilchek, M. and Bayer, E. A. (1989) Avidin–biotin technology ten years on: Has it lived up to its expectations? *Trends Biochem. Sci.* **14,** 408–412.
6. Wilchek, M. and Bayer, E. A. (1988) The avidin–biotin complex in bioanalytical applications. *Anal. Biochem.* **171,** 1–32.

CHAPTER 14

Avidin–Biotin Technology

Preparation of Avidin Conjugates

Edward A. Bayer and Meir Wilchek

1. Introduction

The second step required in the application of avidin–biotin technology is to prepare an appropriate avidin-associated probe or probes for the desired application (for general reviews, *see* refs. 1 and 2). For example, a fluorescent form of avidin can be used for fluorescence microscopy, fluorescence-activated cell sorting, and in some cases, for immunoassay. Likewise, an avidin-enzyme conjugate can be used for immunoblotting, immunoassay, light microscopy, and in some cases, electron microscopy. An immobilized form of avidin can be used for isolation purposes (see Note 1).

Although a wide range of avidin–probe conjugates and complexes can be purchased from dozens of commercial enterprises (and the investigator with poor chemical or biochemical means or background is encouraged to do so), we present in this chapter some simple examples for production of such avidin-associated probes. The procedures described here can, of course, be adapted for the preparation of other types of probes.

2. Materials

2.1. Preparation of Fluorescein-Derivatized Avidin

1. Avidin: 10 mg/mL in 10 m*M* sodium phosphate buffer, pH 7.4 (*see* Note 2).
2. Fluorescein isothiocyanate: 2.5 mg/mL in 0.5*M* sodium bicarbonate, pH 9.5.

From: *Methods in Molecular Biology, Vol. 10: Immunochemical Protocols*
Ed.: M. Manson

3. Sephadex G-50 column, 10 mL bed volume.
4. Sodium phosphate buffer: 10 m*M*, pH 7.4.

2.2. Preparation of Ferritin-Conjugated Avidin (see *Note 3*)

5. Avidin: 15 mg/mL in 5 mL of acetate-buffered saline, pH 4.5.
6. Ferritin, 100 mg/mL.
7. Sodium periodate: 0.1*M*, freshly prepared aqueous solution.
8. Sodium borohydride: 10 mg/mL, freshly prepared in 10 m*M* NaOH.
9. Acetate-buffered saline: 50 m*M* sodium acetate, 0.9% NaCl, pH 4.5.
10. Borate-buffered saline: 0.1*M* sodium borate, 0.9% NaCl, pH 8.5.
11. PBS, pH 7.4.
12. Dialysis tubing.
13. 50-mL Erlenmeyer flask.
14. 2-L Erlenmeyer flask.
15. Ice bath.
16. Magnetic stirrer.

2.3. Preparation of Avidin-Coated Erythrocytes

17. PBS, pH 7.4.
18. Avidin: 1 mg/mL in PBS.
19. Sheep erythrocytes: 2 mL of packed cells.
20. Formaldehyde (2%).
21. Sodium periodate: 10 m*M*, freshly prepared solution in PBS.
22. Sodium azide: 10% aqueous solution (CARE—TOXIC).

2.4. Preparation of Sepharose-Avidin Affinity Resin (see *Note 5*)

23. Sodium bicarbonate solution: 0.1*M*, pH 8.5.
24. Avidin: 50 mg dissolved in 50 mL of sodium bicarbonate solution.
25. Sepharose CL-4B (25 g).
26. Cyanotransfer reagent: N-Cyanotriethylammonium tetrafluoroborate.
27. Triethylamine: 0.2*M* aqueous solution.
28. Distilled water.
29. Acetone: 30% and 60% aqueous solutions.
30. Washing medium: acetone and 0.1*M* HCl (1:1).
31. PBS, pH 7.4.
32. Sodium azide (10%).
33. Ice bath.
34. Sintered glass funnel (coarse).

2.5. Preparation of Avidin-Conjugated Peroxidase (see Note 6)

35. PBS, pH 7.4.
36. Avidin: 4 mg/mL in PBS.
37. Horseradish peroxidase: 8 mg/mL in PBS.
38. Glutaraldehyde (1%).
39. Sephacryl S-200 column (1 × 60 cm), pre-equilibrated with PBS.
40. Dialysis tubing (for retention of M_r 12,000 proteins): boiled in distilled water immediately before use.

2.6. Preparation of Streptavidin-Complexed Alkaline Phosphatase

41. PBS, pH 7.4.
42. Streptavidin: 1 mg/mL in PBS (*see* Note 1).
43. Biotinylated alkaline phosphatase: 500 U/mL in PBS.
44. Bovine-serum albumin: 2% in PBS.

3. Methods

3.1. Preparation of Fluorescein-Derivatized Avidin

1. Add 0.1 mL of the fluorescein isothiocyanate solution to 1 mL of the avidin solution.
2. Let the reaction stand overnight at 4°C.
3. Apply the reaction mixture to the Sephadex column, and elute with sodium phosphate buffer.
4. Collect the first peak, pool the fractions, and store at –20°C in aliquots.

3.2. Preparation of Ferritin-Conjugated Avidin

1. Add the solution of ferritin to the avidin solution.
2. Add 0.66 mL of the stock solution of sodium periodate, and stir magnetically for 30 min in ice.
3. Add the reaction mixture to the dialysis sac, and dialyze for 6 h at 4°C against 2 L of acetate-buffered saline.
4. Transfer the dialysis sac to a 2-L flask containing borate-buffered saline, and dialyze overnight at 4°C.
5. Transfer the contents of the dialysis sac to a 50-mL Erlenmeyer flask, add 0.5 mL of sodium borohydride solution and stir magnetically for 1 h in an ice bath.
6. Dialyze exhaustively (2 L, 3–5 buffer changes) at 4°C against PBS.

7. Remove the contents of the dialysis sac, and centrifuge (100,000*g*) for 3 h in an ultracentrifuge.
8. Discard the supernatant fluids, resuspend the pellet (gently) in 10 mL of PBS, and repeat the centrifugation.
9. Discard the supernatant fluids, resuspend to 50 mL in PBS, sterilize by filtration, and store at 4°C in 1-mL aliquots.

3.3. Preparation of Avidin-Coated Erythrocytes

1. Fix the erythrocytes with formaldehyde (optional) (*see* Note 4).
2. Wash the cells three times by centrifugation (1000*g*, 5 min) with PBS.
3. Resuspend the cells in 20 mL of the sodium periodate solution, and incubate at room temperature for 1 h.
4. Wash the cells once by centrifugation in PBS.
5. Add 5 mL of the avidin solution, and incubate for 1 h at room temperature.
6. Wash three times by centrifugation, resuspend in 20 mL of PBS, add 200 µL of 10% sodium azide solution (for formaldehyde-fixed erythrocytes only), and store at 4°C.

3.4. Preparation of Sepharose-Avidin Affinity Resin

1. Wash the Sepharose by successive filtration in a sintered glass funnel with large volumes (0.5–1 L) of distilled water, 30 and 60% acetone. Resuspend in 50 mL of acetone (60%,), and cool the resin in an ice bath.
2. Add 500 mg of cyanotransfer reagent.
3. Stir the resin (very gently) with a magnetic stirrer, and add dropwise 5 mL of the triethylamine solution.
4. After 2 min, filter the activated gel and wash with 100 mL of cold washing medium. Wash exhaustively with distilled water.
5. Add the avidin solution, and stir overnight at 4°C.
6. Wash the gel first with 0.5 L of sodium bicarbonate, then with a similar volume of PBS, and resuspend in 50 mL of the latter buffer (containing 0.1% sodium azide). Store at 4°C.

3.5. Preparation of Avidin-Conjugated Peroxidase

1. Mix equal volumes of the avidin and peroxidase solutions.
2. Add 50 µL of glutaraldehyde solution per mL of the mixed proteins, and incubate for 3 h at room temperature.
3. Dialyze overnight at 4°C against PBS.
4. Remove the contents of the dialysis sac and centrifuge for 15 min at 10,000*g*.

5. Apply the supernatant to the Sephacryl column, and elute in PBS.
6. Monitor the column spectrophotometrically at 280 nm, collect and pool void volume fractions, sterilize by filtration, and store at 4°C in 100-µL aliquots.

3.6. Preparation of Streptavidin-Complexed Alkaline Phosphatase

1. For every 10 mL of complex required, dilute 50 µL of the streptavidin stock to 5 mL with bovine serum albumin solution. Similarly, dilute 50 µL of the stock solution of the biotinylated enzyme.
2. Mix the two diluted protein solutions rapidly, and allow the resultant mixture to incubate at room temperature for 30 min prior to use.

4. Notes

1. In recent years, streptavidin has largely replaced egg-white avidin in avidin–biotin technology. The main reason for this is that avidin is a glycosylated positively charged protein, and these two properties (the high pI and the presence of sugars) are considered to be responsible for a variety of nonspecific interactions, which may cause false positive results and/or contamination of the experimental system. In contrast, streptavidin is a nonglycosylated neutrally charged molecule. In coming years, however, there may well be a return to the egg-white variety, since avidin can easily be derivatized to neutral pI, and methods for obtaining a nonglycosylated form are now available. Nonglycosylated avidin is currently being developed commercially by Belovo Soc. Coop. (Bastogne, Belgium). Moreover, there is evidence that the use of streptavidin is marred by secondary affinity interactions (not associated with the biotin-binding site) with a variety of cell-based receptors.
2. In some cases, the positively charged properties of egg-white avidin provide an advantage for its use in lieu of streptavidin. This is particularly the case when high levels of derivatization (for example in the synthesis of a fluorescent derivative of avidin) is required. Thus, the high number of free lysines in avidin furnishes an excellent medium for extensive derivatization with simultaneous reduction of the pI to an acceptable level. The use of nonglycosylated avidin for such purposes would counteract the other major source of nonspecific binding, i.e., saccharide-specific interactions.
3. The procedure described for periodate-induced conjugation of ferritin and avidin yields a high ratio of unit-paired conjugates which are especially appropriate for ultrastructural localization using electron microscopy.

4. Erythrocytes can be formaldehyde fixed either before or after periodate-induced coupling of avidin. The preparation is then stable for at least several months if stored in the cold room. Sodium azide (0.05%) can be added for good measure.
5. Under the conditions described in this chapter, using the cyanotransfer reagent for activation of Sepharose, an affinity resin containing about 1 mg of avidin per mL of resin is produced. Higher amounts of avidin can be coupled by simply increasing the concentration of avidin in the solution that is incubated with the activated resin. The properties of the resultant resin are chemically identical to those of CNBr-activated resin. However, the cyanotransfer reagent is not volatile, and, hence, the activation is much safer and can be performed in the absence of a fume hood.
6. Glutaraldehyde-induced conjugation of proteins usually results in the formation of high-mol-wt multimers. The largest fraction appears in the void volume of the Sephacryl column with a trail of lower-mol-wt species. The preparation is suitable for immunoassay studies. The use of preformed streptavidin-complexed enzymes is a convenient, reliable and reproducible alternative to classic covalently crosslinked conjugates. The major advantage in using such complexes is that the signal is often enhanced. Furthermore, stock solutions of streptavidin and many biotinylated enzymes can be stored indefinitely with little influence on their performance. In forming complexes, a compromise must be achieved between the amount of biotinylated enzyme that forms the complex with streptavidin, and the amount of free biotin-binding sites that remain on the streptavidin for subsequent interaction with the biotinylated antibody or antigen. A high level of enzyme is required in the complex to provide a good signal. The most effective ratio of streptavidin for complex formation with a given biotinylated enzyme is determined empirically in optimization experiments. For alkaline phosphatase, a ratio of 2 µg of avidin per unit of enzyme activity of the biotinylated enzyme has been determined to be optimal. This ratio is also identical for egg-white avidin or nonglycosylated avidin.

References

1. Wilchek, M. and Bayer, E. A., eds. (1990) Avidin–Biotin Technology. *Methods in Enzymology*, vol. 184, Academic, Orlando, FL.
2. Wilchek, M. and Bayer, E. A. (1988) The avidin–biotin complex in bioanalytical applications. *Anal. Biochem.* **171**, 1–32.

CHAPTER 15

Immunochemical Applications of Avidin–Biotin Technology

Edward A. Bayer and Meir Wilchek

1. Introduction

1.1. Application in Immunoaffinity Chromatography

The major advantage in the use of avidin–biotin technology for isolation purposes is that an improved capacity for purification of an antigen often results *(1,2)*. The antibody, bound to the column via an avidin–biotin bridge, is less affected by the chemistry of immobilization or by physical interactions (hydrophobic, electrostatic, salting-out effects, precipitation, and so on) with the solid support. Such potential deleterious effects are borne by avidin, which is a highly stable protein, and thus serves as a physicochemical buffer for the antibody.

In certain cases (e.g., isolation of cell-based components), the immunochemical reaction between the biotinylated antibody and the target antigen can be performed first *in situ,* and the isolation of the immunocomplex can be achieved subsequently on the avidin column.

Owing to the strength of the avidin–biotin complex, the immunochemical bond can be preferentially severed, permitting selective elution of the antigen. Variations on this same theme allow the immunoaffinity isolation of other components of the immune system, such as antibodies, complements, blood cell receptors, and the like.

From: *Methods in Molecular Biology, Vol. 10: Immunochemical Protocols*
Ed.: M. Manson ©1992 The Humana Press, Inc., Totowa, NJ

One drawback of the system is that the original avidin-resin cannot be regenerated because of the stability of the avidin–biotin complex *(3)*.

1.2. Applications in Immunocytochemistry

Historically, the extensive application of avidin–biotin technology was initially propagated by its suitability for localization studies *(4)*. In this regard, an underivatized primary antibody can be employed, followed by a biotinylated second antibody or biotinylated protein A. The localization of the latter in a given experimental system can be accomplished by an electron-dense, fluorogenic, or chromogenic form of avidin. For this purpose, ferritin–avidin, avidin–gold, and avidin–enzyme conjugates are the most popular.

In Section 3.2., we present simple procedures for localizing a cell-associated antigen by light or electron microscopy. By simple modification of the protocol, a fluorescent derivative of avidin can be used for fluorescence microscopy.

At the expense of versatility and perhaps sensitivity, one incubation step may be precluded by employing biotinylated primary antibodies. In addition, localization of other components of the immune system can be effected by substituting the primary antibody with an appropriate biotinylated binder.

In recent years, the immunocytochemical approach using avidin–biotin technology has experienced an additional flurry of activity owing to its applicability in pathodiagnostics.

1.3. Applications in Cell Identification and Sorting

Identification and separation of cell populations (particularly lymphocyte subpopulations) is of great importance for further progress in cellular immunology *(5)*. Two major approaches have proved valuable: fluorescence-activated cell sorting and affinity-based physical separation of cells using differential settling or adsorptive properties.

In the first approach, only relatively low numbers of cells can be isolated by cell-sorting procedures. Nevertheless, this method can result in efficient cell separation, particularly if two parameters (e.g., size and fluorescence) are employed.

The second approach is more amenable to scale-up procedures. Thus, agglutinated cells can be separated from nonagglutinated cells, erythrocyte rosettes can be employed, as well as other immobilizing matrices.

Again, avidin–biotin technology has served to increase the sensitivity, versatility, and efficacy of the identification and separation processes. As in all applications of avidin–biotin technology, there is a great variety of strategies that can be instituted. We have chosen to provide two simple exemplary procedures that can be used for the selective retrieval or elimination of a given antibody-bearing cell type.

1.4. Applications in Immunoassay

Immunoassay and immunodiagnostics are fields that have benefited greatly from the mediation afforded by avidin–biotin technology *(6, 7)*. The four biotin-binding sites on avidin coupled with the potential for attaching many biotin groups to an antibody have served to provide an amplified signal in immunoassays. Alternatively, the stability of the avidin–biotin complex can be used for improving the properties of the immobilizing or separating systems. Consequently, improved assay sensitivities have been reported extensively in the literature.

In this section, we provide a procedure for a solid-phase enzyme-linked assay for quantitative determination of either an antibody or an antigen. (For details of the theory of radioimmunoassay and ELISA, *see* vol. 1, Chapters 37 and 38 and also, this vol., Chapters 29, 30, and 35.) In this case, the avidin–biotin complex mediates between a second antibody and an enzyme, resulting in enhanced signal intensities. A second procedure is also described wherein the avidin–biotin complex mediates in the separation stage of a competitive radioimmunoprecipitation assay. Again, these procedures exemplify the varied possibilities of avidin–biotin technology; the components of these assays can be modified in many ways to accomodate a given experimental system or strategy.

1.5. Application in Immunoblotting

During the last decade, immunoblotting has become a versatile and essential technique for any biochemically oriented research laboratory *(6, 7)*. As observed for the other immunochemical techniques described above, mediation by avidin–biotin technology has served to improve signal intensities, to enhance sensitivity levels, and has contributed greatly to the efficacy and versatility of blotting procedures.

2. Materials

2.1. Immunoaffinity Chromatography

1. Phosphate-buffered saline (PBS), pH 7.4.
2. Sample material (up to 10 mg/mL protein) e.g., serum, cell extract, and so on, diluted if necessary in PBS.
3. Sepharose–avidin: ~1.5 mg of avidin per mL of Sepharose.
4. Biotinylated monoclonal or polyclonal antibody: ~2 mg/mL in PBS.
5. Acetic acid (0.1*M*).
6. Column (glass or plastic, containing sintered glass or plastic filter to retain the affinity resin). The size of the column should be sufficient to facilitate application and allow for a small reservoir for effluent, wash, and eluant liquids.

2.2. Immunocytochemistry

2.2.1. Light Microscopy

7. Rehydrated deparaffinized sections or fixed frozen sections.
8. PBS, pH 7.4.
9. Blocking solution, wash solution and diluent: Low-fat (1%) milk, 10% in PBS.
10. Primary monoclonal or polyclonal antibody: 2 μg/mL in PBS.
11. Biotinylated secondary antibody: 20 μg/mL in PBS.
12. Avidin-conjugated peroxidase: 10 μg/mL in PBS.
13. Substrate solution should be prepared fresh as required: Add 1.5 g of nickel ammonium sulfate to 100 mL of acetate-buffered saline (pH 6); stir until the solid is dissolved. Add 50 mg of 3,3'-diaminobenzidine (*see* Note 7). After the substrate dissolves, add 15 μL of 30% hydrogen peroxide. Filter the solution (Whatman No. 1 filter paper).

2.2.2. Electron Microscopy

14. Cell suspension, fixed if desired.
15. Wash solution: PBS, pH 7.4.
16. Blocking solution and diluent: Bovine serum albumin (2%), in PBS.
17. Biotinylated primary antibody: 0.5 mg/mL in diluent.
18. Ferritin-conjugated avidin: 1 mg/mL protein in diluent.

2.3. Cell Identification and Sorting

2.3.1. Fluorescence-Activated Cell Sorting

19. PBS, pH 7.4.
20. Diluent medium: Bovine serum albumin (2%) in PBS. Sodium azide (CARE—TOXIC) may be added to a final concentration of 0.1% if separation of viable cells is not required.
21. Cell sample: Complex mixture of cell types (ca. 2×10^6/mL) suspended in diluent medium.
22. Biotinylated monoclonal or polyclonal antibodies: 0.1 mg/mL, in diluent medium.
23. Fluorescein-derivatized avidin: 10 μg/mL in diluent medium.
24. Formaldehyde: 4% in PBS.

2.3.2. Rosette Formation

25. PBS, pH 7.4.
26. Cell sample: Complex mixture of cell types (ca. 5×10^7/mL) suspended in PBS.
27. Avidin–coated erythrocytes: 10^9/mL suspended in PBS.
28. Biotinylated monoclonal or polyclonal antibodies: 0.1 mg/mL in PBS.

29. Ficoll solution: 24%, 1.090 g/mL final density.
30. Ammonium chloride solution (0.85%).
31. Fetal calf serum: 10%, in PBS.

2.4. Immunoassay

2.4.1. Enzyme-Linked Immunoassay

32. Antigen or capture antibody: 5 µg/mL in coating buffer, which is 0.1*M* carbonate–bicarbonate buffer, pH 9.6, brought to 0.01% sodium azide.
33. PBS, pH 7.4.
34. Blocking solution, wash solution and diluent: Low-fat (1%) milk, 10%, in PBS.
35. Test antigen, primary monoclonal or polyclonal antibody (serial dilutions of unknown or standard samples).
36. Biotinylated secondary antibody: 3 µg/mL, in diluent.
37. Avidin-conjugated peroxidase: 1.0 µg/mL, in diluent.
38. Citrate–phosphate buffer (pH 5): Mix equal vol of 0.1*M* citric acid and 0.2*M* dibasic sodium phosphate, and dilute 1:1 with distilled water.
39. Substrate solution: Add 2.5 mg of 2,2'azino-*bis*(3-ethylbenz-thiazoline-6-sulfonic acid) to 10 mL of citrate–phosphate buffer. Add 10 µL of 30% hydrogen peroxide to this solution. Make fresh as required.

2.4.2. Immunoprecipitation

40. PBS, pH 7.4.
41. Precipitation medium: Polyethylene glycol 6000 (5%) in PBS containing 0.1% sodium azide.
42. Carrier protein solution: Biotinylated bovine serum albumin (0.1 mg/mL) diluted in 2% underivatized bovine serum albumin (in PBS).
43. Radioactive antigen: ^{125}I-labeled (about 2×10^6 cpm/mL) in PBS.
44. Test antigen: serial dilutions of standard or unknown samples in precipitation medium.
45. Biotinylated monoclonal or polyclonal antibody: 0.5 µg/mL in carrier protein solution.
46. Avidin: 0.1 mg/mL in precipitation medium.
47. Microfuge tubes.

2.5. Immunoblotting

48. Antigen-adsorbed dot blots or nitrocellulose blot transfers containing antigenic material separated by SDS-PAGE, isoelectric focusing, and so on (*see* vol. 1, Chapter 12; vol. 3, Chapters 15 and 19; and this vol., Chapter 24).
49. PBS, pH 7.4.

50. Blocking solution: Low-fat (1%) milk, 10%, in PBS.
51. Biotinylated monoclonal or polyclonal antibody: 3 µg/mL in blocking solution.
52. Streptavidin-complexed alkaline phosphatase (for preparation and effective concentration, *see* Chapter 14, Section 3.6.).
53. Substrate solution: Add 10 mg of naphthol AS-MX phosphate (sodium salt) to a solution containing 30 mg of Fast Red dissolved in 100 mL of 0.1*M* Tris-HCl, pH 8.4.

3. Methods

3.1. Immunoaffinity Chromatography

1. Pour the Sepharose–avidin into an appropriate column and wash with PBS.
2. Apply the biotinylated antibody (about 1 mg per mL of resin).
3. Wash the column first with PBS and then with 5 column vol of 0.1*M* acetic acid. Wash again with PBS, monitoring the effluent fractions until the A_{280} reaches a minimum value (preferably less than 0.01).
4. Apply the sample solution to the column.
5. Wash with PBS until the A_{280} reaches a minimum value.
6. Elute the sequestered material with acetic acid.
7. Examine eluted fractions for the presence of biological material or activity. Pool and dialyze appropriate fractions.

3.2. Immunocytochemistry

3.2.1. Immunocytochemistry for Light Microscopy

1. Incubate the slide with the tissue sections for 1 h at room temperature with blocking solution.
2. Incubate for 2 h at room temperature with primary antibody solution, using a volume sufficient to cover the sample without drying out. Rinse the slide 3–5 times by dipping briefly in wash solution.
3. Incubate with biotinylated second antibody for 30 min at room temperature, applying an appropriate volume as described in Step 2. Rinse the slide as before.
4. Incubate with avidin-conjugated peroxidase for 30 min at room temperature. Rinse as before.
5. Incubate the slide with freshly prepared substrate solution for 5 min. Wash with PBS, and then rinse well with distilled water.

6. Counterstain for cell structure by an appropriate procedure (*see* this vol., Chapter 10).

3.2.2. Immunocytochemistry for Electron Microscopy

1. Wash the cell suspension by centrifugation (500*g*, 15 min, 4°C), and incubate in blocking solution for 1 h at room temperature. Wash again.
2. Resuspend the pellet in the solution containing the biotinylated antibody, and incubate for 1 h at room temperature. Wash by centrifugation.
3. Resuspend the sample in the solution containing ferritin-conjugated avidin; incubate for 1 h at room temperature.
4. Wash the cells in PBS, postfix in 2% glutaraldehyde, and process for electron microscopy.

3.3. Cell Identification and Sorting

3.3.1. Fluorescence-Activated Cell Sorting

1. Incubate 1 mL of washed cell sample for 30 min at 4°C with 50 µL of biotinylated antibody solution. Wash the samples three times by centrifugation (500*g*, 15 min, 4°C) with diluent medium.
2. Resuspend the cells in 0.9 mL of diluent medium, and add 0.1 mL of fluorescein-derivatized avidin solution. Incubate at 4°C for 30 min. Wash three times with PBS. Keep cells on ice.
3. Fix the cells with formaldehyde (unless separation of viable cells is desired).
4. Analyze, separate, and collect cell subpopulations by cell cytometry and sorting (*see* this vol., Chapter 41).

3.3.2. Rosette Formation

1. Wash the cell sample with PBS by centrifugation (500*g*, 10 min).
2. Resuspend in the original volume of biotinylated antibody solution, and incubate for 1 h at 37°C. Wash twice by centrifugation.
3. Mix equal volumes of the cell sample and the erythrocyte suspension.
4. View rosette formation in the light microscope.
5. Separate rosetted cells by layering the cell sample on a 24% Ficoll solution; centrifuge the mixed cell preparation (1000*g*) for 30 min at 4°C. Remove nonrosetting cells from the interface and wash by centrifugation in PBS.
6. If unfixed avidin-coated erythrocytes are used, lyse rosetted erythrocytes by resuspending the pellet in ammonium chloride solution (1 mL for

each mL of original suspension). Add 9 vol of 10% fetal calf serum, wash, and resuspend the separated antigen-bearing cell subpopulation.

3.4. Immunoassay

3.4.1. Enzyme-Linked Immunoassay

1. Add to the microtiter plates 100 µL/well of either antigen or capture antibody. Store the plates at 4°C until use.
2. Wash the plates once, add blocking solution (250 µL/well), and incubate for 1 h at 37°C. Wash the plates once again.
3. Incubate with primary antibody or test antigen solution (100 µL/well) for 2 h at 37°C. Wash three times. For plates containing the capture antibody that have been incubated with test antigen, repeat this step now using the primary antibody.
4. Incubate with biotinylated secondary antibody solution (100 µL/well) for 1 h at 37°C. Wash three times.
5. Treat with avidin-conjugated peroxidase solution (100 µL/well) and allow to stand for 30 min at room temperature. Wash three times.
6. Incubate with substrate solution, and read the absorption of each well periodically at 420 nm.

3.4.2. Immunoprecipitation

1. Add successively to the microfuge tubes (briefly vortex once after each addition): the test antigen solution (200 µL), the biotinylated antibody solution (50 µL), and the radioactive antigen solution (10 µL). Incubate overnight at 37°C.
2. Add 10 µL of the avidin solution to the tubes and incubate for 1 h at room temperature.
3. Centrifuge, discard supernatant fluids (appropriately for ^{125}I), wash, and count the radioactivity in the pellets.

3.5. Immunoblotting

1. Incubate the blots in blocking solution for 1 h at 37°C. Rinse three times with PBS.
2. Add the biotinylated antibody solution. Incubate for 2 h at 37°C with gentle rocking, ensuring uniform distribution of solution on the surface of the blot. Rinse three times with PBS.
3. Treat with the streptavidin-complexed alkaline phosphatase solution for 30 min at room temperature. Rinse three times with PBS.
4. Add the substrate solution. When the desired level of stain is reached, rinse, and store the blot in water.

4. Notes

4.1. Immunoaffinity Chromatography

1. Use of avidin–biotin technology in immunoaffinity chromatography can lead to improved performance of the affinity column, which translates both into yields of the target material and stability of the column *(1,2)*. The presence of free biotin-binding sites of the avidin-resin allows secondary interaction with biotinylated antibody or its complexes with avidin that may have leaked from the affinity resin.
2. Use of a long-chain biotin-containing reagent (e.g., biotinyl ε-aminocaproyl *N*-hydroxysuccinimide ester) in producing the biotinylated antibody results in a more stable affinity column.
3. The avidin-containing resin is stable to storage at 4°C for many years in the presence of azide. (Sodium azide is a potent poison and should be treated accordingly.) Biotinylated antibody (and other proteins) can also be stored almost indefinitely at –20°C. Thus, the preparation of the affinity resin can be performed in several minutes.
4. In order to circumvent nonspecific interactions with lectins (e.g., if cell-associated antigens are being purified), resins containing streptavidin or nonglycosylated avidin can be employed in place of native egg-white avidin. Resins containing avidin or nonglycosylated avidin can be acetylated or succinylated, in order to preclude undesired ion-exchange interactions.

4.2. Immunocytochemistry

4.2.1. Light Microscopy

5. Owing to its convenience and versatility, the method is especially appropriate for pathohistochemical analysis of large numbers of potentially diseased tissue samples. A series of different monoclonal and/or polyclonal antibodies can be used in a sequence of samples. Only the primary antibody has to be altered for a given sample; the other steps (successive introduction of biotinylated second antibody, avidin-conjugated probe, substrate, wash solutions) are identical.
6. In this chapter, we do not address the question of collection, storage, and sectioning of tissues. Frozen tissues (e.g., derived from surgical biopies), plastic- or paraffin-embedded tissue (bearing denaturation-resistant antigenic determinants), and, of course, cell suspensions can be analyzed with equivalent facility.
7. Diaminobenzidine is considered to be carcinogenic, and its handling thus deserves appropriate respect and caution. Gloves and face mask

should be worn during its weighing and addition to solution.

8. The avidin–biotin technique is especially useful in double-labeling experiments. This can be done using two different enzymes in the final step, combined with two different antibody species (monoclonal and/or polyclonal). One of the antibodies can be biotinylated and the other immunochemical reaction can be based on a different procedure (e.g., employing an antibody–enzyme conjugate). Alternatively, both antibodies can be biotinylated and the secondary reaction with the respective avidin–enzyme conjugate can be performed on different sides of an impermeant (e.g., plastic-embedded) tissue section.
9. Stringent controls [e.g., employing native (nonbiotinylated) antibody in the primary step or biotin-blocked avidin–enzyme conjugate in the secondary step] should also be carefully implemented.
10. Other avidin–enzyme conjugates that convert solid substrates to colored precipitates can be used. Likewise, fluorescent derivatives of avidin can be used, and the sample can be viewed with a fluorescent microscope; two different fluorophores can be employed in double-labeling experiments (*see* this vol., Chapters 11 and 42). Preformed complexes containing avidin and biotinylated enzyme can also be used in place of the covalently coupled conjugate.
11. In order to avoid nonspecific binding of native egg-white avidin with cell surface lectins, streptavidin– or nonglycosylated avidin–enzyme conjugates or complexes can be employed in its stead.

4.2.2. Electron Microscopy

12. Other electron-dense markers can be used instead of ferritin–avidin conjugates. Enzyme-conjugated avidin has been used in the past, but the product that precipitates often does so after diffusion, and the resolution afforded using such conjugates is consequently reduced. More recently, colloidal gold conjugated with either avidin or streptavidin has been extensively employed for electron microscopic analysis. The resolution of these particles is superior to that of ferritin and is particularly useful for detecting intracellular antigens and other target molecules (*see also* Chapters 16–19).
13. The use of the avidin–biotin system enables a separation of the primary and secondary steps. Previous employment of an antibody-marker for direct detection was problematic since such conjugates are very large and subject to a variety of irrelevant interactions that interfere with the desired immunochemical binding. In contrast, the biotinylated antibody can first be incubated with the cell or tissue sample; following this initial interaction, the sample can be fixed and incubated subsequently with

the avidin–containing marker. This last step can be performed even after embedding and sectioning of the cells or tissue.

14. As in all protocols involving avidin–biotin technology, controls consisting of native (nonbiotinylated) antibody in the primary step or biotin-blocked avidin–enzyme conjugate in the secondary step should always be employed. For localization of intracellular material, it is especially important to carry out strict controls. It should be remembered that biotin is a naturally occurring vitamin that comprises a covalently attached prosthetic group of cell-based enzymes that require its presence for their respective activities. Thus, a low-level positive response could easily represent the localization of resident biotin-requiring enzymes rather than the desired target molecule.

4.3. Cell Identification and Sorting

4.3.1. Fluorescence-Activated Cell Sorting

15. The four binding sites on avidin in conjunction with the capacity to introduce many biotin moieties per antibody molecule have resulted in enhanced sensitivity and reduced background in flow cytometric experiments. The limit of detection using the procedure described here is about 5000 to 10,000 antigenic determinants per cell. This can be significantly improved using polymeric fluorescent forms of avidin (e.g., avidin–complexed fluorescent microspheres).

4.3.2. Rosette Formation

16. The percent of cells undergoing rosette formation can be quantified by empirically determining (microscopically) the number of target cells that have associated with a given number (e.g., four) of erythrocytes. The cells can be stained with brilliant cresyl blue for clarity.
17. The method is appropriate for many cell types. In some cases, the density of the Ficoll medium must be altered to accommodate unusually high or low densities of either the target cell type or a contaminant in the original mixture.
18. Special care must be taken to avoid secondary interactions when cells of the immune system are examined. In such cases, it is advisable to use biotinylated antibody fragments that lack the Fc portion.
19. Unlike flow cytometry in which multiple parameters (i.e., size vs presence of antigenic determinant) can be assessed, the information and/or separation afforded by rosetting are based on a single parameter (presence or lack of antigen). Nevertheless, rosetting is appropriate for scale-up procedures and large quantities of the desired cell type can be isolated using this approach.

20. It should be kept in mind that, after separation, positively selected cells are still labeled with biotinylated antibodies that are complexed with avidin, and probably retain residual components of the erythrocyte membrane (following its lysis by ammonium chloride).

4.4. Immunoassay

4.4.1. Enzyme-Linked Immunoassay

21. The two protocols described here are but examples of the many possibilities for immunoassay. Many different variations are possible. The approach is also suitable for competitive immunoassay using a biotinylated antigen standard. Homogeneous immunoassays have also been described. Although we have given a protocol using a biotinylated second antibody, the primary antibody can be biotinylated and the avidin–enzyme probe can then be used immediately afterwards. Biotinylated protein A or protein G can be used instead of the secondary antibody. Of course, other avidin-conjugated enzymes, other probes (fluorescent, chemiluminescent, radioactive, and so on), or avidin complexed with the biotinylated probes, can be substituted for the avidin-conjugated peroxidase given as an example in this section. The entire protocol can be reversed, and avidin–biotin mediation can be incorporated in the capture system: avidin- or streptavidin–immobilized plates or beads can serve as a highly efficient capture system for immobilizing a desired biotinylated antibody or antigen.
22. Plates containing the capture antibody can be stored at 4°C in the presence of azide for a period of at least 6 mo. They should be sealed hermetically in order to prevent evaporation of the solution.
23. The optimal concentration of antibody or antigen for coating plates varies for any given preparation and should be determined empirically for each system. Likewise, other components of the assay system should be pretested one at a time in model experiments. The concentrations provided here are based on the reagents that were prepared in our own laboratory, and may be used as a starting point for similar assays.
24. Whereas polyclonal antibody preparations usually retain their binding properties following biotinylation, the susceptibility of a given monoclonal antibody to biotinylation is impossible to predict. Thus, it is imperative that preliminary experiments be carried out to determine the residual activity following the biotinylation step.

4.4.2. Immunoprecipitation

25. The protocol described here is a solution-phase competitive radioimmunoassay. The results can be graphed as percent inhibition vs amount

of test antigen. The procedure is especially convenient for determining affinity constants and epitope specificities for monoclonal antibodies. Protocols can be modified to accommodate nonradiolabeled probes (e.g., avidin-conjugated enzymes). An alternative and complementary approach can be taken that consists of biotinylating the antigen rather than the antibody. The advantage of using avidin–biotin mediated precipitation (over procedures involving a second antibody or protein A) is that a general method for selective and quantitative precipitation is now available. For a more thorough treatment of immunoprecipitation methods using avidin–biotin technology, *see* ref. 8.

4.5. Immunoblotting

26. For incubations in all blotting experiments, we generally use between 0.5–1 mL of the required solution for every cm^2 of the blot.
27. The stated concentration of biotinylated antibody has been found to be effective in many of our studies; nevertheless, for any given immunochemical antibody–antigen pair, a range of concentrations of the biotinylated antibody should be examined in terms of signal intensity, background levels, and extent of nonspecific binding. For this purpose, dot blots are especially convenient.
28. The Fast Red reagent in combination with the precipitable substrate provides a deep red stain on the blot for positive reactions. Many other "Fast" colors (blue, yellow, green, violet, and so on) are available and can be substituted in the protocol; the choice is up to the personal taste of the individual investigator. Double-labeling experiments are also possible by using a different enzyme combined with a different-colored substrate (*see also* this vol., Chapter 11). Radioactive, fluorescent, or chemiluminescent derivatives of avidin or streptavidin can also be used instead of the streptavidin-complexed enzyme described here. As in all of the avidin–biotin-based applications, the versatility, efficiency, and convenience for blotting technology are remarkable.

References

1. Bayer, E. A. and Wilchek, M. (1990) The application of avidin–biotin technology for affinity-based separations. *J. Chromatography* **510,** 3–11.
2. Bayer, E. A. and Wilchek, M. (1990) Avidin column as a highly efficient and stable alternative for immobilization of ligands for affinity chromatography. *J. Molec. Recog.* **3,** 115–120.
3. Wilchek, M. and Bayer, E. A. (1989) A universal affinity column using avidin–biotin technology, in *Protein Recognition of Immobilized Ligands* (Hutchens, T. W., ed.), Alan R. Liss, New York, pp. 83–90.

4. Bayer, E. A., Skutelsky, E., and Wilchek, M. (1979) The avidin–biotin complex in affinity cytochemistry. *Methods Enzymol.* **62,** 308–315.
5. Wilchek, M. and Bayer, E. A. (1984) The avidin-biotin complex in immunology. *Immunol. Today* **5,** 39–43.
6. Wilchek, M. and Bayer, E. A. (1988) The avidin–biotin complex in bioanalytical applications. *Anal. Biochem.* **171,** 1–32.
7. Bayer, E. A. and Wilchek, M. (1978) The avidin–biotin complex as a tool in molecular biology. *Trends Biochem. Sci.* **3,** N237–N239.
8. Wagener, C., Krüger, U., and Shively, J. E. (1990) Selective precipitation of biotin-labeled antigens or monoclonal antibodies by avidin for determining epitope specificities and affinities in solution-phase assays. *Methods Enzymol.* **184,** 518–529.

CHAPTER 16

Preparation of Gold Probes

Julian E. Beesley

1. Introduction

Colloidal gold probes are widely used in the biological sciences for both light and electron microscopy. A gold probe is an electron dense sphere of gold coated with an immunologically active protein. There are two aspects to be considered when making the probe.

1. Gold spheres are produced by the chemical reduction of gold chloride. This happens in three stages *(1)*. Initially, the reduction of Au^{3+} produces a supersaturated molecular Au solution. Nucleation is initiated when the concentration of Au increases and the gold atoms cluster and form nuclei. Particle growth proceeds with the deposition of molecular gold on the nuclei. The theoretical size of the gold is inversely proportional to the cube root of the number of nuclei formed if it is assumed that the conversion of Au^{3+} to Au is complete and that the concentration of gold remains constant *(2)*.

   The speed of reduction of the gold chloride determines how many gold nuclei are formed. In a closed system, this will determine the final size of the gold. White phosphorus or sodium or potassium thiocyanate are fast reducers and are used for producing small, 2–12 nm gold spheres *(3)*, whereas sodium citrate is a slow reducing agent for producing relatively large, 15–30 nm spheres. In this manner gold sols of diameters useful for electron microscopy, between 2 and 15 nm can be prepared.

From: *Methods in Molecular Biology, Vol. 10: Immunochemical Protocols*
Ed.: M. Manson ©1992 The Humana Press, Inc., Totowa, NJ

Slot and Geuze *(1)* improved the sodium citrate method for reducing gold salt by the addition of tannic acid to the reaction. The reaction rate, and hence the size of the gold, could be controlled by the amount of tannic acid added, and monodisperse gold sols, ranging in size from 3 to 17 nm could be produced.

2. Gold spheres are coated with the desired protein. The process is complex and only partially understood. It relies on the negative charge of the gold interacting with the positive charge of the protein. The technique of complexing proteins with colloidal gold was first described in detail by Geoghegan and Ackerman *(4)*. Complexing a protein with colloidal gold depends greatly on the pH of the medium. To form stable complexes, it should be 0.5 pH units basic to the isoelectric point of the protein in question. Almost any protein may be complexed with gold spheres. The most common for immunocytochemistry, are antibodies, either primary or secondary, proteins A and G, streptavidin, and lectins. Proteins are often expensive or in short supply, and to avoid wastage, a titration is performed to determine the exact quantity of protein required to coat the gold spheres in the sol. The titration is carried out by adding successive dilutions of protein to a constant volume of gold. Sodium chloride is added. In those tubes in which there is sufficient protein to stabilize the gold against the flocculating effect of the sodium chloride there is no color change but in those in which the amount of protein is insufficient, the gold flocculates and changes color from red to blue. The minimum quantity of protein necessary to stabilize the gold sol can be calculated and used to prepare the required quantity of probe. The minimum quantity of gold can be considered synonymous with the amount needed to coat the gold spheres to produce an efficient probe. Adding more than this may produce a probe of high initial activity but this is unstable *(5,6)*.

Experience has shown that the desired routine sizes are 5, 10, and 15 nm. All these are useful for transmission electron microscope studies and the 5 nm gold is recommended for light microscopy. Gold spheres of 30 nm are occasionally used for scanning electron microscope immunocytochemistry. A 1 nm probe is available and might be useful for both light and electron microscopy after silver enhancement.

2. Materials

1. Gold chloride crystals.
2. 1% Aqueous trisodium citrate. $2H_2O$.
3. 25 m*M* and 0.2*M* potassium carbonate.
4. 0.1*M* hydrochloric acid.

5. 10% Aqueous sodium chloride.
6. 1% Tannic acid (from Aleppo nutgalls, code 8835, supplied by Mallinckrodt, St. Louis, MO).
7. The protein to be complexed with the gold. It is essential that the isoelectric point of the protein is known.
8. Ultracentrifuge and 10-mL centrifuge tubes.
9. All distilled water to be used should be double-distilled and filtered through a 0.45-μ millipore filter.
10. All glassware should be thoroughly cleaned and siliconized.
11. Electron microscope facilities with Formvar/carbon-coated grids.
12. 1% Aqueous polyethylene glycol (Carbowax 20, Union Carbide).
13. Phosphate buffered saline (PBS): 0.1*M* Phosphate, 0.15*M* NaCl, pH 7.4.
14. PBS containing 0.2 mg/mL of PEG.
15. Sodium azide (toxic, *see* Chapter 6, Section 2.3., Step 18).

3. Methods

There are three stages in the production of gold probes: (1) production of gold spheres, (2) estimation of the amount of protein to be added to the gold, and (3) making the required amount of probe.

3.1. Production of the Gold Spheres (1)

1. Freshly prepare a gold solution from an ampule of gold chloride by adding 1 mL of a 1% aqueous gold chloride solution to 79 mL of distilled water.
2. Prepare the reducing mixture with 4 mL of 1% trisodium citrate. $2H_2O$, 2 mL of 1% tannic acid, 2 mL of 25 m*M* potassium carbonate and distilled water to make 20 mL.
3. Warm the solutions to 60°C and quickly add the reducing mixture to the gold solution while stirring. The temperature is critical at this stage. Evidence of sol formation is the red color of the mixture.
4. After the sol has formed, heat the mixture to boiling, and cool. According to Slot and Geuze *(1),* the quantities stated here should produce 4 nm (+/– 11.7%) particles. For 6 nm (+/– 7.3%) particles, add 0.5 mL of tannic acid and 0.5 mL of potassium carbonate to the sodium citrate. The potassium carbonate counteracts the pH effect of the tannic acid. Below 0.5 mL, the tannic acid has no effect on the pH and may be omitted. Therefore, for 8.2 nm (+/– 6.9%) particles, add 0.125 mL tannic acid to the sodium citrate, and for 11.5 nm (+/– 6.3%) particles, add 0.03 mL tannic acid. The sol forms within seconds if a high amount of tannic acid has been added or will take up to 60 min if the tannic acid has been omitted.

The nominal values for the sizes of the gold are those given above. In practice, these tend to vary slightly. Before the gold probes are used for electron microscopy, it is advisable to check the size range, especially if multiple immunolabeling is to be carried out.

5. Dry a small aliquot of the gold onto a Formvar/carbon-coated 400-mesh copper grid.
6. Wash away any salts in the preparation by floating the grid, specimen-side down, on distilled water.
7. Measure approx 100 gold particles and calculate the mean size range. This exercise will also show if the gold particles are clumped. The majority, at least 80–85% should be single particles. If the size distribution is unacceptably high, the preparation can be purified by centrifugation over a continuous sucrose or glycerol density gradient *(7)*.

3.2. Titration to Determine the Minimum Amount of Protein to Stabilize the Gold Sol (8)

1. Adjust the pH of the gold sol to 0.5 pH units above the isoelectric point of the protein to be complexed. Care should be taken when adjusting the pH since nonstabilized colloidal gold will plug the pore of the electrode *(9)*. Take an aliquot of a few milliliters of the gold and add five drops of 1% aqueous polyethylene glycol, before measuring the pH. Make the necessary adjustments to the pH and repeat until the required pH is obtained. Do not return these aliquots to the remaining colloidal gold sol. Add 0.1*M* HCl to lower the pH or add 0.2*M* potassium carbonate to raise the pH, each of these being carried out with vortexing.
2. Measure five aliquots of 0.5 mL of gold sol.
3. Prepare five aliquots of serially diluted protein in distilled water and add one of these, while shaking, to each of the 0.5-mL gold sol aliquots.
4. After 1 min, add 0.1 mL of 10% aqueous sodium chloride to each tube. Where there is excess protein in the tubes the sol will not change color, but in those tubes where there is insufficient protein to stabilize the gold, flocculation will have occurred and the liquid will be blue. The correct concentration of protein is the minimal amount that will inhibit flocculation. Horisberger *(10)* suggests that for accurate determination of color change, a spectrophotometric assay should be used.

3.3. Production of the Colloidal Gold Probe

Once the minimal amount of protein necessary to stabilize a given quantity of gold sol is known, any amount of gold probe can be produced. For general laboratory use, this is 10 mL.

1. Dissolve the required amount of protein in 0.1–0.2 mL of distilled water in a centrifuge tube and add 10 mL of the gold sol.
2. After 2 min, add 1 mL of 1% aqueous polyethylene glycol solution to stabilize the gold probe *(8)*.
3. Centrifuge the mixture at a speed depending on the size of the gold complex (15 nm at 60,000*g* for 1 h at 4°C *[8]*, 12 nm at 5000*g* at 4°C *[1]*, 5–12 nm at 105,000*g* for 1.5 h at 4°C *[8]*, 5 nm at 125,000*g* for 45min at 4°C *[1]*, and 2–3 nm at 105,000*g* for 1.5 h at 4°C *[8]*).

   The pellet formed consists of two phases *(8)*. There is a large loose part, which is the protein–gold complex. In addition, there is a tight, dense pellet on the side of the tube, which contains agregated gold particles and gold particles that have not been fully stabilized.
4. Resuspend the loose part of the pellet in 1.5 mL of PBS containing 0.2 mg/mL of polyethylene glycol. This can be stored for up to one year at 4°C. If necessary, 0.5mg/mL of sodium azide may be added to prevent small organisms from growing in the probe. It is suitably diluted before use.

The technique described above is the basis for the preparation of colloidal gold probes and the principles of probe production are identical for each type of probe. Those who are interested should refer to Slot and Geuze *(1)* for preparation of protein A–gold probes, Roth *(3)*, for production of antibody–gold complexes, Tolson et al. *(11)* for production of the avidin–gold complex, Horisberger *(10)* for production of the lectin–gold complex, and Bendayan *(12)* for production of the enzyme–gold complex.

4. Notes

1. There are many ways of testing the probe, but the most convincing is on a known positive sample. Therefore, this could be a histological section, an EM specimen, or a dot-blot. Estimation of the concentration of the probe by optical density measurements is a good method to standardize the concentration of probes from one batch to another, but in addition, it is always preferable to test the performance of the probes on known positive samples.
2. Many of the gold probes are available commercially. It is reasonably easy to make good quality probes in the laboratory, and the investigator has to consider nonscientific factors, such as time and cost. A good compromise is to purchase the reagents that will be in constant demand and make the required quantity of those that are not readily available.

References

1. Slot, J. W. and Geuze, H. J. (1985) A new method of preparing gold probes for multiple-labelling cytochemistry. *Eur. J. Cell Biol.* **38,** 87–93.
2. Frens, G. (1973) Preparation of gold dispersions of varying size: Controlled nucleation for the regulation of the particle size in monodisperse gold suspensions. *Nature: Physical Science* **241,** 20–22.
3. Roth, J. (1983) The colloidal gold marker system for light and electron microscopic cytochemistry, in *Techniques in Immunocytochemistry,* Vol. 2 (Bullock, G. R. and Petrusz, P., eds.), Academic, London, pp. 216–284.
4. Geoghegan, W. D. and Ackerman, G. A. (1977) Adsorption of horseradish peroxidase, ovomucoid and anti-immunoglobulin to colloidal gold for the indirect detection of concanavalin A, wheat germ agglutinin and goat anti-human inmunoglobulin G on cell surfaces at the electron microscope level: A new method, theory and application. *J. Histochem. Cytochem.* **25,** 1182–1200.
5. Geoghegan, W. D. (1985) The adsorption of IgG and IgG fragments to colloidal gold: Molecular orientation. *J. Cell Biol.* **101,** 85a.
6. Geoghegan, W. D. (1986) The adsorption of rabbit IgG to colloidal gold: Molecular orientation. (Proc. Histochem. Soc. America.) *J. Cell Biol.* **32,** 1360.
7. Slot, J. W. and Geuze, H. J. (1981) Sizing of Protein-A colloidal gold probes for immunoelectron microscopy. *J. Cell Biol.* **90,** 533–536.
8. Roth, J. (1982) The protein A-gold (pAg) technique: A qualitative and quantitative approach for antigen localisation on thin sections, in *Techniques in Immunocytochemistry,* Vol. 1 (Bullock, G. R. and Petrusz, P., eds.), Academic, London, pp. 107–33.
9. Lucocq, J. M. and Roth, J. (1985) Colloidal gold and colloidal silver—metallic markers for light microscope histochemistry, in *Techniques in Immunocytochemistry,* Vol. 3 (Bullock, G. R. and Petrusz, P., eds.), Academic, London, pp. 203–236.
10. Horisberger, M. (1985) The gold method as applied to lectin cytochemistry in transmission and scanning electron microscopy, in *Techniques in Immunocytochemistry,* Vol. 3 (Bullock, G. R. and Petrusz, P., eds.), Academic, London, pp. 155–178.
11. Tolson, N. D., Boothroyd, B., and Hopkins, C. R. (1981). Cell surface labelling with gold colloidal particles: The use of avidin and staphylococcal protein A-coated gold in conjunction with biotin and Fc-bearing ligands. *J. Microsc.* **123,** 215–226.
12. Bendayan, M. (1985) The enzyme-gold technique: A new cytochemical approach for the ultrastructural localisation of macromolecules, in *Techniques in Immunocytochemistry,* Vol. 3 (Bullock, G. R. and Petrusz, P., eds.), Academic, London, pp. 179–201.

CHAPTER 17

Immunogold Probes for Light Microscopy

Julian E. Beesley

1. Introduction

Light microscope immunocytochemistry was initiated by the classical work of Coons et al. *(1,2)*, who developed the immunofluorescent technique for antigen localization. All other immunocytochemical techniques are based on the same philosophy, but use different microscopically dense markers. For instance, Avrameas and Uriel *(3)* and Nakane and Pierce *(4)* described the use of the enzyme peroxidase as a dense marker for immunocytochemistry, and this was later developed by Sternberger *(5)*, who described the sensitive peroxidase–antiperoxidase techniques. The initial immunoenzyme techniques have been further expanded by the use of alkaline phosphatase as a marker (*6, see also* this vol., Chapter 10). These techniques, including the original immunofluorescent technique, are all routinely used for immunohistochemistry. Recently, the use of colloidal gold probes as immunocytochemical markers has been described for electron immunocytochemistry *(7)*, and these are now proving to be of use at the light microscope level.

A colloidal gold probe is a gold sphere, usually between 1 and 15 nm in diameter, coated with an immunological protein. As in the immunoenzyme techniques, the immunological protein could be one of a wide range of proteins. It could be a primary antibody, for use in a direct one-step system, or a secondary or tertiary antibody, for indirect labeling. It could be protein A, protein G, or, if biotinylated antibodies are used, streptavidin could be coupled to the gold. Immunolabeling is carried out by incubating the antigen with

From: *Methods in Molecular Biology, Vol. 10: Immunochemical Protocols*
Ed.: M. Manson ©1992 The Humana Press, Inc., Totowa, NJ

the primary antibody–gold complex in the direct technique, or primary antibody followed by the gold conjugate in the indirect technique. The natural pink color of the gold probes can be seen *(8)* on lightly stained material, although the color of the gold probes is generally masked by the dense staining of the specimen. Gold probes may be observed without further treatment by dark field illumination *(9)*. Recently, the silver enhancement technique has been described *(10,11)* for improved detectability of gold probes (*see also* this vol., Chapter 19). Gold immunolabeling is carried out and the immunolabeled specimen is flooded with a silver solution in the presence of a reducing agent. The gold spheres act as nuclei on which the silver ions precipitate as metallic silver, thereby increasing the size of the probe. The probe eventually becomes sufficiently large to be seen easily as a black deposit on densely stained specimens. This technique permits all the routine stains to be carried out, and in addition, the immunolabeling remains visible and permanent. Initially, the silver solutions had to be made by the investigator, but more recently, light insensitive kits have become commercially available.

It is important to have the maximum number of gold nuclei for optimum signal. It is therefore advisable to immunolabel with the 5 nm gold *(12)*. Ultrasmall gold, 1 nm in diameter, is now available and may eventually prove to be superior to the 5 nm particles.

The silver enhancement technique may be used with the immunoenzyme techniques for multiple immunolabeling *(13)*. There is, in addition, increased interest in producing different colors during silver enhancement and these may well be of use in the future for multiple immunolabeling experiments. The silver enhancement technique has also found use in scanning electron microscopy, in which the antigens are labeled with small gold probes for great sensitivity, and are enhanced for detection by back-scattered electron detectors.

The technique may be carried out on any suitably fixed antigen. For light microscopy this is usually a smear of isolated cells, such as a blood smear, or on wax-embedded or frozen sections of tissue. More recently, the technique has been used for *in situ* hybridization techniques.

2. Materials

1. An antibody of known specificity and species. It does not matter whether the antibody is monoclonal or polyclonal.
2. A gold probe, 5-nm diameter, coated with an immunological protein specific to the primary antibody. If the primary antibody is from a rabbit, gold coated with protein A or antirabbit serum may be used. If the primary antibody is from mouse, gold coated with protein G or an antibody

raised against the correct species of IgG is used. If the antibody is biotinylated, a streptavidin-gold complex should be used. These will be referred to as the gold probe. In the context of the final result, it does not matter which probe is used as long as it reacts with the primary antibody.

3. Lugol's iodine.
4. 2.5% Sodium thiosulfate.
5. Phosphate buffered saline (PBS): 0.01*M* sodium phosphate, 0.5*M* NaCl, pH 7.2, as suggested by Slot and Geuze *(14)*.
6. Heat-inactivated serum from the second antibody species (not for use with the protein A–gold technique).
7. 1% Bovine serum albumin in PBS (BSA–PBS).
8. Commercial silver enhancement kit.
9. Tap water.
10. 2.5 and 5% sodium thiosulfate.
11. Mayer's hematoxylin and Merkoglas mountant.
12. Dehydrating alcohol.
13. Xylene.
14. Gum acacia (500 g/L in distilled water).
15. Trisodium citrate. $2H_2O$.
16. Citric acid.
17. Hydroquinone: Freshly prepared (0.85 g/15 mL).
18. Silver lactate: Freshly prepared (0.11 g/15 mL).
19. Tris-buffered saline (TBS): 0.05*M* Tris in isotonic (0.9%) saline, pH 7.6.
20. Trypsin: 0.1% in TBS.
21. Calcium chloride.
22. Neutral buffered formalin.

3. Methods

3.1. Paraffin Sections

1. Fix the specimen in neutral buffered formalin and embed in wax. Cut 3–5 μm sections.
2. Dewax the sections in xylene and rehydrate through alcohols to water.
3. Immerse in Lugol's iodine (5 min), and remove the iodine by rinsing thoroughly in 2.5% sodium thiosulfate, followed by washing in PBS. There is some controversy surrounding the use of Lugol's iodine, some workers claiming that it is effective in increasing the stain concentration, others maintaining that it can safely be omitted. Its efficacy for individual applications should be tested.

4. Wipe excess liquid from the area around the sections and flood the sections with 5% heat-inactivated normal serum from the second antibody species (20 min). If protein A–gold probes are to be used, flood the sections with BSA–PBS rather than normal serum.
5. Tip the slides and wipe away the normal serum from around the section and incubate at room temperature with specific antibody diluted in BSA–PBS for 1 h. The concentration of the specific antibody required will be determined by the titer of the antibody. This is found either empirically by carrying out a series of dilutions and determining which gives the best signal to noise ratio, or alternatively, many suppliers recommend dilutions of their antibodies. It is important to keep the sections covered with antibody throughout the incubation. Usually between 50 and 200 μL of antibody is sufficient. During the longer incubations, the slides are prevented from drying out by carrying out the incubations in a moist chamber, either a specially designed commercially available chamber or a plastic lunch box lined with moist tissue paper.
6. Rinse the slides in BSA–PBS by flooding several times.
7. Cover the sections with gold probe suitably diluted in BSA–PBS (1 h). The concentration of gold probe can be found empirically by a series of dilutions, or by following the manufacturer's recommended dilutions, usually in the region of 1:40.
8. Wash the slides thoroughly in BSA–PBS for 5 min and wash again in distilled water for 5 min.
9. If a commercial silver enhancement kit is used, make up enhancer immediately before use according to the manufacturer's instructions, and with respect to times and lighting conditions. If a silver solution is mixed in the laboratory (Section 3.4.), immerse sections in this and develop while monitoring with a microscope in a darkroom with a Safelight 5902 or F904.
10. When the reaction has developed sufficiently, according to the investigator's preference, wash the slides thoroughly in tap water for 10 min.
11. To prevent the silver intensification from fading, fix the reaction product with 5% sodium thiosulfate for 5 min and wash with water.
12. Counterstain as required (1 min with Mayer's hematoxylin).
13. Wash in water.
14. Dehydrate in alcohol and clear in xylene.
15. Mount in a synthetic mountant, such as Merkoglas.
16. The antigen/antibody complexes will appear black when viewed with transmitted light. If dark field observations are carried out, the gold

probes will be yellow. If epipolarization microscopy is used *(15)* the gold/ silver probes will appear silvery.

3.2. Frozen Sections

Follow the above procedure from Step 3. If immunolabeling is weak, the sections may be permeabilized by preincubation for 5 min in 0.1% trypsin in TBS containing 0.1% calcium chloride at 37°C. The section may be incubated with normal serum without washing.

3.3. Cell Suspensions

De Waele et al. *(9)* recommend immunolabeling isolated cells while in suspension. This necessitates centrifugation after each step until a cytocentrifuge preparation is made prior to contrasting. A preferred method in this laboratory is to prepare cytocentrifuge preparations and to immunolabel them as for frozen sections thereby avoiding several centrifugation steps.

3.4. Preparation of a Silver Solution (10)

The silver solution for enhancing the gold probes is prepared by mixing the following reagents in the following order and using immediately. All solution are made in distilled water.

1. 7.5 mL of gum acacia (50 g/L) prepared by stirring overnight and filtering through gauze. Stock solutions may be kept frozen in aliquots.
2. 10 mL of citrate buffer at pH 3.0. This consists of 23.5 g of trisodium citrate. $2H_2O$, 2.5 g of citric acid. $1H_2O$, and 100 mL of distilled water.
3. 15 mL of freshly prepared hydroquinone.
4. 15 mL of freshly prepared silver lactate.

Keep the silver lactate and the final mixture containing the silver lactate dark by wrapping the containers within foil and use in the darkroom.

4. Notes

1. Problems with immunolabeling usually fall into two categories. Either there is no immunolabeling, or the labeling is so high that there is so much background it is difficult to make out the specific labeled sites. Both Beltz and Burd *(16)* and Beesley *(17)* deal with these problems. Problems can be subdivided into:
   a. Personal: Has the operator made an error in the technique?
   b. No antigen: The antigen may have been damaged during processing or there may be no antigen in the sample.
   c. No antibody: The antibody may have been destroyed during preparation or there may be no specific antibody in the serum.

   d. No gold probes: The gold probes may have been destroyed during storage.
   e. Wrong buffers may have been used.

   These problems are at first sight daunting but usually the remedy is quite simple.

2. No immunolabeling. It is always advisable to run a positive control with each experiment to confirm that the reagents are functioning properly. If there is no immunolabeling on the experimental sections, the concentration of the antibody may be too low, or indeed too high. In the latter case, there is no room for the second antibody to bind, therefore reducing the signal *(16)*. It is possible that the antigen has been masked by the thickness of the section. Cut thinner sections or permeabilize the sections with a detergent, such as Triton X-100, or carry out proteolysis to unmask the antigen *(18)*. A common problem is the fixation regime. If the fixation is too harsh, the antigenicity of the specimen will be lost, whereas if the fixation regime is too light antigens and other material may leach from the tissue.
3. Too much labeling. An unacceptably high background may be caused by a very poor antibody. It is always wise to ascertain the titer and specificity of an antibody before immunolabeling and estimate the expected results. If a suitable antibody has been chosen, the concentration of immunological reagents may be too high or there may be nonspecific attachment of reagents to the tissue. The former is remedied by carrying out a series of dilutions of the reagents and silver development times and selecting concentrations that produce high signal to low noise. High background caused by the nonspecific attachment of reagents to the specimen is reduced by including 0.5% Triton X-100 in the buffers used for diluting the reagents. A preincubation of the specimen with 1% BSA and 1% gelatin in PBS *(17)* will block some of these nonspecific sites and will also, if necessary block reactive electrostatic sites on the gold probes *(19)*. Beltz and Burd *(16)* recommend that the addition of up to 0.5*M* sodium chloride should reduce nonspecific labeling by preventing ionic interactions between the sera and the tissue. They warn also that high salt can interfere with low affinity antigen–antibody binding, and this technique should be used with care. Beltz and Burd *(16)* additionally recommend that if there is trouble with background immunolabeling, the antiserum can be absorbed with fresh tissue from the same host that does not contain the antigen in question. If these fail, they suggest selection of another antibody against the same antigen.

4. Methodological comments. Immunocytochemistry is a routine technique, but nonetheless, care must be taken to achieve optimal results. Pay particular care to the freshness of reagents, especially buffers, which will become contaminated with bacteria that may interfere with imunolabeling reagents. During the immunolabeling, care should be taken to prevent the specimens from drying. During the drying process, the reagents concentrate, leading to unacceptable concentrations and high background. Immunolabeling at room temperature is normal, but beware of carrying this out in direct sunlight on the laboratory bench. In addition, the silver solutions are light-sensitive and should be protected from direct sunlight.

   A common problem with silver staining is the presence of dark spots over the tissue. This may be remedied by reducing the enhancement times. If it is too much of a problem, possibly caused by metal ions in the tissue, one of the immunoenzyme techniques should be considered.

References

1. Coons, A. H., Creech, H. J., and Jones, R. N. (1941) Immunological properties of an antibody containing a fluorescent group. *Proc. Soc. Exp. Biol. Med.* **47,** 200–202.
2. Coons, A. H. and Kaplan, M. H. (1950) Localization of antigen in tissue cells. *J. Exp. Med.* **91,** 1–13.
3. Avrameas, S. and Uriel, J. (1966) Methode de marquage d'antigenes et d'anticorps avec des enzymes et son application en immunodiffusion. *C. R. Acad. Sci., Paris, Ser. D.* **262,** 2543–2545
4. Nakane, P. R. and Pierce, G. B., Jr. (1966) Enzyme-labelled antibodies: Preparation and application for the localization of antigens. *J. Histochem. Cytochem.* **14,** 929–931.
5. Sternberger, L. A. (1979) *Immunocytochemistry,* 2nd Ed., John Wiley, New York.
6. Mason, D. Y. (1985) Immunocytochemical labelling of monoclonal antibodies by the APAAP immunoalkaline phosphatase technique, in *Techniques in Immunocytochemistry,* Vol. 3 (Bullock, G. R. and Petrusz, P., eds.), Academic, London, pp. 25–42.
7. Faulk, W. R. and Taylor, G. M. (1971) An immunocolloid method for the electron microscope. *Immunochemistry* **8,** 1081–1083
8. Roth, J. (1982) Applications of immunocolloids in light microscopy: Preparation of protein A–silver and protein A–gold complexes and their applications for localization of single and multiple antigens in paraffin sections. *J. Histochem. Cytochem.* **30,** 691–696.
9. De Waele, M., De Mey, J., Moeremans, M., De Brabander, M., and Van Camp, B. (1983) Immunogold staining method for the light microscopic detection of leukocyte cell surface antigens with monoclonal antibodies. *J. Histochem. Cytochem.* **31,** 376–381.
10. Holgate, C. S., Jackson, P., Cowen, P. N., and Bird, C. C. (1983) Immunogold silver staining: A new method of immunostaining with enhanced sensitivity. *J. Histochem. Cytochem.* **31,** 938–944.

11. Danscher, G. and Norgaard, J. O. R. (1983) Light microscopic visualisation of colloidal gold on resin embedded tissue. *J. Histochem. Cytochem.* **31,** 1394–1398.
12. Hoefsmit, E. C. M., Korn, C., Blijleven, N., and Ploem, J. S. (1986) Light microscopical detection of single 5 and 20 nm gold particles used for immunolabelling of plasma membrane antigens with silver enhancement and reflection contrast. *J. Microsc.* **143,** 161–169.
13. De Mey, J., Hacker, G. W., De Waele, M., and Springall, D. R. (1986) Gold probes in light microscopy, in *Immunocytochemistry: Modern Methods and Applications,* 2nd Ed. (Polak, J. M. and Varndell, I. M., eds.), J. Wright and Sons, Bristol, pp. 71–88.
14. Slot, J. W. and Geuze, H. J. (1984) Gold markers for single and double immunolabelling of ultrathin cryosections, in *Immunolabelling for Electron Microscopy* (Polak, J. M. and Varndell, I. M., eds.), Elsevier, Amsterdam, pp. 129–142.
15. De Waele, M., De Mey, J., Renmans, W., Labeur, C., Jochmans, K., and Van Camp, B. (1986) Potential of immunogold–silver staining for the study of leukocyte subpopulations as defined by monoclonal antibodies. *J. Histochem. Cytochem.* **34,** 1257–1263.
16. Beltz, B. S. and Burd, G. D. (1989) *Immunocytochemical Techniques: Principles and Practice.* Blackwell, Oxford.
17. Beesley, J. E. (1989) Colloidal gold: A new perspective for cytochemical marking. *Royal Microscopical Society Handbook,* 17. Oxford Science Publications, Oxford.
18. Finley, J. C. W. and Petrusz, P. (1982) The use of proteolytic enzymes for improved localization of tissue antigens with immunocytochemistry, in *Techniques in Immunocytochemistry,* Vol. 1 (Bullock, G. R. and Petrusz, P., eds.), Academic, New York, pp. 239–250.
19. Behnke, O., Ammitzboll, T., Jessen, H., Klokker, M., Nilausen, K., Tranum-Jensen, H., and Olsson, L. (1986) Non-specific binding of protein-stabilised gold sols as a source of error in immunocytochemistry. *Eur. J. Cell Biol.* **41,** 326–338.

CHAPTER 18

Immunogold Probes in Electron Microscopy

Julian E. Beesley

1. Introduction

Electron microscopy permits the detailed study of cell relationships within tissues and organelles within cells. Electron immunocytochemistry is the high resolution study of antigens within cells and their relation to cell ultrastructure. Fixation to achieve optimal fine structural detail for electron microscopy is exactly that which damages antigens with respect to reaction with specific antibody. Cell preparation for electron microscopy is therefore a compromise between retaining sufficient antigenicity while preserving the cell ultrastructure.

The specimen is prepared to expose antigens at the surface of the sample, and these samples are successively incubated with primary antibody, a wash, and a secondary antibody that is complexed with an electron dense marker. Blocking steps are included to reduce nonspecific attachment of the reagents to either the sample or the background.

Until 1980, peroxidase was the marker of choice, but since then, the use of colloidal gold has increased and now, is almost universally used for electron immunocytochemistry. Colloidal gold is ideal for electron microscopy. It is particulate and very dense, therefore, it can be identified on heavily stained biological tissue. Because it is small, it will not obscure the fine structure of the sample, and it can be prepared in several different sizes to be used for multiple immunolabeling experiments and quantification.

From: *Methods in Molecular Biology, Vol. 10: Immunochemical Protocols*
Ed.: M. Manson ©1992 The Humana Press, Inc., Totowa, NJ

Some experiments require that antigenic information is obtained from the surface of the cell, others are interested in intramembrane antigens or intracellular antigens, or indeed the relationship of one antigen-bearing cell with another. The specimen can be prepared by one of several different methods to expose antigens to obtain the desired antigenic information from the sample *(1)*.

a. The preembedding technique allows external antigens to be localized. The sample, usually whole cells, is incubated with the immunological reagents before dehydration, embedding, and sectioning.
b. In contrast, for the postembedding technique, the sample is embedded and sectioned before immunolabeling. This permits the localization of internal antigens.
c. The immunonegative stain technique is used for the localization of external antigens on viruses, bacterial pili and any small objects that may be dried onto an electron microscope grid and immunolabeled *in situ.*
d. The immunoreplica technique is a high resolution technique for the localization of antigens on the plasma membrane of cultured cells. After immunolabeling, the cells are replicated with carbon and platinum to reveal immunolabeling on the cell surface.
e. The freeze-fracture immunolabeling technique can be used in two ways. In the first, the cells are immunolabeled and freeze-fractured. This shows immunolabeling of the outer surface of the cell membrane superimposed on a high resolution replica of the inner surface of the membrane. In the second, the cells are fractured before immunolabeling. This permits immunolabeling to be correlated with large areas of the fractured faces.
f. Finally, when relationships between cells expressing particular surface antigens are required, specimens are immunolabeled before preparation for scanning electron microscope observation when immunolabeling on the surface of the cells in the tissue is observed.

The process of immunolabeling for each of these different methods is similar. The difference between the techniques is in preparation of the antigen and the contrasting method.

2. Materials

1. Antibody of known specificity.
2. Colloidal gold probe of the required size (10 nm diameter for routine use), which will react with the primary antibody.
3. Phosphate-buffered saline (PBS): 0.01*M* Phosphate, 0.15*M* NaCl, pH 7.2, as recommended by Slot and Geuze *(2)*.

4. 1% Gelatin in PBS.
5. 0.02*M* Glycine in PBS.
6. 1% Bovine serum albumin in PBS (BSA–PBS).
7. 1% Glutaraldehyde in PBS.
8. Distilled water.
9. Materials for embedding the samples in the pre- and postembedding techniques.
10. Contrasting reagents for samples (*see* individual techniques).
11. 4% Formaldehyde freshly prepared from paraformaldehyde powder, in 0.1*M* cacodylate buffer, pH 7.2 or 1% glutaraldehyde in 0.1*M* cacodylate buffer, pH 7.2.
12. Graded series of ethanol (70%, 90%, 100%).
13. 2.3*M* Sucrose.
14. 200-Mesh Butvar/carbon-coated grids.
15. 1% Osmium tetroxide.
16. 2% aqueous uranyl acetate.
17. 1% uranyl acetate made up in 70% methanol.
18. Neutral uranyl acetate. To prepare neutral 2% uranyle acetate, mix equal quantities of 4% aqueous uranyl acetate and 0.3*M* oxalic acid and adjust the pH to 7.2–7.4 with 10% ammonium hydroxide *(3)*.
19. Reynold's lead citrate stain *(4)*.
20. 1% sodium phosphotungstate.
21. Freon 22.
22. Liquid nitrogen.
23. Sodium hypochlorite.
24. 40% chromic acid.

3. Method

Some techniques require the specimen to be fixed. Fixation for electron microscopy depends very much on the antibody being used. If the antibody is a monoclonal, fixation is in 4% cacodylate-buffered formaldehyde. Keep the specimen in this fixative until further processing. Polyclonal antibodies are able to withstand much greater fixation and so, 1% cacodylate buffered glutaraldehyde for 1 h is used as the fixative. This preserves much more ultrastructure than formaldehyde. The sample should be transferred to cacodylate buffer until further processing.

3.1. The Postembedding Technique

This technique is the most popular of the immunolabeling techniques. All immunolabeling steps are carried out at room temperature, and the various reagents are placed dropwise on a piece of Parafilm in a Petri dish. If required, the chamber can be kept moist by inclusion of several pieces of damp filter paper.

3.1.1. Specimen Preparation

1. Ascertain which primary antibody is to be used and fix the specimen with either freshly prepared 4% formaldehyde or with 1% glutaraldehyde.
2. Dehydrate the specimen in a graded series of ethanol and embed in the resin of choice. This could be LR White *(1)*. If a more sensitive system is required, LR Gold or Lowicryl *(5)* should be used. For greatest sensitivity, cryoprotect the specimen in 2.3*M* sucrose for 1 h before freezing in liquid nitrogen and cut ultrathin frozen sections that are collected and thawed on a droplet of 2.3*M* sucrose before mounting on 200 mesh Butvar/carbon-coated grids and immunolabeling *(6,7)*.

3.1.2. Immunolabeling

1. Float the grids, section side down on 1% gelatin in PBS for 10 min. The gelatin adsorbs nonspecifically to nonimmunological sticky sites on the surface of the section. This prevents antibody attachment, thereby reducing nonspecific background.
2. Float the grid on 0.02*M* glycine in PBS for 3 min. This will block any free aldehyde groups in the tissue and prevent the antibodies from being fixed nonimmunologically on the tissue.
3. Rinse the sections with BSA–PBS for 2 min.
4. Incubate the sections with an appropriate dilution of specific antibody diluted with BSA–PBS for 1 h. The concentration of antibody is found by immunolabeling at different concentrations until a satisfactory signal:noise ratio is achieved. Suitable starting dilutions for monoclonal antibodies are 1:5, 1:10, and 1:20. For polyclonal antibodies, 1:10, 1:50, and 1:100 are recommended.
5. Rinse excess reagents from the section by floating the grid for 4× 1 min washes on BSA–PBS.
6. Incubate the sections with a suitable dilution of colloidal gold probe, diluted with BSA–PBS. This dilution is found empirically, by testing a number of dilutions. A good starting dilution is 1:20 for a 10 nm gold probe.
7. Wash excess reagents off the sections by floating the grid, 4× 1 min, on PBS (no BSA).
8. Fix the reagents by floating the grid for 1 min on 2.5% glutaraldehyde in PBS.
9. Wash the sections thoroughly with water, 4× 1 min.

10. Stain resin sections with 1% methanolic uranyl acetate and lead citrate according to routine electron microscopy procedures. If ultrathin frozen sections are being used, stain with 2% aqueous uranyl acetate, neutral uranyl acetate, and embed in methyl cellulose before examination *(7)*.

3.2. The Preembedding Technique

The preembedding technique is used to localize antigens on the surface of isolated cells, either prokaryotes or eukaryotes. If it is necessary to store the cells before immunolabeling, they must be fixed as in Section 3.1.1. If a very sensitive method is required, fixation may be omitted before immunolabeling.

1. Centrifuge the cells lightly into a pellet and resuspend this in the reagents, following the schedule from Step 3 in Section 3.1.2. It is important to resuspend the cell suspensions thoroughly in each reagent, otherwise the cells in the center of the clump will not be immunolabeled because of poor penetration of reagents *(1)*.
2. After glutaraldehyde fixation (Step 8), fix the cells further with 1% osmium tetroxide, saturated uranyl acetate, dehydrate in an ascending series of ethanol (70%, 90%, 100%), and embed in epoxy resin. Ultrathin sections of the block stained with 1% methanolic uranyl acetate and lead citrate will reveal immunolabeling on the outer surface of the cells.

3.3. The Immunonegative Stain Technique

A suspension of viruses, bacterial pili, or even isolated cell organelles, in distilled water is dried onto the electron microscope grid and immunolabeled from Step 3 in Section 3.1.2. After immunolabeling (Step 7), the specimen is contrasted with a negative stain, such as 1% sodium phosphotungstate. If a fixative is included (Step 8), the specimens are difficult to stain. For very small specimens, it is advisable to use the 5 nm gold probes. Care should be taken with highly pathogenic samples. If necessary, they should be killed with glutaraldehyde or formaldehyde and tested for viability before immunolabeling.

3.4. The Immunoreplica Technique

The immunoreplica technique *(8)* is used when it is necessary to detect antigenic sites on the plasma membrane of cultured cells. The cells are cultured on cover slips, and are fixed as described above depending on the antibody in question, and immunolabeled *in situ* from Step 1 in Section 3.1.2. After immunolabeling (Step 9), they are further fixed with 1% osmium tetroxide and are dehydrated in a graded series of ethanol (70%, 90%, 100%), critically point dried and replicated with a layer of carbon and platinum before examination with the transmission electron microscope. Large areas of the replicated plasma membrane remain intact for observation. Colloidal gold probes are probably the only probes of sufficient density that can be detected on these surfaces.

3.5. The Immunofreeze-Fracture Techniques

These techniques were developed and have been pioneered by Pinto da Silva et al. *(9)*.

3.5.1. The Label-Fracture Technique (10)

Isolated cells, usually unfixed, may be immunolabeled with antibody and gold probe as in Steps 3–9 of Section 3.1.2. and freeze-fractured by routine techniques *(11)*. Briefly, the tissue is cryoprotected with a suitable agent, such as 30% glycerol, and rapidly frozen in Freon 22 cooled with liquid nitrogen. The specimen is fractured in a freeze-fracture plant and the surfaces replicated. These replicas are thawed and biological debris removed by cleaning with sodium hypochlorite and 40% chromic acid, followed by thorough washing in distilled water. After mounting on uncoated electron microscope grids, the replicated fracture faces can be observed simultaneously with immunolabeling on the outer surfaces of the cells.

3.5.2. The Fracture-Label Technique

Alternatively, the specimen may be fractured before immunolabeling *(9)*. The sample is fixed depending on which antibody is being used, cryoprotected, frozen rapidly, and ground into small pieces under liquid nitrogen in a homogenizer. The small pieces are thawed in glycerol and immunolabeled as in Steps 3–8 of Section 3.1.2. At this stage, they may be processed to resin for ultrathin sectioning to observe immunolabeling on the fractured faces (the thin section fracture-label technique), or they may be dehydrated, critical point-dried and replicated, before examination (the critical point drying fracture-label technique) for observation of replicas of the labeled surface of fractured tissue.

3.6. The Immunoscanning Technique

Colloidal gold immunolabeling is suitable for scanning electron microscopy *(12)*. This yields a further useful dimension to the technique for observing external antigens. Specimens excised from the animal or cells grown in culture are suitably fixed in either 4% cacodylate buffered formaldehyde or 1% cacodylate buffered glutaraldehyde, and immunolabeled as in Steps 3–8 of Section 3.1.2. After fixation with 2.5% glutaraldehyde, the samples are further fixed with 1% osmium tetroxide and prepared for scanning electron microscope observation by any of the routine techniques, such as the osmium tetroxide/thiocarbohydrazide technique *(13)*. Until recently, the resolution of this technique was limited to that achievable with the scanning electron microscope and only relatively large 30 nm gold probes could be used. De Harven et al. *(14)* have combined the use of small gold probes with the silver enhancement technique, which gives high density of immunolabeling combined with relatively high resolution and this should be of importance in the future.

4. Notes

1. Colloidal gold probes are the most popular of all the immunolabeling techniques for electron immunocytochemistry. Since individual gold probes can be easily identified with the electron microscope, there is increased interest in multiple immunolabeling and quantitative studies. The technique is so popular that there are many different nuances in immunolabeling technique. The methods given in this chapter are those that have proved to be satisfactory in this laboratory.
2. The choice of the size of gold to be used is often questioned. Small 5 nm gold particles are surrounded by fewer protein molecules than the 15 nm particles. Therefore, for a given number of closely spaced antigens, it can be assumed that there will be more 5 nm gold particles than 15 nm particles, since each 15 nm gold probe is able to saturate more antibodies than the 5 nm probe. There will also be less steric hindrance when using the 5 nm gold compared with the 15 nm probe. Therefore, it is advisable to use the smallest probe consistent with the magnification required to detect the structures of interest. For thin section studies (Sections 3.1., 3.2., and 3.5.), a gold probe of 10 nm is a useful size. When immunolabeling small virus particles, such as polio (Section 3.3.), the size of the probe could be reduced to 5 nm. For general work the 15 nm probe is slightly large and the level of immunolabeling is low.
3. For multiple immunolabeling experiments, it is important that there should be no overlap of sizes of the gold probe. It is advisable to use the 5 nm gold in conjunction with the 15 nm probe. There is no difficulty in differentiating these two-sized probes. The 5 and the 10, or the 10 and the 15 nm probes could be used, if the size range of the probes were sufficiently small, but it has been my experience in triple immunolabeling with the 5, 10, and 15 nm probes, that it is sometimes confusing to separate the 5 from the 10 and the 10 from the 15 nm probes.
4. The postembedding technique (Section 3.1.) is the most popular of all the immunolabeling techniques. The preembedding technique (Sec-

tion 3.2.) and the immunonegative stain technique (Section 3.3.) are also reasonably popular. The remaining techniques are specialized and although the information they provide is extremely useful, they are not widely used.

5. There is a continuing debate about which method of specimen preparation is optimum for the postembedding technique (Section 3.1.). There are reports of the technique working with routinely fixed and embedded tissue, although these are exceptional. Methacrylate and LR White have been used for ambient temperature embedding, but there is now a great interest in the low temperature techniques using LR Gold and Lowicryl K4M. Advantages of these techniques are that at the low temperature, there is very little leaching of cell components, since, during dehydration at temperatures down to –40°C, proteins precipitate and are effectively stabilized within the tissue. There is also interest in techniques in which the unfixed sample is frozen before being dried and embedded in a resin *(15)*. Another further avenue is to freeze the sample and cut frozen sections, which are thawed at room temperature before immunolabeling *(6)*. The deciding factor in which technique to use is the abundance of the antigen and its sensitivity to fixation. If the antigen is abundant and withstands fixation, sensitivity is not usually a problem. Techniques and demands of immunocytochemistry are becoming increasingly stringent, and the investigator is often required to identify relatively few sensitive antigens in a tissue. It would be advisable in these instances to consider the use of cryotechniques for antigen preparation.
6. Any of the gold probes may be conveniently used for immunolabeling provided they link with the primary antibody. Slight differences are seen between immunolabeling with the protein A-gold probes and the antibody–gold complexes. If the antibody–gold complex is used, up to 10 conjugated gold probes may attach to a single Fc component of the primary antibody, thereby producing labeling in clusters. Protein A possesses one binding site for the Fc region, and therefore, clumps of Protein A-gold probes are not observed. Protein A-gold may therefore be of more use in quantitative studies (although many other factors such as steric hindrance and binding several antibodies with one probe, must be considered). A 1-nm gold probe has been reported and this contains several gold particles attached to one antibody. It may be of use for increasing the signal for electron microscope immunocytochemistry. It is a little small for routine use, but this could be remedied by use of the silver enhancement techniques.

7. Immunolabeling for electron microscopy is theoretically identical to immunolabeling for light microscopy. The problems discussed in Chapter 17 describing light microscope immunolabeling are entirely relevant to electron immunocytochemistry.

References

1. Beesley, J. (1989) Colloidal gold: A new perspective for cytochemical marking. *Royal Microscopical Society Handbook* 17, Oxford Science Publications, Oxford, UK.
2. Slot, J. W., and Geuze, H. J. (1984) Gold markers for single and double immunolabelling of ultrathin cryosections, in *Immunolabelling for Electron Microscopy* (Polak, J. M. and Varndell, I. M., eds.), Elsevier, Amsterdam, pp. 129–142.
3. Tokuyasu, K.T. (1978) A study of positive staining of ultrathin frozen sections. *J. Ultrastruct. Res.* **63,** 287–307.
4. Reynolds, E. S. (1963) The use of lead citrate at high pH as an electron-opaque stain in electron microscopy. *J. Cell Biol.* **17,** 208–212.
5. Carlemalm, E., Garavito, R. M., and Villiger, W. (1982) Resin development for electron microscopy and an analysis of embedding at low temperature. *J. Microsc.* **126,** 123–143.
6. Griffiths, G., McDowall, A., Back, R., and Dubochet, J. (1984) On the preparation of cryosections for immunocytochemistry. *J. Ultrastruct. Res.* **89,** 68–78.
7. Tokuyasu, K. T. (1986) Immunocryoultramicrotomy. *J. Microsc.* **143,** 139–149.
8. Mannweiler, K., Hohenberg, H., Bohn, W., and Rutter, G. (1982) Protein A-gold particles as markers in replica immunocytochemistry: High resolution electron microscope investigations of plasma membrane surfaces. *J. Microsc.* **126,** 145–149.
9. Pinto da Silva, P., Barbosa, M. L. F., and Aguas, A. P. (1986) A guide to fracture label: Cytochemical labelling of freeze-fractured cells, in *Advanced Techniques in Biological Electron Microscopy* (Koehler, J. K., ed.), Springer-Verlag, Berlin, pp. 201–227.
10. Pinto da Silva, P. and Kan, F. W. (1984) Label-fracture: A method for high resolution labeling of cell surfaces. *J. Cell Biol.* **99,** 1156–1161.
11. Robards, A. W. and Sleytr, U. B. (1985) Low temperature methods in biological electron microscopy, in *Practical Methods in Electron Microscopy,* Vol. 10 (Glauert, A. M., ed.), Elsevier, Amsterdam.
12. Hodges, G. M., Southgate, J., and Toulson, E. C. (1987) Colloidal gold—a powerful tool in S.E.M. immunocytochemistry: An overview of bioapplications. *Scanning Microscopy* **1,** 301–318.
13. Murphy, J. A. (1980) Non-coating techniques to render biological specimens conductive/1980 update. *Scanning Electron Microscopy* **1,** 209–220.
14. De Harven, E., Soligo, D., and Christensen, H. (1990) Double labelling of cell surface antigens with colloidal gold markers. *Histochemical J.* **22,** 18–23.
15. Dudek, R. W., Varndell, I. M., and Polak, J. M. (1984) Combined quick-freeze and freeze-drying techniques for improved electron immunocytochemistry, in *Immunolabelling for Electron Microscopy* (Polak, J. M. and Varndell, I. M., eds.), Elsevier, Amsterdam, pp. 235–248.

CHAPTER 19

Electron Microscopic Silver Enhancement for Double Labeling with Antibodies Raised in the Same Species

Kurt Bienz and Denise Egger

1. Introduction

In immunoelectron microscopy (IEM), simultaneous labeling of two or more antigens on the same section is desirable for many applications. If the antibodies (Ab) to be used are raised in the same species, as is usually the case with monoclonal antibodies (MAb), the difficulty arises that the labeled secondary, antispecies Ab used in the first labeling step traps the primary Ab directed against the second antigen, thus leading to a nonspecific signal for the second antigen.

We report here a method ("EM-silver enhancement," ref. *1*) to overcome this problem. This procedure increases the size of the gold marker by a predeterminable amount, thereby inactivating the antispecies Ab present on the gold grain, but fully retaining the immunoreactivity of the section. For IEM double labeling, therefore, the EM-silver enhancement (B in Fig. 1) has to be performed after the first labeling step (A in Fig. 1) to render the section ready for a second immunocytochemical reaction (C in Fig 1) with primary Ab from the same species and with the same (small) gold marker as in the first labeling step.

The physical development employed in the silver enhancement procedure was originally used in photographic work, including electron microscopic autoradiography *(2,3)*, and is widely applied today in intensifying

From: *Methods in Molecular Biology, Vol. 10: Immunochemical Protocols*
Ed.: M. Manson ©1992 The Humana Press, Inc., Totowa, NJ

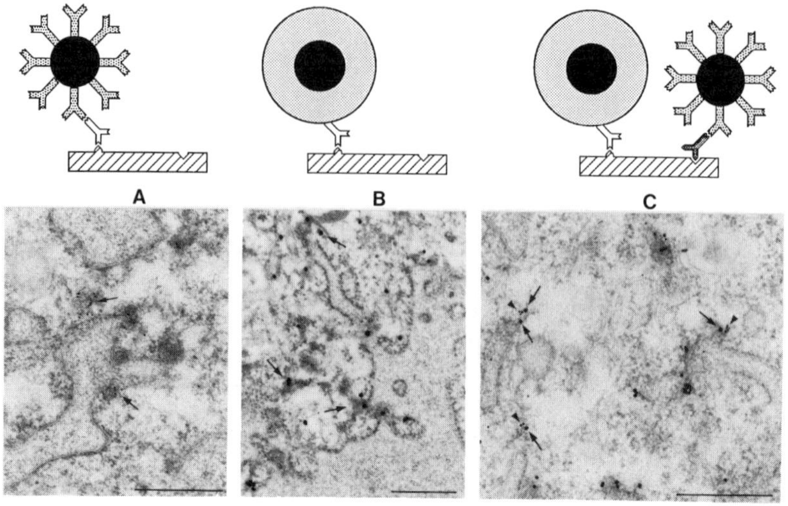

Fig. 1. Schematic representation and electron microscopic aspect of the IEM double labeling method. Panel A: The first antigen is labeled with the primary Ab (white) and a secondary antispecies Ab (hatched), tagged with a small (e.g., 5–10 nm) gold grain (black). The EM-picture above shows a viral replication complex in a poliovirus-infected HEp-2 cell, labeled (arrows) with a mouse MAb directed against the viral protein 2C and a goat- antimouse (GAM) Ab tagged with 10 nm gold (Janssen Auroprobe EM-grade, GAMIgG G10). Bar: 500 nm. Panel B: The EM-silver enhancement inactivates the antispecies Ab and enlarges the gold grain by the deposition of a silver layer (shaded area) on its surface. EM-picture: same as panel A, gold grains (arrows) enlarged approx threefold by the silver enhancement. Bar: 500 nm. Panel C: After the silver-enhancement, a second antigen in the same section can be labeled by an Ab (dark hatched) raised in the same species as was the Ab (white) against the first antigen. The gold labeled antispecies Ab is the same as used in the first labeling step. EM-picture: double-labeled poliovirus infected cell. First label was with anti-2C Ab, gold grains enlarged by the silver enhancement (arrows) as in panel B. Second label (small grains, arrowheads) was with MAb against the viral protein VPl and the same gold marker GAM G10 as in panel A. Bar: 500 nm.

immunogold stains in light microscopic preparations or on Western blots (immunogold–silver stain, IGSS; refs. *4,5. See also* vol. 3, Chapter 34). In essence, the method employs a photographic developer and dissolved silver ions. The photographic development reduces the silver ions to silver atoms on the surface of the gold particles in the immunocytochemical preparation. Thus, it leads to an autocatalytic increase in size of the gold label during the developing process.

It may be noted that, besides IEM double labeling, the EM-silver enhancement *per se* is also useful when IEM preparations are to be photographed at low magnifications. The immunocytochemical labeling can be done with 5 or 10 nm gold grains, which provide the best labeling efficiency, and the size of the marker can afterwards be reproducibly adjusted at will.

2. Materials

1. IEM specimens: Use sections of material, fixed and embedded (e.g., in Lowicryl K4M or LR-White) in order to retain antigenicity (*see* Note 1). It is essential to use gold grids, as the chemicals involved in the procedures react with copper or nickel. The use of negatively stained, unfixed material (subcellular fractions, viruses, and so on) is not recommended, as the EM-silver enhancement affects the structure of such specimens.
2. Antibodies: The type of primary Abs is determined by the antigens to be detected. The secondary antispecies Ab should ideally be labeled with 5 or 10 nm gold and good quality Abs can be obtained commercially from several suppliers (*see* Note 2).
3. TBS–BSA buffer (washing buffer): 0.9% NaCl and 0.1% bovine serum albumin in 20 m*M* Tris, adjusted with HCl to pH 8.2.
4. Blocking buffer: Normal serum of the species in which the secondary Ab was raised, is used diluted to 5% in TBS–BSA to block nonspecific binding sites.
5. To dilute the primary Abs, blocking buffer is used that contains only 1% serum.
6. The developer for EM-silver enhancement ("Agfa-Gevaert-developer," ref. *2*) is freshly made up by dissolving 0.075 g of Metol (4-(methylamino)-phenolsulfate, Fluka, Buchs, Switzerland), 0.05 g of sodium sulfite (anhydrous), and 0.02 g of potassium thiocyanate in 10 mL of distilled water. The final pH is 6.3.
7. Silver ions: Ilford L4 nuclear research emulsion, gel form.
8. Safelight illumination: Ilford filter S902.
9. Glutaraldehyde (2%) in phosphate-buffered saline is used to fix the sections.
10. Sections are stained with a 4% aqueous solution of uranyl acetate.

3. Methods

3.1. Immunocytochemical Labeling of First Antigen

All steps are performed at room temperature.

1. Float the grids with the sectioned cells for 10 min on distilled water.
2. Then block nonspecific binding sites by floating for 20 min on blocking buffer.

3. Follow this by incubation on the primary Ab (e.g., hybridoma culture supernatant, diluted appropriately in blocking buffer containing 1% serum) for 1 h.
4. Wash the grids by floating them on TBS–BSA in a watch glass on a gyrotory shaker for 2 × 5 min.
5. Then incubate them for 1 h on the gold-labeled secondary Ab, diluted in TBS–BSA according to the manufacturer's instructions.
6. Rinse again twice on TBS–BSA as above and "jet-wash" with distilled water.

3.2. EM-Silver Enhancement

Perform all steps in the darkroom with safelight illumination turned on.

1. To prepare the developing solution (physical developer), weigh 100 mg of L4 emulsion (silver donor) into a watch glass and add 2.5 mL of the freshly prepared "Agfa-Gevaert" developer. Alternatively, put an estimated amount of L4 emulsion into a light tight container in the darkroom, weigh outside the darkroom before adding the amount of developer required to obtain the proportion of 40 mg of emulsion per mL of developer.
2. Stir the freshly prepared physical developer with a small magnetic stirring bar for 5 min. Only part of the emulsion will dissolve; for reproducible results, keep the stirring time constant and record the temperature of the developer (preferably 20–21°C).
3. Place the gold-labeled sections, moistened with distilled water, on the surface of the developer and stir continuously for 4–6 min.
4. To stop the action of the developer, wash the grids by quickly dipping them several times in distilled water. The final size of the silver-coated gold grain depends on the amount of emulsion used, the stirring time, during which some of the emulsion is dissolved, the temperature of the developer, and the developing time (*see* Note 3).

3.3. Immunocytochemical Labeling of the Second Antigen

The grid is now ready for repeating the immunocytochemical labeling, using a primary Ab directed against another antigen. The procedure is performed exactly as described in 3.1.

3.4. Staining of Sections

1. To avoid loss of gold-labeled Ab during staining with the acidic uranium acetate, fix the sections with glutaraldehyde for 15 min.

2. Then rinse them with distilled water and stain with uranium acetate for 30 min at 37°C. They may then further be stained with a conventional lead stain (*see* Note 4).

4. Notes

1. The IEM double labeling method described was performed with cells fixed in 2% paraformaldehyde and 0.04% glutaraldehyde and embedded in Lowicryl K4M or LR-White. Other fixation and embedding procedures should work equally well, provided that the specimen resists the chemicals used in the silver enhancement.

   Some embedding media, such as Lowicryl, evaporate to a certain extent under the electron beam. Thus, when the section is monitored after the first labeling step, extensive irradiation should be avoided, otherwise, evaporated methacrylate might become deposited in the vicinity of the irradiated area. In such regions of the section, the gold grains are no longer accessible to the photographic developer.

   If the sections have been dried during the procedure, e.g., for viewing in the EM, they must be rehydrated with water before proceeding to the next step to avoid increased background.
2. When labeling the first and second antigen, the same gold-coupled, secondary Ab can be employed. Owing to enlargement of the first label, first and second labels can easily be distinguished. It is, however, necessary that the colloidal gold, used to label the secondary Ab, is very uniform in diameter, so that, after enhancement, no overlapping in size between the two labels occurs. Although we did not test it, the method should also work for gold-labeled protein A instead of an antispecies secondary Ab.
3. To allow a clear-cut distinction between the two labeled antigens after silver enhancement, the final size of the developed silver grains should be rather homogeneous and accurately predictable. The following parameters influence size, variation in size, and shape of the final grain:
   a. Developer: Metol, a slow-working fine grain developer, produces silver grains of uniform size and of round or slightly oval form. Hydrochinone, a coarse grain developer, which is used widely in light microscopic immunocytochemistry, leads, in the EM, to silver grains of irregular size and outlines. The same holds true or the IntenSE M-procedure (formerly Janssen, now Amersham).
   b. Source of silver ions: Silver nitrate, lactate, bromide, and chloride were tested, but it was found that only pieces of Ilford L4

emulsion as silver ion donor, in combination with the Metol developer as described in 3.2. above, lead to compact silver grains of predeterminable, uniform size (coefficient of variation less than 10%). We do not know why the L4 emulsion is superior to all other silver donors tested, as its exact composition is not known to us.

The useful concentration of emulsion is in the range of 20–50 mg of emulsion per mL of developer. It should be noted, however, that the number of emulsion pieces influences the developing speed, because the silver halide in the emulsion is only slowly soluble in the sodium sulfite incorporated in the developer, and the larger surfaces of several small emulsion pieces accelerate the dissolution of the silver halide. Similarly, the time of stirring the developer, before the section is put on, influences the concentration of silver ions, and thus, the developing speed.

If the gold grains need only to be enlarged to be more easily visible in low magnification work, silver bromide (10–20mg/mL) can be substituted for L4 emulsion, or the commercial IntenSE M-procedure can be performed. As mentioned above, both methods yield very heterogeneously sized grains, so that they can not be used for double labeling experiments.

c. Developing time and temperature: The grain size increases linearly with the developing time, doubling its diameter at 20–21°C in approx 4 min *(1)*. For double labeling, it is convenient to obtain a two to threefold increase in size of the gold grain. Smaller grains may be hard to distinguish from the unenlarged grains of the second labeling step, and larger grains obscure underlying details. Changes in temperature influence the developing speed considerably. Note in this respect, that magnetic stirrers may give off heat.

4. For sufficient contrast, most specimens will require lead staining after the uranium acetate stain. Although all conventional lead stains can be used, we found staining with Millonig's lead hydroxide for 2 min under N_2 best for Lowicryl embedded material.

References

1. Bienz, K., Egger, D., and Pasamontes, L. (1986) Electron microscopic immunocytochemistry: Silver enhancement of colloidal gold marker allows double labeling with the same primary antibody. *J. Histochem. Cytochem.* **34**, 1337–1342.

2. Kopriwa, B. M. (1975) A comparison of various procedures for fine grain development in electron microscopic radioautography. *Histochemistry* **44,** 201–224.
3. Bienz, K. (1977) Techniques and applications of autoradiography in the light and electron microscope. *Microsc. Acta* **79,** 1–22.
4. Holgate, C. S., Jackson, P., Cowen, P. N., and Bird, C. C. (1983) Immunogold-silver staining: New method of immunostaining with enhanced sensitivity. *J. Histochem. Cytochem.* **31,** 938–944.
5. Moeremans, M., Daneels, G., van Dijck, A., Langanger, G., and De Mey, J. (1984) Sensitive visualization of antigen–antibody reactions in dot and blot immune overlay assays with immunogold and immunogold/silver staining. *J. Immunol. Meth.* **74,** 353–360.

CHAPTER 20

Quantitative and Qualitative Immunoelectrophoresis

General Comments on Principles, Reagents, Equipment, and Procedures

Anne Laine

1. Introduction

The immunoelectrophoretic methods presented in Chapters 21 to 23 are all based on the electrophoretic migration of antigens in antibody-containing agarose and the specific immunoprecipitation of antigens with the corresponding antibodies. These methods exploit the property of immunoglobulins of remaining essentially stationary during electrophoresis at pH 8.6 (buffer pH most often used) and the property of most of the proteins to migrate in these conditions. Precipitating antibodies are required for these methods. For each antibody/antigen system, an individual precipitate is formed. The area enclosed by the precipitate is directly proportional to the concentration of antigen applied. These principles may, therefore, be used in a standardized procedure for immunological identification and quantitation of antigens and/or antibodies.

The rocket technique or electroimmunodiffusion (EID), which allows a precise quantitation of antigen, is described in Chapter 21, the crossed immunoelectrophoresis (CIE), which is a qualitative and quantitative method, is dealt with in Chapter 22, and the crossed immunoaffinoelectrophoresis (CIAE), which is more specially designed for the detection of biospecific

From: *Methods in Molecular Biology, Vol. 10: Immunochemical Protocols*
Ed.: M. Manson

interaction between macromolecular components and which constitutes a prediction method for the experiments of preparative separation, is covered in Chapter 23.

2. Materials

Several compositions for buffer with a pH about 8.6 have been published *(1)*, most of them containing barbital. The most commonly used is barbital buffer.

1. Electrophoresis buffer A: 20.6 g of barbital-Na and 4 g of barbital dissolved in 5 L of deionized water with magnetic stirring (this requires at least 2 h at room temperature).

   A barbital-glycine/Tris buffer has also been recommended *(1,2)*. It is obtained by mixing equal volumes of buffer 1 and buffer 2 prepared as follows. Buffer 1: 65 g of barbital-Na and 10.35 g of barbital dissolved in 5 L of deionized water; buffer 2: 281 g of glycine and 226 g of Tris dissolved in 5 L of deionized water. One volume of this buffer diluted with 4 vol of deionized water is most often used.
2. Electrophoresis buffer B: For crossed immunoaffinoelectrophoresis, another buffer is used according to Bøg-Hansen et al. *(3)*. The composition is: 44.8 g of Tris, 23 g of barbital, 0.555 g of calcium lactate, and 0.065 g sodium azide (very toxic if swallowed, contact with acids liberates very toxic gas) dissolved in 5 L of deionized water with magnetic stirring (this requires at least 3 h at room temperature). Store between 4 and 8°C. A cold buffer may be used instead of cooling the electrophoresis apparatus unless very high voltage (>15 V/cm in the gel) is used.
3. Agarose: This can be obtained from various sources (e.g., L'Industrie Biologique Française [France], Pharmacia [Sweden], and Litex [Denmark]). Its strength and low electroendosmotic properties are of particular importance. A 1% agarose solution in electrophoresis buffer (w/v) is generally used. Several batches of agarose need to be tested in order to find one that has the right properties for the proteins under investigation.
4. Washing solutions: 0.15*M* NaCl is used for washing out the nonprecipitated proteins from the gel, and deionized water for removing the NaCl from the gel before drying.
5. Staining solution: Coomassie Brilliant blue is a convenient and sensitive dye: 5 g of Coomassie Brilliant blue, 450 mL of 95% ethanol, 450 mL of deionized water, and 100 mL of acetic acid (glacial). Dissolve the dye in ethanol first and then add water and acetic acid.

6. Destaining solution: 450 mL of 95% ethanol, 450 mL of deionized water, and 100 mL of acetic acid (glacial).
7. Electrophoresis apparatus: It may have a built-in cooling system, but this is not essential if cold buffer (4–8°C) is used, except when very high voltage is applied (>15 V/cm in the gel for more than 1 h). The buffer vessels should have a volume of 1 L or more to ensure a sufficient capacity of the buffer system (*see* Note 2).
8. Power supply: Rectified current with stabilized voltage is used. A total output of 70–100 V corresponds to 1–3 V/cm, whereas 200–250 V gives 8–10 V/cm in the gel.
9. Voltmeter: It is used to check the voltage in the gel and enables a rapid discovery of anode/cathode confusion.
10. Glass plates: Various sizes are possible. For instance, we use 90 × 110 × 1.5 mm for electroimmunodiffusion and the first dimension of the crossed immunoelectrophoresis, and 100 × 100 × 1.5 mm for the second dimension of the crossed immunoelectrophoresis and for crossed immunoaffinoelectrophoresis. The plate dimensions must suit the individual electrophoresis apparatus. The glass plates must be cleaned with hot water and carefully dried with clean absorbent paper.
11. Wicks: The buffer-gel connections can be made of paper (Whatman No. 1). The wicks have to be exactly the width of the gel and are 6 cm long. For low voltage, two layers suffice, for high voltage, three layers are necessary. They are wetted with electrophoresis buffer just before being placed on the gel (*see* Note 3).
12. Horizontal table: It is important to make a uniform gel and to ensure a smooth uniform thickness. Thus, gel casting on a glass plate must be carried out using a horizontal table leveled with a spirit level before pouring the agarose. If a gel thinner than 1.5 mm is prepared, the use of a thermostated horizontal table (about 50°C) may facilitate pouring of the agarose. Glass plates preheated to about 50–60°C are also recommended.
13. A boiling water bath, hot plate, or microwave oven is needed to dissolve the agarose.
14. Water bath (50–56°C): The agarose solution must be cooled to 50–56°C before the antiserum is added to avoid protein denaturation. But the exact temperature at which the agarose remains fluid depends on the type of agarose that is used (56°C for agarose from Litex, 50°C for agarose from L'Industrie Biologique Française, for example).
15. Gel puncher and template: A gel plug is punched out with a metal punch and then sucked away with another punch (smaller) connected

to a suction apparatus, such as a water jet pump. Avoid too strong suction that will crack the gel surrounding the hole.

A template with holes in line in a plexibridge was designed and recommended by Weeke *(1)*. It allows punching of holes reproducibly side by side with uniform distances between them.

16. Razor blades: To cut out the gel and to transfer the gel from plate to plate (in crossed immunoelectrophoresis technique), a blade at least 2 × 11 cm is used.
17. Hair drier: It is time-saving to use the warm air stream from a hair drier in order to rapidly dry the agarose gel to a fine film before staining.

3. Methods

Some general procedures for the techniques detailed in the following three chapters are described here.

3.1. Preparation of Agarose and Gel Casting on a Glass Plate

1. Prepare 1% (w/v) agarose in electrophoresis buffer in a boiling water bath with continuous stirring until it is clarified (too much boiling turns the solution yellow) (*see* Note 4).
2. Divide the agarose into aliquots (in tubes), corresponding to the amount used for each plate and place the tubes to be mixed with antiserum in the water bath (50–56°C). When the agarose is used without antibodies, it can be poured immediately onto the glass plate since it does not need to be cooled.
3. Spread the molten gel carefully and very rapidly over the whole plate surface; when the tube is almost empty, move it very quickly to the corners of the plate in order to cover the plate completely (*see* Note 5).

   Adequate gelling requires from 5–15 min, depending on the type of agarose used.

3.2. Removal of the Nonprecipitated Proteins After Electrophoresis

After electrophoresis, precipitation peaks can often be detected directly in the gel. Nevertheless, staining of the gel provides more precise results and it is a good way of keeping a record.

Nonprecipitated proteins that would otherwise be strongly stained must be removed from the gel. The procedure depends on the amount of antiserum present in the gel; if antiserum is under 1% (v/v) in the gel, then washing is performed directly *(see* Step 4 *below)*, when over 1% (v/v), pressing of the gel is necessary. It is performed according to Laurell *(4)* as follows.

1. Place the plate with the gel on a clean glass plate (bigger than the gel plate), and cover the gel with a layer of filter paper wetted with water (avoid air bubbles).
2. Place several layers of soft absorbent paper on top of the filter paper and put one glass plate (500 g) on the top for 20 min.
3. Thereafter, carefully remove the filter paper (it may be stuck on the agarose and should be wetted with deionized water before removing). The pressing procedure can be repeated after replacing the filter paper if the gel contains antiserum at more than 5% v/v. The gel is now a thin film.
4. Wash the gel by placing the glass plate with the gel or the pressed gel in 0.15*M* NaCl for 30 min and in deionized water for 15 min. Do not put too much liquid in the bowl and hold the plate horizontally when immersing it in the liquid, otherwise, the gel may float off the plate.

3.3. Drying and Staining

1. After washing, cover the gel plate with a layer of filter paper wetted with water (to avoid air bubbles) and dry in hot air from a hair drier or in an oven. Drying is complete when the filter paper begins to lift from the plate. Drying the gel too fast or with too much heat will produce cracking or cause the filter paper to stick on the gel (*see* Note 6).
2. When the gel is completely dried to a fine film, place it in the staining solution for 10 min.
3. Destain with at least two or three changes of destaining solution.
4. Then press the plate on filter paper and allow to dry with the gel uppermost.

The staining of the gel film will fade if unprotected from light for too long (several months). However, another similar operation will restore the staining.

4. Notes

1. It is better to use fresh buffer for each experiment. Contamination of the buffer with proteins takes place during electrophoresis, and proteins that are recognized in the following antigen–antibody system may cause trouble. Moreover, pH changes take place during the electrophoresis with lowering pH on the anodic side and increasing pH on the cathodic side.
2. The electrophoresis apparatus must be carefully washed after each experiment to avoid growth of bacteria and fungi.
3. The wicks become contaminated by proteins from both samples and

antisera (especially when these are made monospecific by absorption procedures) and must be changed after each electrophoresis.

4. Agarose solution can be stored at 4°C and remelted in a boiling water bath when required. Do not remelt the agarose more than once, since it may be denatured.
5. Weeke *(1)* recommends coating the glass plate before pouring the agarose onto it in order to obtain a better contact between the electrophoretic gel and the glass plate. In fact, coating of glass plates is only necessary if long washings are needed after electrophoresis.
6. An alternative method for drying the agarose is to cover the plate with a piece of wet filter paper and leave overnight.
7. Sensitivity of the immunoelectrophoretic methods may be increased by using radioactive *(5)* or enzyme *(6)* labeling of antigen or antibody.

References

1. Weeke, B. (1973) General remarks on principles, equipment, reagents, and procedures. *Scand. J. Immunol.* **2(Suppl. 1)**, 15–35.
2. Bjerrum, O. J. and Lundahl, P. (1973) Detergent-containing gels for immunological studies of solubilized erythrocyte membrane components. *Scand. J. Immunol.* **2(Suppl. 1)**, 139–143.
3. Bøg-Hansen, T. C., Bjerrum, O. J. and, Ramlau, J. (1975) Detection of biospecific interaction during the first dimension electrophoresis in crossed immunoelectrophoresis. *Scand. J. Immunol.* **4(Suppl. 2)**, 141–147.
4. Laurell C. B. (1972) Electroimmunoassay. *Scand. J. Clin. Lab. Invest.* **29(Suppl. 124)**, 21–37.
5. Norgaard-Petersen, B. (1973) A highly sensitive radioimmunoelectrophoretic quantitation of human α-fetoprotein. *Clin. Chim. Acta.* **48,** 345,346.
6. Kindmark, C. and Thorell, J. (1972) Quantitative determination of individual serum proteins by radioelectroimmunoassay and use of ^{125}I-labelled antibodies (application to C-reactive protein). *Scand. J. Clin. Lab. Invest.* **29(Suppl. 124)**, 49–53.

CHAPTER 21

Rocket Immunoelectrophoresis Technique or Electroimmunodiffusion

Anne Laine

1. Introduction

The rocket immunoelectrophoresis technique or electroimmunodiffusion (EID) *(1)* is a simple, fast, and reproducible technique for quantitation of a single protein, and is also applicable in a protein mixture. Several unknown samples can be analyzed on a single plate. Known reference solutions have to be included in each plate. To obtain an accurate quantitation, the proteins in the reference solutions and in the unknown samples have to be physicochemically and immunologically identical. The samples are applied in wells punched out of an agarose gel containing the corresponding monospecific antiserum. One-dimensional electrophoresis is performed and rocket-shaped precipitates are formed. The quantitation is based on measuring the height of the precipitate peak.

This method is very flexible since several factors influencing the height of the precipitates are modifiable: antigen concentration (sample dilution), antibody concentration, field strength, gel batch (electroendosmosis), or time of electrophoresis. To reduce the consumption of antibodies, it is possible to pour antibody-containing agarose on the middle part of the plate only *(see below)*, and agarose without antibody on the edges of the plate is used to carry the paper wicks.

The size heterogeneity of antigen-carrying molecules has a smaller influence on EID than on single radial diffusion *(2)*. On the contrary, the charge heterogeneity has a greater influence on EID than on single radial diffusion.

From: *Methods in Molecular Biology, Vol. 10: Immunochemical Protocols*
Ed.: M. Manson ©1992 The Humana Press, Inc., Totowa, NJ

Furthermore, simultaneous quantitation of two proteins can be performed when the corresponding antisera are mixed in the gel. This requires that the precipitate morphology for the two proteins be different enough to allow satisfactory identification *(3)*.

The electrophoresis can be performed with high voltage (8–10 V/cm in the gel) for about 2–4 h or with low voltage (1.5–2 V/cm) overnight. With low voltage the results are better, and this is convenient for all proteins, even those that migrate slowly.

2. Materials

General equipment and reagents are as described in Chapter 20.

1. Glass plates: 90 × 110 × 1.5 mm.
2. Electrophoresis buffer A.
3. 1% (w/v) agarose in electrophoresis buffer.
4. Monospecific antiserum (for concentration, *see* Note 1).
5. Reference solution (*see* Note 2).

3. Methods (*see also* Chapter 20)

1. Prepare a 1.5-mm thick gel by pouring 15 mL of agarose on a glass plate (90 × 100 × 1.5 mm) placed on the horizontal table. After gelling has occurred, remove agarose from the middle part of the plate (45 × 90 mm) as indicated in Fig. 1a and replace by 7.5 mL of agarose (kept in a tube in the water bath 50–56°C) to which the monospecific antiserum has been added (*see* Notes 1 and 3).
2. Dilute the samples and the reference solution in the electrophoresis buffer (*see* Notes 1 and 2).
3. When the agarose is set, punch out from the gel a linear row of 2.5-mm diameter wells (5- or 7-mm distant from each other) and 0.5 cm in from the separation between agarose and antibody-containing agarose (Fig. 1a).
4. Place the glass plate with the gel on the electrophoresis apparatus. The connection to the buffer in the buffer vessels is made by means of the paper wicks (two layers).
5. Apply a current of 1.5 V/cm to the gel and add the buffer-diluted samples to the wells with the current on, to avoid diffusion (warning: electrical hazard, even if the voltage is low), by means of a syringe (1–10 µL from Hamilton) (*see* Notes 4 and 5). A 2.5-mm well diameter is suitable for 3–5 µL vol in a 1.5-mm deep gel. Check the voltage across the wicks with the voltmeter. The electrophoresis is performed at 1.5 V/cm overnight.

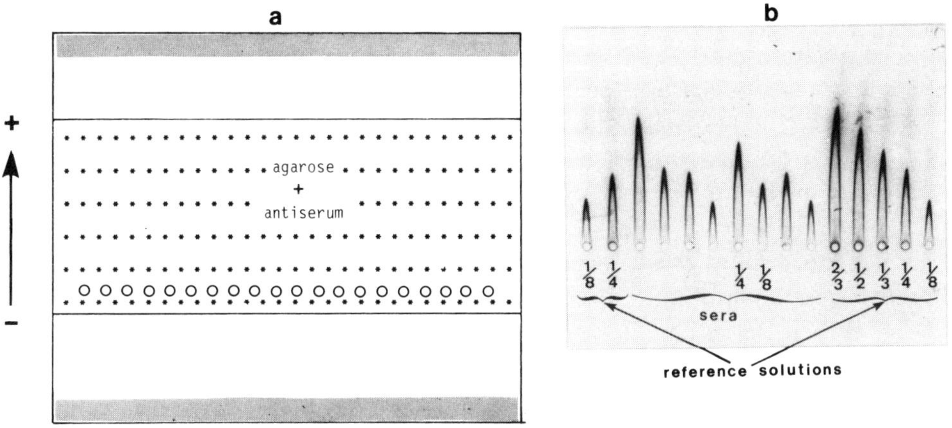

Fig. 1. (a) Template for the electroimmunodiffusion technique. * * * indicates area to be replaced; ▩ paper wick positions; ○ sample well. (b) α_1-antichymotrypsin in human serum. Antiserum: 1% rabbit antihuman α_1-antichymotrypsin (Dakopatts Ltd) corresponding to 1.5 µL of antiserum/cm^2 gel area. Reference solutions: 3 µL of dilutions of Plasma-Serum-Standard from Behringwerke (1/8, 1/4, 1/3, 1/2, 2/3, v/v in buffer). Unknown samples: 3 µL of serum diluted with buffer A (for example, 1/8 and 1/4 (v/v)).

6. After electrophoresis, press the gel (if the antibody amount in the gel is higher than 1% [v/v]), wash, dry, stain, and destain as described in Chapter 20. As an example, the quantitation of α_1-antichymotrypsin in human sera is shown in Fig. 1b.
7. Measurement of the peak height may be carried out using millimeter squared paper, the gel side of the plate facing the paper. Measure from the top of the wells to the rocket tips (this is easier to measure than from the center of the wells as recommended by other authors *[4]*). Draw the calibration curve (abscissa: protein concentrations, ordinate: peak heights) (*see* Note 2).

   Measure the peak heights of the unknown samples and read the concentrations on the calibration curve (*see* Notes 6–11). The concentrations in the unknown samples are, of course, calculated by multiplying the concentration in the diluted samples by the dilution factor.

4. Notes

1. The antiserum concentration used in the gel may be varied widely. It is determined empirically against an antigen dilution series. To begin with, 1% antiserum (v/v) in the gel is advisable.

2. At least four dilutions of the reference solution are needed to construct a standard curve. Reference solutions are commercially available (e.g., Protein-Standard-Serum or Protein-Standard-Plasma from Behringwerke).
3. If the monospecific antiserum is not too expensive or too precious, one can use antiserum-containing agarose to cover the whole plate, thus avoiding having to cut away one part of the gel and replace it. In this case, 15 mL of agarose (kept in the water bath) are mixed with the antiserum (the amount of antiserum added is then twice as much as for 7.5 mL of agarose) and immediately poured onto the glass plate.
4. Precision in the delivery of the sample volume is of major importance. Hamilton microsyringes are often used instead of automatic dispensing pipets.
5. It is necessary to rinse the syringe very carefully with buffer (at least five times) and with the next sample between two sample applications.
6. If the relationship between the concentration of the protein in the dilutions and the height of the precipitate does not give a straight line, it is often possible to obtain a straight line by further dilution of the samples and the use of less antiserum. But the limits of sensitivity for the rocket technique are determined by the formation of visible precipitin lines and this depends on the protein under study (for example, a weaker precipitin line is obtained for human serum α_1-antichymotrypsin than for human serum α_1-proteinase inhibitor in the same concentration range *[5]*).
7. The reproducibility of the method is best when the peak height is between 10 and 50 mm. If peaks are higher than 50 mm, either the antiserum concentration must be increased or the samples to be tested and the reference solutions must be further diluted. The reverse applies if the peaks obtained are smaller than 10 mm (accurate measurement in this case is uncertain). The rocket heights of the samples to be tested must lie within the curve constructed by plots obtained from lowest to highest reference solutions.
8. Validity of the method requires that charge and size of the antigen-carrying molecules in both samples to be tested and reference solutions are controlled. These molecules have to be physicochemically and immunologically identical. Precipitate morphology provides a clue as to whether these conditions are met: The same morphological appearance is necessary.
9. Precision. Standard errors between 2 and 5% are usually obtained.
10. If only partial identity exists between the protein in the unknown samples and that in the reference solution, errors in quantitation

may occur. Inspect critically and carefully the precipitate morphology in each experiment.

11. A more exact quantitation may be carried out by comparison of the areas enclosed by the precipitates, but this is more time-consuming.
12. Antigen molecules with cathodic electrophoretic mobility (also different from that of the antibodies) may also be quantified by EID, but the electrophoresis must be performed from the anode to the cathode.
13. The resolved peaks can be transferred electrophoretically to a nitrocellulose membrane and immunostained, as described by Hansen *(6)*.
14. An excellent review has been published by Laurell and McKay *(7)*. These authors paid particular attention to the different types of precipitate morphology that can be encountered with EID. They specified that the shape of precipitin loops, density of precipitates, sharpness of the precipitin line, and formation of adjacent peaks, all provide clues regarding physicochemical differences that may occur, not only between the samples and the reference solution, but also between the samples. Accurate quantitative analysis is not possible when precipitates of individual samples demonstrate abnormal appearance. If all samples produce the same type of deviation from the reference solution, then results can be given as relative rather than absolute values.

References

1. Laurell, C. B. (1966) Quantitative estimation of proteins by electrophoresis in agarose gel containing antibodies. *Anal. Biochem.* **15,** 45–52.
2. Mancini, G., Carbonara, A. O., and Heremans, J. F. (1965) Immunochemical quantitation of antigens by single radial immunodiffusion. *Immunochemistry* **2,** 235–254.
3. Krøll, J., Jensen, K. A., and Lyngbye, J. (1970) Quantitative immunoelectrophoresis as routine analysis, in *Methods in Clinical Chemistry,* Vol. 1. Karger, Basel/München/Paris/New York, pp. 131–139.
4. Laurell, C. B. (1972) Electroimmunoassay. *Scand. J. Clin. Lab. Invest.* **29(Suppl. 124),** 21–37.
5. Davril, M., Laine, A., and Hayem, A. (1987) Studies on the interactions of human pancreatic elastase 2 with human α_1-proteinase inhibitor and α_1-antichymotrypsin. *Biochem. J.* **245,** 699–704.
6. Hansen, S. A. (1988) Immunostaining of rocket immunoelectrophoresis precipitates too weak for identification following staining with Coomassie Brilliant Blue. *Electrophoresis* **9,** 101–102.
7. Laurell, C. B. and McKay, E. J. (1981) Electroimmunoassy, in *Methods in Enzymology* (Langone, J. J. and Van Vunakis, H., eds.), Academic, London, pp. 339–369.

CHAPTER 22

Crossed Immunoelectrophoresis

Anne Laine

1. Introduction

Ressler first described in 1960 *(1)* a form of immunoelectrophoresis now called crossed immunoelectrophoresis (CIE) or two-dimensional immunoelectrophoresis, which was later improved by Laurell *(2)*, Clarke and Freeman *(3)*, and Weeke *(4)*, among others. CIE is superior to the classical immunoelectrophoresis, according to Grabar and Williams *(5)*, particularly in providing better resolution and quantitative capabilities. It combines the electrophoretic separation of the sample proteins in agarose gel with electrophoresis at right angles to the initial separation in an antibody-containing agarose. Each protein separated during the first dimension forms a separate precipitation peak during the second dimension. Moreover, the area under any protein peak is directly proportional to the concentration of that protein in the analyzed sample and inversely proportional to the concentration of antibody to that protein in the antiserum used.

Today, CIE is a technique that has been considerably expanded and improved and many applications and modifications have been described (*see [6]* for a review). A CIE pattern looks complicated to the beginner, but regular users rapidly become used to recognizing individual antigens. This technique is not complex to use and interpret. A very important application of CIE is its precise quantitation of a large number of antigens simultaneously.

Thus, CIE is an effective tool that may be used in at least two ways, depending on whether quantitation or electrophoretic resolution is the purpose of the study. It is, for example, well suited to protein studies on

From: *Methods in Molecular Biology, Vol. 10: Immunochemical Protocols*
Ed.: M. Manson

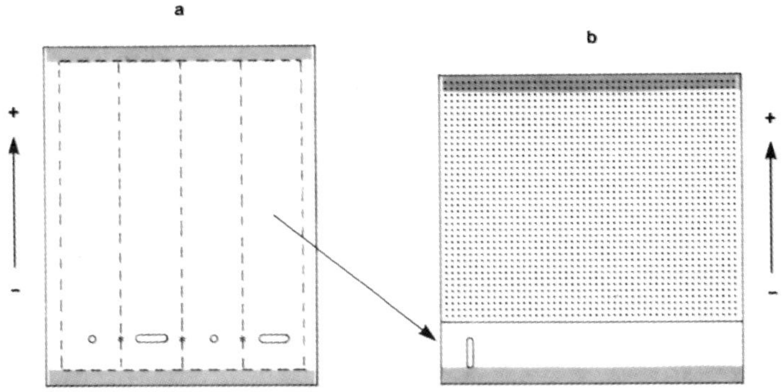

Fig. 1. Templates for crossed immunoelectrophoresis. (a) First dimension: After the first dimension, 5 mm of gel along the edges is discarded. O or ⊂⊃:sample well or slit. (b) Second dimension: The first dimension strip is transferred to the bottom of the plate and the remaining area is filled with 12 mL of antiserum-containing agarose. * bromophenol blue marker spot; ▒ paper wick positions; ::: agarose containing antiserum.

hereditary polymorphism, microheterogeneity, or complex formation with other proteins (an example is shown in ref. 7).

2. Materials

General equipment and reagents are the same as described in Chapter 20.

1. Glass plates: 90 × 110 × 1.5 mm for the first dimension; 100 × 100 × 1.5 mm for the second dimension.
2. Electrophoresis buffer A.
3. Agarose 1% (w/v) in electrophoresis buffer divided into aliquots of 15 mL or 12 mL.
4. Concentrated bromophenol blue solution: 1 mg/mL in electrophoresis buffer A.
5. Antiserum: mono-, oligo-, or plurispecific.
6. A predrawn template (such as the one shown in Fig. 1a) protected with a sticky plastic sheet.

3. Methods

1. Pour 15 mL of heated 1% agarose onto a 90 × 110 × 1.5 mm glass plate placed on the horizontal table.
2. After gelling, punch out four wells or four slits (depending on the volumes of the samples to be applied) as indicated in Fig. 1a (*see* Note 1).

This is easily done by placing the glass plate with the gel on the template. Place the glass plate with the gel on the electrophoresis apparatus.

3. Wet the wicks (three layers) with the electrophoresis buffer and place them over the 90-mm edges of the plate.
4. Apply samples by means of a Hamilton syringe.
5. Apply concentrated bromophenol blue solution by means of a Pasteur pipet on the same line as the sample wells but between them (*see* Fig. 1a), in order to follow the migration in the plate.
6. For the first dimension electrophoresis, apply 10 V/cm across the gel (*see* Note 2). It takes one hour for the marker dye (which is free) to cross the plate. Then stop the electrophoresis.
7. Cut away a 5-mm broad edge all around the gel with the razor blade. This portion will have suffered from heating owing to the high voltage.
8. Divide the remaining part of the gel into four strips (100 × 20 mm) (Fig. 1a). Transfer each strip with the razor blade to a 100 × 100 × 1.5 mm glass plate placed on the horizontal table (Fig. 1b) (*see* Notes 3–5).
9. Mix 12 mL of agarose (brought to the boil to clarify it and then kept in the 50–56°C water bath) with the antiserum (*see* Note 1, Chapter 21) and pour onto the remaining part of the plate adjacent to the strip (Fig. 1b).
10. When the agarose has set, place the plate on the electrophoresis apparatus with the first dimension strip at the cathode end. Connect the gel with the buffer by means of fresh paper wicks (two layers).
11. Carry out electrophoresis in the second dimension overnight with 1.5 V/cm in the gel from the cathode to the anode.
12. At the end of the electrophoresis, press the gel plate, wash, dry, stain, and destain as described in Chapter 20.
13. Identification of the individual precipitates in a pattern may be carried out immunochemically since a large number of monospecific antisera are now commercially available and can easily be used for identification purposes (*see* for example, Note 9 on "CIE with a trap-gel"). The tandem CIE also detailed in Note 10 can be another way to identify a precipitate if pure known antigen is available.
14. Quantitation of an individual antigen can be performed by measuring the area enclosed by the precipitate. Area measurement can be performed by planimetry (by means of a planimeter), by drawing the precipitate outlines on paper, cutting out the drawing and weighing it, or by multiplying height and width of the precipitate peak obtained if the peaks are almost symmetrical. This area is compared with the area determined for

a known amount of a reference antigen, using the same antiserum and the same technique. A major technical contribution to the precision of antigen quantitation by CIE has been made by adding a quantitative reference marker like carbamylated transferrin, which migrates near the albumin *(3,4)*.

4. Notes

1. When the electrophoretic mobility of the proteins is not known, the sample can be applied in the middle of the strip and the strip placed in the middle of the second dimension plate, the antibody-containing agarose is then poured equally on both sides of the strip on the glass plate.
2. The time and the voltage indicated for electrophoresis are suitable for most proteins. When slowly moving components are studied or when a microheterogeneity is to be detected, the appropriate time for the first dimension electrophoresis should be determined in initial trial experiments. In the second dimension, once the precipitation peaks are formed, they are quite stable to further electrophoresis.
3. To transfer a first dimension gel strip, the razor blade is gently pushed under the strip until about one-half of the strip is supported. Place the free part of the gel strip on the 100 × 100 × 1.5 mm plate and, with another razor blade, push the strip from the first razor blade, making sure that it is not broader than the glass plate (this could cause the agarose for the second dimension to pour off the glass plate).
4. Do not push the strips from the first dimension plate one behind the other, even if it seems easier than transferring them with the razor blade. Proteins could remain on the plate and contaminate the following gel strip. This could cause artifacts in the second dimension electrophoresis (lines across the gel if the contaminating proteins precipitate with the antiserum used).
5. To avoid the transfer of the first dimension strip, which is a tricky process, it is easier to carry out the first and second dimension electrophoreses on the same 100 × 100 × 1.5 mm plate. After the first dimension, the gel strip (100 × 20 mm) containing the migrated sample is left, whereas the remaining part of the gel is removed and replaced by antiserum-containing agarose.
6. Components that move toward the cathode can be investigated in this technique by performing the first dimension electrophoresis from the anode to the cathode. Under these conditions, a pyronin solution can be used as a marker dye. The strip is placed on the anodic side of the

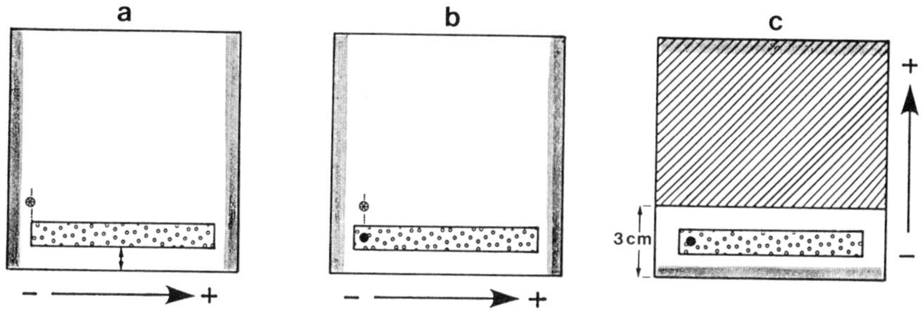

Fig. 2. Templates for the three steps for trap gel crossed immunoelectrophoresis. (a) preelectrophoresis; (b) first dimension; (c) second dimension. ∴ strip to be removed and replaced by 1.2 mL of agarose containing monospecific antiserum; /// agarose containing polyspecific antiserum; ▩ paper wick positions; * bromophenol blue marker spot; ● sample well.

second dimension plate and the electrophoresis performed from the anode to the cathode.

7. If peaks go off the top of the agarose plate, either the amount of proteins applied in the first dimension must be reduced or the antiserum concentration increased in the second dimension. If the peaks obtained are too small, the reverse applies.
8. Simple improvements of the CIE method have been described: line immunoelectrophoresis *(8)*, intermediate gel technique *(9)*, selective precipitation of the antigen in the first dimension *(10)*, among others. A procedure we improved in our laboratory *(11)* and the tandem CIE from Krøll *(12,13)* are described below in more detail.
9. We developed a procedure for the identification of proteins by CIE with a trap-gel containing monospecific antiserum *(11)*. This procedure can be used with all commercially available antisera, even when made monospecific by absorption procedures and thus containing antigen excesses.

   Heated 1% agarose (15 mL) is poured onto a glass plate (100 × 100 × 1.5 mm). After gelling of the agarose, a gel strip (1 × 8 cm) is removed as indicated in Fig. 2a. The resultant trough is refilled with 1.2 mL of agarose (kept in the 50–56°C water bath) containing the monospecific antiserum (15–50 µL approx).

   A preelectrophoresis is performed at 10 V/cm in the gel (this allows the removal of most of the excess antigens that would otherwise disturb the pattern during the second dimension electrophoresis) for at least 1 h until the bromophenol blue marker spot has crossed and left the plate.

A well is punched out in the antiserum containing agarose as indicated in Fig. 2b. The sample is then applied. In other respects, the procedure is as for CIE, except that 10.5 mL of antiserum-containing agarose are used for the second dimension instead of 12 mL. Pressing of the gel is avoided because of the fragility of the first dimension strip. During the first dimension, the trap-gel blocks an individual protein as a rocket. The corresponding peak is reduced or disappears in the second dimension, when compared with the pattern obtained in normal CIE. There is immediate unequivocal identification of a specific antigen.

10. A modification of the CIE technique, tandem CIE *(12,13)*, makes possible a direct correlation of immunologically identical fractions in different samples by producing partial fusion of related immunoprecipitate patterns.

    Application wells or slits are punched out in the same strip in the first dimensional electrophoresis gel with an intercenter distance of 8 mm. After absorption of the samples in the gel, the wells should be carefully sealed with agarose before electrophoresis, since otherwise, the separation pattern of the cathodic sample is disturbed during its electrophoretic passage around the anodic well. In other respects, the procedure is as for CIE.

    Identification of the precipitates can be done as follows. Absorption of a protein fraction in one of two identical serum samples can be done by adding 1–5 µL of a monospecific antiserum to one of the two application wells, or a solution of purified protein can be applied in one of the two application wells, whereas the sample containing the complex antigen mixture is applied in the other well. The precipitation peak corresponding to this purified protein can be identified as part of the only double peak that appears in the pattern.

References

1. Ressler, N. (1960) Two dimensional electrophoresis of protein-antigens with an antibody-containing buffer. *Clin. Chim. Acta* **5,** 795–800.
2. Laurell, C. B. (1965) Antigen antibody crossed electrophoresis. *Anal. Biochem.* **10,** 358–361.
3. Clarke H. G. M. and Freeman T. (1967) Quantitative immunoelectrophoresis method (Laurell electrophoresis), in *Protides of the Biological Fluids,* 14th Colloquium, 1966 (Peeters, H., ed.), Elsevier, Amsterdam, pp. 503–509.
4. Weeke, B. (1970) Carbamylated human transferrin used as a reference in the Laurell crossed electrophoresis. *Scand. J. Clin Lab. Invest.* **25,** 161–163.
5. Grabar, P. and Williams, C. A. (1953) Méthode permettant l'étude conjuguée des

propriétés électrophorétiques et immunochimiques d'un mélange de protéines: Application au sérum sanguin. *Biochim. Biophys. Acta* **10,** 193.

6. Emmett, M. and Crowle, A. J. (1982) Crossed immunoelectrophoresis: Qualitative and quantitative considerations. *J. Immunol. Methods* **50,** R65–R83.
7. Davril, M., Laine, A., and Hayem, A. (1987) Studies on the interactions of human pancreatic elastase 2 with human α_1-proteinase inhibitor and α_1-antichymotrypsin. *Biochem. J.* **245,** 699–704.
8. Krøll, J. (1968) Quantitative immunoelectrophoresis. *Scand. J. Clin. Lab. Invest.* **22,** 112–114.
9. Svendsen, P. J. and Axelsen, N. H. A. (1972) A modified antigen-antibody crossed electrophoresis characterizing the specificity and titre of human precipitins against *Candida albicans. J. Immuncl. Methods* **1,** 169–176.
10. Platt, H. S., Sewell, B. M., Feldman, T., and Souhami, R. L. (1973) Further modifications of the two-dimensional electrophoretic technique of Laurell. *Clin. Chim. Acta* **46,** 419–429.
11. Laine, A., Ducourouble, M. P., and Hannothiaux, M. H. (1987) Identification of proteins by crossed immunoelectrophoresis with a trap-gel. *Anal. Biochem.* **161,** 39–44.
12. Krøll, J. (1968) On the immunoelectrophoretical identification and quantitation of serum proteins. *Scand. J. Clin. Lab. Invest.* **22,** 79–81.
13. Krøll, J. (1969) Immunochemical identification of specific precipitin lines in quantitative immunoelectrophoresis patterns. *Scand. J. Clin. Lab. Invest.* **24,** 55–60.

CHAPTER 23

Crossed Immunoaffinoelectrophoresis

Anne Laine

1. Introduction

The crossed immunoaffinoelectrophoresis technique (CIAE) combines the principle of biospecific interaction with the principle of identification of proteins by immunoprecipitation in CIE. Biospecific interaction of macromolecular components during electrophoresis was first described by Nakamura et al. *(1)*. Lectins, or plant agglutinins, are proteins that react with carbohydrate groups with high specificity. Very rapidly, a combination of CIE and affinity electrophoresis with lectins was developed for identification and characterization of glycoproteins. Among the lectins, concanavalin A (Con A) is the most commonly used. Originally, Con A was introduced into an intermediate gel as immobilized Con A bound to Sepharose or free Con A *(2)*. Glycoproteins can be partially characterized with respect to the number of lectin binding sites per molecule (e.g., ref. *3*). Bøg-Hansen et al. *(4)* modified the procedure by introducing the lectin into the first dimension gel. Con A is electrophoretically immobile under the experimental conditions used for CIAE. The procedure can be used with other lectins, but their electrophoretic mobility must be checked beforehand. This procedure allows detection and separation of microheterogeneous forms of a glycoprotein. The degree of retardation during the first dimension electrophoresis in the gel with lectin is an expression of the affinity between the glycoprotein and the lectin. Higher affinity means stronger binding, which in turn means a higher degree of retardation.

From: *Methods in Molecular Biology, Vol. 10: Immunochemical Protocols*
Ed.: M. Manson ©1992 The Humana Press, Inc., Totowa, NJ

The second dimension gel contains the antiserum. The proteins that do not interact with the lectin in the first dimension gel keep their electrophoretic mobility, and their precipitation patterns are similar to those observed in CIE without lectin in the first dimension.

An affinity precipitate is formed during the first dimension electrophoresis (with a rocket-shape) between the lectin and the glycoproteins that strongly interact with it. Dissociation of this lectin–glycoprotein complex can be achieved by adding to the second dimension gel a carbohydrate (a displacer) specific for the lectin used *(5)*. As a consequence of the cathodic (electroendosmotic) flow of carbohydrate molecules, the lectin–glycoprotein complex is dissociated and glycoprotein–antibody precipitation can then be obtained in the second dimension gel, even for the glycoprotein molecules that have a very strong affinity for the lectin. Quantitation of the peak areas corresponding to each microheterogeneous form of a glycoprotein can be performed by planimetry, for instance, by weighing a paper tracing of each pattern.

Significant variations in the relative ratios of the microheterogeneous forms separated by CIAE have been detected in various diseases for several glycoproteins, particularly acute phase glycoproteins (*6–11,* among others).

Moreover, the analytical experiments in CIAE are thought to constitute a prediction method for preparative separation experiments *(2)*.

2. Materials

General equipment and reagents are as described in Chapter 20.

1. Glass plates: 100 × 100 × 1.5 mm.
2. Electrophoresis buffer B.
3. Buffer for dissolving Con A: 0.05*M* Tris-HCl, pH 7.6, containing 0.15*M* NaCl, 1 m*M* each of $MnCl_2$, $MgCl_2$, and $CaCl_2$ and 0.02% NaN_3 (*see* Note 1).
4. Con A.
5. 1% (w/v) Agarose in electrophoresis buffer B.
6. Bovine serum albumin (5 mg/0.1 mL) in concentrated bromophenol blue solution (1 mg/mL in electrophoresis buffer B).
7. Mono- or polyspecific antiserum.
8. α-Methyl-D-glucoside (from Sigma, for example).

3. Methods

1. The first and second dimensions are carried out on the same 100 × 100 × 1.5-mm glass plate. Pour 15 mL of heated 1% agarose onto the glass plate placed on the horizontal table.

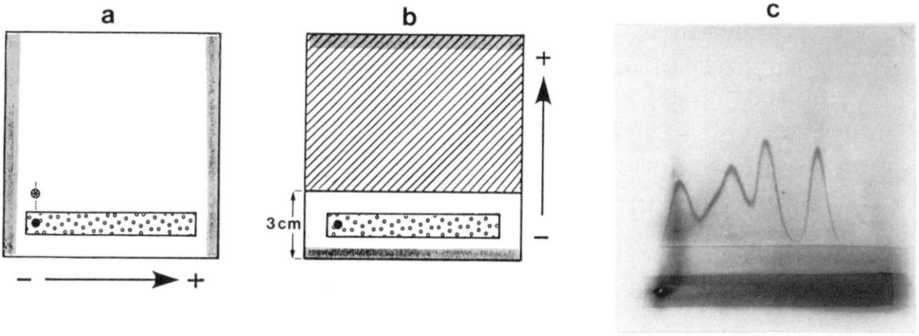

Fig. 1. Templates for the crossed immunoaffinoelectrophoresis technique. (a) first dimension; (b) second dimension. ⁘ strip to be removed and replaced by 1.1 mL of agarose + 200 µL of Con A solution; /// agarose containing antiserum; ▓ paper wick positions; * marker spot (bovine albumin in concentrated bromophenol blue solution); ● sample well; (c) crossed immunoaffinoelectrophoresis pattern of human α_1-antichymotrypsin in normal serum.

2. Prepare the Con A solution just before use. For each plate, 2 mg of Con A are required, dissolved in 20 µL of the buffer in 2.3. above. Add the dissolved Con A to 180 µL of electrophoresis buffer B. For several plates, prepare only one Con A solution and divide into 200 µL aliquots in small tubes (*see* Note 2).
3. After the agarose has gelled, remove a strip (1 × 8 cm) as indicated in Fig. 1a).
4. Refill the resultant trough with 1.1 mL of agarose (kept in the 50–56°C water bath) to which has just been added the 200 µL of Con A solution.
5. After gelling, punch out a well in the Con A containing-agarose as indicated in Fig. 1a.
6. Place the glass plate with the gel on the electrophoresis apparatus. Wet the wicks (three layers) with the electrophoresis buffer and place over the edges of the plate, then apply the sample to the well (for example, 3 µL of human serum).
7. Apply a small drop of the bovine albumin solution prepared in concentrated bromophenol blue solution by means of a Pasteur pipet above the sample well as indicated in Fig. 1a, in order to follow the migration in the plate.
8. Carry out the first dimension electrophoresis with 10 V/cm in the gel (controlled with the voltmeter). It takes 80 min for the colored bovine albumin to cross the plate. This electrophoresis time was convenient in our studies of the microheterogeneity of human serum α_1-antichymotrypsin (*see* Note 5).

9. As indicated in Fig. 1b, leave a gel strip (30 × 100 mm in area) containing the first dimension electrophoresis, and remove the remaining 70 × 100 mm gel.
10. Add 10.5 mL of agarose (kept in the 50–56°C water bath) to 200 mg of α-methyl-D-glucoside prepared in another tube heated in the water bath. Then add the antiserum and carefully mix the tube contents (*see* Notes 4 and 6).
11. With the plate on the horizontal table, pour the resultant mixture very quickly onto the remaining part of the plate adjacent to the strip.
12. When gelled, place the plate in the electrophoresis apparatus, with the first dimension strip at the cathode end. Connect the gel with the buffer by means of fresh paper wicks (two layers).
13. Carry out electrophoresis in the second dimension overnight with 2 V/cm in the gel, from the cathode to the anode.
14. After electrophoresis, wash, dry, and stain and destain the gel plate as described in Chapter 20. The appearance of several peaks for a single antigen indicates heterogeneity of the carbohydrate part of the molecule (Fig. 1c) *(12)*.

4. Notes

1. When preparing the buffer to dissolve Con A, the pH must be adjusted to 7.6 before addition of $MnCl_2$, since at pH greater than pH 8.0, the solution turns brown because of the formation of Mn oxide.
2. The amount of Con A used here (250 μg/cm^2) is sufficient to block the reacting glycoproteins contained in 3 μL of normal serum *(12)*, as well as in 3 μL of inflammatory serum in which acute phase glycoproteins are increased *(11)*.
3. It is important to preheat the tubes containing the Con A solution before adding the 1.1 mL of agarose, otherwise, agarose will gel in the tube.
4. The agarose added with α-methyl-D-glucoside and antiserum must be very carefully and rapidly mixed immediately before being poured onto the glass plate to properly dissolve the sugar, but not to have the agarose gelling.
5. The time quoted for the first dimension electrophoresis is suitable for proteins that migrate relatively fast, like α_1-antichymotrypsin. For glycoproteins with a low electrophoresis mobility, this time should be lengthened and the appropriate time for the first dimension run should be determined in initial trial experiments.

6. The amount of antiserum required in the second dimension in CIAE is lower than that for the same sample in CIE, depending on the number of individual peaks and on the repartition of these peaks. For example, for α_1-antichymotrypsin studied in our laboratory, when 3 µL of normal serum are applied, we need 75 µL of specific antiserum to obtain a precipitation peak that is 4–6 cm high in CIE, whereas only 20 µL is sufficient in CIAE to obtain four equivalent peaks 4 cm high.

References

1. Nakamura, S. (1966) *Cross Electrophoresis. Its Principle and Applications.* Igaku Shoin, Tokyo and Elsevier, Amsterdam.
2. Bøg-Hansen, T. C. (1973) Crossed immunoaffinoelectrophoresis. A method to predict the result of affinity chromatography. *Anal. Biochem.* **56,** 480–488.
3. Bøg-Hansen, T. C. and Brogren, C. H. (1975) Identification of glycoproteins with one and with two or more binding sites to Con A by crossed immunoaffinoelectrophoresis. *Scand. J. Immunol.* **4(Suppl. 2),** 135–139.
4. Bøg-Hansen, T. C., Bjerrum, O. J., and Ramlau, J. (1975) Detection of biospecific interaction during the first dimension electrophoresis in crossed immunoelectrophoresis. *Scand. J. Immunol.* **4(Suppl. 2),** 141–147.
5. Salier, J. Ph., Faye, L., Vergaine, D., and Martin, J. P. (1980) True and false glycoprotein microheterogeneity observed with lectin crossed affinoelectrophoresis of inter-alpha-trypsin-inhibitor. *Electrophoresis* **1,** 193–197.
6. Nicollet, I., Lebreton, J. P., Fontaine, M., and Hiron, M. (1981) Evidence for alpha-l-acid glycoprotein populations of different pI values after concanavalin A affinity chromatography. Study of their evolution during inflammation in man. *Biochim. Biophys. Acta* **668,** 235–245.
7. Wells, C., Bøg-Hansen, T. C., Cooper, E. H., and Glass, M. R. (1981) The use of concanavalin A crossed immunoaffinoelectrophoresis to detect hormone-associated variations in α_1-acid glycoprotein. *Clin. Chim. Acta* **109,** 59–67.
8. Raynes, J. G. (1982) Variations in the relative proportions of microheterogeneous forms of plasma glycoproteins in pregnancy and disease. *Biomedicine* **36,** 77–86.
9. Bowen, M., Raynes, J. G., and Cooper, E. H. (1986) Changes in the relative amount of individual microheterogeneous forms of serum α_1-antichymotrypsin in disease, in *Lectins* (Bøg-Hansen, T. C., ed.), Walter de Gruyter, Berlin-New York, pp. 403–411.
10. Damgaard, A. M., Heegard, P. M., Hansen, J. E., and Bøg-Hansen, T. G. (1986) The microheterogeneity of the acute phase reactant alpha-l-antichymotrypsin in testicular and colorectal cancer, in *Protides of the Biological Fluids,* 34th Colloquium, 1986 (Peeters, H., ed.), Elsevier, Amsterdam, pp. 449–452.
11. Hachulla, E., Laine, A., and Hayem, A. (1988) α_1-Antichymotrypsin microheterogeneity in crossed immunoaffinoelectrophoresis with free concanavalin A: A useful diagnostic tool in inflammatory syndrome. *Clin. Chem.* **34,** 911–915.
12. Laine, A., Hachulla, E., and Hayem, A. (1989) The microheterogeneity of serum α_1-antichymotrypsin revealed by interaction with concanavalin A in crossed immunoaffinoelectrophoresis and in affinity chromatography. *Electrophoresis* **10,** 227–233.

CHAPTER 24

Immunodetection of Proteins by Western Blotting

Colin J. Henderson and C. Roland Wolf

1. Introduction

The technique of protein immunoblotting, more commonly known as "Western" blotting, was first described a decade ago by Towbin et al., *(1)*, using electric current to transfer polypeptides from polyacrylamide gels to nitrocellulose, although several groups had previously reported the use of capillary forces to effect the transfer of DNA ("Southern" blotting *[2]*), RNA ("Northern" blotting *[3–5]*), and protein *(6, 7)* from agarose and polyacrylamide gels.

Essentially, Western blotting *(8)* involves the separation of polypeptides in gels by a variety of means (denaturing or nondenaturing one- or two-dimensional, isoelectric focusing), followed by the electrophoretic transfer of the separated polypeptides to an immobilizing matrix or membrane, the most commonly used membrane being nitrocellulose. In this way, a "replica" of the polypeptide separation profile from the gel electrophoresis step is created on the membrane, which can then be probed with antisera or other ligands, such as oligonucleotides or lectins, to identify specific polypeptide(s). Thus, Western blotting is a powerful technique for identifying specific proteins in a complex mixture, and may be used, for example, to investigate the regulation of protein expression, protein phosphorylation, or the identification of DNA-binding proteins.

From: *Methods in Molecular Biology, Vol. 10: Immunochemical Protocols*
Ed.: M. Manson ©1992 The Humana Press, Inc., Totowa, NJ

There are many advantages in transferring the polypeptides to such a membrane:

1. The membrane is more easily handled than the original gel (especially true for the nylon-based membranes);
2. The polypeptides immobilized on the membrane are more accessible to probing with various reagents (i.e., antibodies);
3. The membrane with its immobilized polypeptides can be stored for some time prior to probing, and in some cases, it may be possible to reprobe the same membrane several times with different reagents;
4. More than one replica can be made from a single gel; and
5. Quantitative changes in protein expression can be measured, subject to certain limitations.

This chapter will describe in detail the procedure for Western blotting of polypeptides and proteins separated on a denaturing polyacrylamide gel system. This procedure is used routinely in our laboratory for the analysis of polypeptides from a variety of subcellular fractions of whole tissue and cell lines, and has evolved over a number of years in the hands of several people. Many different immunoblotting procedures are currently available; details of variations from the ^{125}I-protein-A method described here are given in the Notes section, as are brief amendments covering electrotransfer from two-dimensional and isoelectric focusing systems. A detailed description of sodium dodecyl sulfate polyacrylamide electrophoresis (SDS-PAGE) is not appropriate for this chapter, and the reader is referred to vol. 1, Chapter 6, and refs. *9–14*. For details of the production of polyclonal and monoclonal antisera, *see* Chapters 1–6 in this vol.

2. Materials

1. Vertical electrophoresis apparatus, preferably with cooling facility (commercial or otherwise).
2. Blotting apparatus (commercial or otherwise). This normally consists of a plastic tank and lid, with platinum electrodes wired on either side of the tank, a plastic holder for the gel/membrane "sandwich," and a set of nylon pads.
3. Stock acrylamide solution (30%): Dissolve 58 g of acrylamide and 2 g of *N,N'*-methylene *bis*-acrylamide in 200 mL of distilled water. Filter and store at 4°C, protected from light (*see* Note 1).
4. Separating gel buffer: 1.5*M* Tris-HCl, 0.5% (w/v) sodium dodecyl sulfate (SDS), pH 8.8. Store at room temperature.

5. Stacking gel buffer: 0.5*M* Tris-HCl, 0.5% (w/v) SDS, pH 6.8. Store at room temperature.
6. Running buffer: 0.5*M* Tris-HCl, 0.5*M* glycine, 1% (w/v) SDS. Make fresh from a 10X stock solution stored at room temperature.
7. *N,N,N′,N′*-Tetramethylene-ethylenediamine (TEMED): Store at 4°C.
8. Ammonium persulfate (AMPS): Dissolve 1 g in 100 mL of distilled water. Make fresh.
9. Sample preparation solution: To 10 mL of stacking gel buffer, add 2 g of SDS, 5 mL of 2-mercaptoethanol, 10 mL of glycerol, and 5 mg of bromophenol blue. Make up to a final vol of 100 mL with distilled water. Store at room temperature.
10. 5X Sample preparation solution: To 50 mL of stacking gel buffer, add 10 g of SDS, 25 mL of 2-mercaptoethanol, 25 mL of glycerol, and 25 mg of bromophenol blue. Store at room temperature.
11. Transfer buffer: 20 m*M* disodium hydrogen phosphate, 20% (v/v) methanol. Make fresh.
12. 10X Tris-buffered saline (10X TBS): Dissolve 90 g of sodium chloride and 60 g of Tris in 1 L of distilled water, and adjust the pH to 7.9. Store at room temperature.
13. Tris-buffered saline with Tween (TBST): Dilute 10X TBS to 1X with distilled water and add Tween-20 to a final concentration of 0.05% (v/v). Store at room temperature.
14. Nitrocellulose membrane: 0.45 μm pore size, in sheets or rolls.
15. Blocking solution: Dissolve 3 g of low-fat dried milk in 100 mL of TBST. Make fresh.
16. First antibody—rabbit (polyclonal) antisera or mouse (monoclonal) antibody to protein(s) of interest, diluted in TBST. Dilution of antibody depends on the titer (*see* Note 2). Make fresh, and store frozen at –20°C. The diluted antibody can be reused up to 10 times.
17. Second antibody—donkey antirabbit (or antimouse) IgG, linked to horseradish peroxidase (HRP), diluted 1:1000 in TBST. Make fresh, but diluted antibody can be stored frozen (–20°C) and reused approx 5 times.
18. HRP-substrate solution: Dissolve 120 mg of 4-chloro-1-naphthol in 40 mL of methanol, add 200 mL of TBS and 80 μL of 30% (v/v) hydrogen peroxide. Make immediately prior to development of the Western blot.
19. ^{125}I-protein A: Obtained commercially in buffered aqueous solution, specific activity >30 mCi/mg, stored at 4°C with lead shielding. Use 5

Table 1
Preparation of SDS-PAGE Gels

Final acrylamide concn., %	Stock 30% acrylamide, mL	Resolving gel buffer, mL	dH_2O mL	1% AMPS mL	TEMED µL
		Separating gel			
6	7.4	9.25	18.4	2	20
9	11.1	9.25	14.7	2	20
12	14.8	9.25	11.0	2	20
15	18.5	9.25	7.3	2	20
18	22.2	9.25	3.6	2	20
		Stacking gel			
4.5	1.5	2.5	5.7	0.3	10

µCi in 50 mL of TBST/membrane, prepared fresh. ^{125}I-protein A should be dispensed from behind adequate lead shielding in a designated Radiation Handling Area. Gloves should be worn at all times, and precautions taken to minimize exposure to ^{125}I both during and after labeling of the blot.

20. Autoradiography cassettes: With intensifying screens.
21. X-ray film: For example, Kodak X-Omat AR5.

3. Methods

3.1. Gel Preparation and Running

1. Prepare a polyacrylamide separating gel of appropriate concentration, according to Table 1, in a vertical electrophoresis apparatus. After the separating gel has set, cast a 4.5% polyacrylamide stacking gel on top (*see* Table 1), with a comb to form sample wells.

   A detailed methodology for SDS-PAGE is given in ref. *1* and vol. 1, Chapter 6 (*see* Note 2).
2. Prepare the denatured protein sample for loading onto the gel by adding 1 vol of sample preparation solution to 1 vol of sample at a protein concn. of approx 3 mg/mL, giving a final protein concn. of 1.5 mg/mL in the denatured sample. For very dilute samples (<0.25 mg/mL), use 1 vol of 5X sample preparation solution to 4 vol of sample. Boil for 5 min at 100°C, cool, and load into the sample wells on the gel. The amount of protein will vary according to source, i.e., approx 5–50 µg/track for whole cell extracts or subcellular fractions, 0.1–1 µg/track for purified proteins (*see* Note 3).

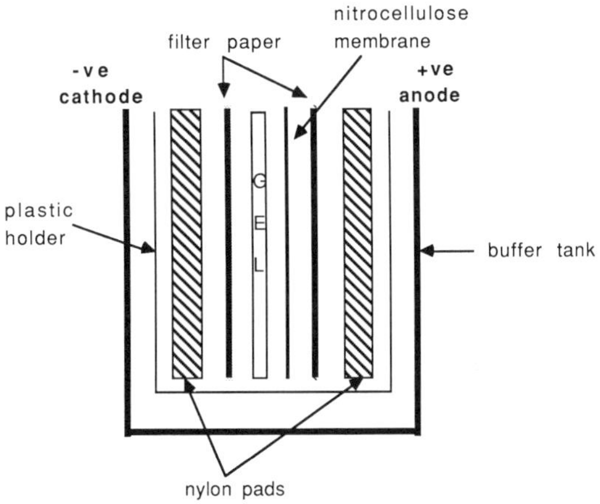

Fig. 1. Assembly of blotting apparatus

3. Carry out electrophoresis at 50–60 mA constant current/gel until the sample has traversed the stacking gel; during this time, the samples "stack," becoming concentrated into a very tight band, affording maximal resolution and separation of the polypeptides in the separating gel. Then continue electrophoresis at 25–30 mA/gel until the bromophenol blue dye front is approx 1–2 cm from the end of the gel (*see* Note 4).
4. Remove the gel assembly from the vertical gel apparatus, separate the glass plates, and cut off the stacking gel with a sharp scalpel blade. Also cut off the separating gel beyond the dye front, and mark the gel uniquely for future orientation by cutting the corner(s) appropriately.

3.2. Assembly of the Gel/Membrane "Sandwich" (see Fig. 1)

1. For each gel, cut a piece of nitrocellulose and two pieces of Whatman 3MM filter paper to a size approx 0.5 cm bigger in each dimension than the gel.
2. Open the plastic holder for the gel/membrane "sandwich" in a basin containing sufficient transfer buffer to cover it. Place a nylon pad flat into the holder, followed by a piece of filter paper and then the nitrocellulose, ensuring there are no air bubbles caught between the layers (*see* Note 5).
3. Carefully lay the gel on top of the nitrocellulose, in the same orientation as it was run in the gel apparatus, and cover with the other piece of filter paper.

4. Lay another nylon pad on top, and carefully close the blotting clamp. Gently squeeze the assembled clamp while holding it under the transfer buffer, to exclude air bubbles, and place the clamp in the blotting apparatus, *ensuring the nitrocellulose is lying between the anode and the gel.*
5. Fill the tank with transfer buffer (*see* Note 6), sufficient to cover the gel/membrane "sandwich" without touching the lid electrodes, and place a stirring bar in the bottom. Carry out the transfer at a constant current of 250 mA overnight, stirring the transfer buffer slowly to assist the distribution of heat generated by electrotransfer (*see* Notes 7–9).

3.3. Development of the Western Blot

1. Remove the plastic holder from the apparatus, open it on a flat surface, and carefully remove the nylon pad and top filter paper and discard.
2. Cut away any excess nitrocellulose from around the gel with a sharp scalpel blade; the nitrocellulose should be marked in an identical manner to the gel.
3. Carefully peel away the gel from the nitrocellulose (*see* Note 10). Place the nitrocellulose membrane in a plastic container with sufficient TBST to adequately cover the membrane (50–100 mL). All subsequent reactions/washings are carried out in this container, on an orbital shaker at room temperature.
4. Wash the nitrocellulose twice with TBST for 5 min, and then incubate with the blocking solution for 1–2 h (*see* Note 11).
5. Wash the membrane again with TBST (3× 10 min) before incubating it with the first antibody for 1 h (*see* Note 2).
6. After washing with TBST (3× 10 min), place the membrane in the second antibody solution for 1 h (*see* Notes 12 and 13).
7. After washing with TBST (3× 10 min), add the HRP-substrate solution. The immunoreactive polypeptides are detected by the appearance of purple bands on the membrane.
8. Once the reaction has stopped, or after 10 min, whichever is the shorter, wash the nitrocellulose in distilled water, and shake for 15 min.
9. Place the membrane in TBST (50 mL), add 5 μCi (0.185 *M*Bq) ^{125}I-protein A, and incubate for 30–60 min, with agitation. Up to four membranes may be treated in this way simultaneously; with more membranes, it may be necessary to increase the volume of TBST and ^{125}I-protein A to maintain sensitivity.
10. After this time, wash the membrane exhaustively with TBST, until no more radioactivity is eluted, blot dry, and cover in cling wrap, before

placing in an autoradiography cassette with X-ray film, and exposing overnight at –70°C. Develop the film the next day, and reexpose the membrane as necessary, depending on the strength of the signal.

4. Notes

1. The concentration of acrylamide used in the resolution of the polypeptides is determined empirically, and depends on the source of the protein and the molecular masses of the polypeptides to be identified. As an approximate guide, for polypeptides of >100 kDa, a separating gel of 5–7% acrylamide should be employed. For polypeptides between 50–100 kDa, 7–9% and for <50 kDa, the acrylamide concentration in the separating gel should be 12–15%. Acrylamide and *bis*-acrylamide are cumulative neurotoxins, and care should be exercised in their handling.
2. The dilution of the primary antibody is dependent on the nature (i.e., monoclonal or polyclonal) and the titer of the antiserum, and should be determined empirically for each individual antiserum. The dilution usually falls between 1:500–1:1000; however, when testing a new antibody preparation, dilutions between 1:50–1:10,000 should be tried. Antibody dilutions can be stored frozen (–20°C) and reused approx 10 times, depending on antibody.
3. The amount of protein loaded on the gel will depend on several factors—the relative abundance of the protein(s) of interest and the titer and specificity of the antibody, for example. Up to 200 μg/track may be loaded, although at this level, high background on development of the blot may be a problem. The smaller the volume loaded onto the gel, the better the resolution will be; in practice, 5–10 μL is an optimal loading volume, although up to 50 μL is possible without completely comprising polypeptide separation.
4. Water-cooling of the vertical electrophoresis apparatus is optional, but separation and resolution of the polypeptides on an SDS/polyacrylamide gel is enhanced by dissipating the heat generated in running the gel. Additionally, the proteins should be run slowly through the resolving gel, 20–25 mA/gel constant current, to prevent "smiling" of the separated polypeptides, where the bands toward the edges of the gel tend to turn upward.
5. Good contact between the gel and the nitrocellulose is essential for efficient electrotransfer, and care should be taken to ensure that all air bubbles are removed, as these will prevent uniform current flow, and thus interfere with the blotting process.

6. The use of methanol in the transfer buffer increases the binding capacity of nitrocellulose for protein, and stabilizes the geometry of the gel during transfer. If methanol is omitted, the gel will swell during transfer, causing a distortion of the bands on the membrane—this may be overcome by presoaking the gel in the transfer buffer for 15–30 min prior to carrying out the transfer. Methanol does, however, decrease the elution efficiency of proteins from SDS/polyacrylamide gels, and therefore, electrotransfer should be carried out for >12 h in order to effect transfer of high molecular mass (M_r) polypeptides (>100 kDa).
7. Generally, elution of the polypeptides from the gel is dependent on their M_r, with low M_r polypeptides being almost entirely eluted after 2 h, but 6–12 h being necessary for higher M_r polypeptides *(8)*. This slow, inefficient transfer of high M_r polypeptides is overcome by the use of 0.1% (w/v) SDS in the transfer buffer, which facilitates transfer of high M_r polypeptides, without affecting their adsorption to the nitrocellulose. However, SDS increases the power requirements of the blotting process, thereby necessitating efficient cooling of the apparatus.
8. The process of protein interaction with nitrocellulose apparently involves hydrophobic effects, although the pore size of the nitrocellulose is also important, with a smaller pore size (0.1 µm) having a higher capacity for binding proteins, but concomitantly resulting in a higher background *(16)*. However, low M_r proteins tend to bind with low affinity to nitrocellulose, thereby being lost during transfer or processing of the membrane. It is therefore important to be aware of the alternative matrices available for Western blotting. These include aminobenzyloxymethyl (ABM) and aminophenylthioether (APT)-cellulose, which require activation to their diazo forms (DBM and DPT, respectively). There is an initial electrostatic interaction between the negatively charged proteins and the positively charged diazonium groups on the DBM- and DPT-cellulose, followed by the formation of an irreversible covalent linkage via the azo derivatives, preventing loss of polypeptides during transfer/processing. However, the coarseness of these cellulose membranes results in a loss of resolution compared to other matrices.

   Nylon-based membranes, such as Genescreen™, Zetaprobe™, or Nitroscreen West™, offer the advantage of mechanical strength and durability (often important in view of the number of steps involved in the processing of the membrane) along with a much greater binding capacity for proteins, thereby allowing the omission of methanol from the transfer buffer. However, the greater protein-binding capacity may lead to increased nonspecific binding, and therefore an increased background. Gloves should be worn at all times when handling membranes.

9. Water cooling of the blotting apparatus is optional, and in our experience unnecessary, at the levels of current and blotting times outlined in this chapter. However, longer transfer times and/or higher currents may require cooling, either via water coils (available for most commercial apparatus), precooling of the transfer buffer, or carrying out the transfer in a cold room. If the transfer buffer is allowed to become too hot, the gel will become distorted, causing problems with the transfer.
10. Following transfer, the gel may be stained for residual protein, in order to check the efficiency of the operation. Several methods are available, which vary in their sensitivity (*see* refs. *9–14*).
11. The process of "blocking" the membrane after transfer, sometimes referred to as "quenching," is essential to block the unoccupied protein-binding sites on the membrane. Failure to carry out this step effectively will result in the primary and secondary antibodies binding to these sites, leading to a reduction in the sensitivity of the blotting process and unacceptably high background "noise." Blocking can be achieved by several means, the most common being gelatin, bovine serum albumin (fraction V), or low-fat dried milk powder, all usually employed at 3% (w/v). The major drawback with the first of these is that, even at room temperature with gelatin of a low bloom number (i.e., 60), the gelatin can partially set and stick to the membrane, interfering with subsequent steps in the processing. For this reason, gelatin is not recommended. Bovine serum albumin is expensive to use in the quantities and volumes required for large numbers of blots, thus the current method of choice is that of low-fat dried milk powder, which is inexpensive, easy to make up, and is very effective (*see also* vol. 3, Chapter 30). The concentration of the blocking reagent can be increased (i.e., to 10% [w/v]); this will reduce background when long autoradiographic exposures of Western blots (up to several weeks) are necessary, for example, when a particular protein is expressed at a very low level.

    The presence of 0.1% (v/v) of polyoxyethylene sorbitan monolaurate (Tween-20) in all buffers during the processing of the membrane assists the prevention of nonspecific binding.
12. For some analyses, it may be desirable to reprobe the same membrane with another antibody (i.e., "erase" the original signal), or probe identical "copies" of a membrane with different antibodies. However, it should be pointed out that both have drawbacks, and wherever

possible, subject to sufficient protein being available, separate gels should be run and blotted.

A recent method of "erasing" membranes is described in ref. *17* and this vol., Chapter 25. Another method *(18)* employs 8*M* urea, 100 m*M* 2-mercaptoethanol, and 5 mg/mL bovine serum albumin at 60°C. Alternatively, a nylon-reinforced nitrocellulose membrane Nitroscreen West™, from DuPont, may be stripped using a very acidic (pH 2.2) buffer; the membrane must not be allowed to dry out at any stage. The manufacturers claim to be able to strip and reprobe the same membrane at least five times without a significant loss of sensitivity.

The danger of protein loss from the membrane under the conditions described above should be taken into account when using these methods.

Multiple replicas may be made by putting several layers of membranes (reportedly up to 10) in the blotting apparatus, capitalizing on the phenomenon whereby protein that is not absorbed to the first nitrocellulose membrane will pass through to the second, and so on *(19)*. Alternatively, several transfers may be made from a single gel by replacing the membrane every 2–3 h *(15,20)*. Both these methods result in each replica being slightly different from the others, each being biased toward a subset of polypeptides, as the rate of elution of polypeptides from the gel is approximately proportional to their molecular mass. Thus, at best, transfer is only qualitative in these cases.

A further alternative is to run the protein sample of interest in a "trough" across the separating gel, blot and cut the membrane into many strips, each having an identical protein separation profile, which can then be probed with different antibodies. A commercial apparatus, Decaprobe™, from Hoefer Scientific Instruments, achieves essentially the same result, without cutting the membrane, by sealing off areas of the membrane from each other, thus allowing different antisera to be used on the same membrane simultaneously.

13. The most commonly employed detection system is probably that of horseradish peroxidase-conjugated second antibody, in which the enzyme converts a colorless substrate (such as *O*-dianisidine, 3,3-diamino benzidine or 4-chloro-1-naphthol) to a colored product, which then becomes irreversibly bound to the membrane. The best substrate, both in terms of chemical safety and sensitivity of detection, is 4-chloro-1-naphthol, which will routinely detect less than 100 pg of protein *(21)*.

    Alternatives to the peroxidase have been developed, including the use of ^{125}I-protein A *(8)*, with a sensitivity level of 1–2 ng of protein; biotinylated second antibody and streptavidin-acid phosphatase, with a detection limit of <20 ng protein *(22)*, and alkaline phosphatase-luciferin *(23)*, which is reported to detect 5–50 pg in a bioluminescence-enhanced system (*see also* this vol., Chapters 15, 26–28). The method of choice in our laboratory is ^{125}I-protein A, applied following the peroxidase step and detection of the polypeptides on the blot. Use of this ligand allows enhancement of the colored signal on an almost continuous scale, simply by using longer autoradiography exposure times. The generation of such an autoradiographic record permits quantitation of the immunoreactive polypeptides by densitometry. It is important to determine that such quantitation is linear, by including a series of known amounts of standard proteins on the blot for calibration purposes.

Protein A does not bind to all subclasses of IgG, nor to IgG from all species. As the majority of primary antibodies are from mice or rabbits, and secondary antisera tend to be raised in either donkeys or rabbits, the IgG of which all bind strongly to protein A, this does not constitute a serious problem. However, if antiserum from another species is used, its binding to protein A should be checked. Recently, an alternative, protein G, has become commercially available, that binds to a wider range of IgG subclasses from a larger number of species.

It must be emphasized, however, that the major determinant of the sensitivity and quality of Western blots is the titer of the primary antiserum used.

14. The transfer of polypeptides from flat-bed isoelectric focusing (IEF) gels may also be effected in a manner similar to that described in this chapter, with the following modifications. After IEF is complete, the gel should be equilibrated for 15–30 min, with shaking, in a transfer buffer of 0.7% acetic acid, to allow the removal of carrier ampholytes, and urea if running a denaturing IEF gel. The transfer should be set up as described, except that the nitrocellulose should be placed between the gel and the cathode, as the acidic transfer buffer gives most polypeptides a net positive charge and they will therefore migrate toward the cathode. However, it is prudent to check that polypeptides are not being lost to the anode by placing a piece of nitrocellulose on that side during the first transfer, and developing this membrane with the cathodal one.
15. Electrotransfer from two-dimensional SDS-PAGE gels may also be carried out with minor alterations to the protocol listed here. Again, it is important to soak the gel in transfer buffer prior to transfer. This will overcome any problems with swelling of the gel, particularly associated with acrylamide gradient gels, often employed in the second dimension separation to facilitate optimal resolution.

16. The recent advent of semidry electrophoretic transfer (*see* vol. 3, Chapter 29) allows the rapid transfer of several gels simultaneously, in a stack of up to 6 gels/membrane "sandwiches," each "sandwich" being separated from the others by a layer of dialysis membrane to prevent transfer of polypeptides from one "sandwich" to the next. The technique uses relatively low power and small amounts of transfer buffer, and the electrodes consist of graphite plates; several types of apparatus are now commercially available. However, in our experience, it is often difficult to exclude all air bubbles from the gel/membrane "sandwich" and to maintain even electrode contact throughout the transfer.
17. "Mini"-gel and blotting apparatus are also now commercially available. These kits offer levels of resolution and detection of polypeptides approximately equivalent to the larger kits, at a fraction of the cost in materials and in time. It should be noted, however, that the blotting step involves a very high transfer current; because of their high conductivity, it is essential that phosphate-containing transfer buffers are not employed. An alternative transfer buffer consists of Tris (25 m*M*), glycine (192 m*M*), and 20% (v/v) methanol. A cooling system is also essential.

References

1. Towbin, H., Staehlin, T., and Gordon, J. (1979) Electrophoretic transfer of proteins from polyacrylamide gels to nitrocellulose sheets: Procedure and some applications. *Proc. Natl. Acad. Sci. USA* **76,** 4350–4354.
2. Southern, E. M. (1975) Detection of specific sequences among DNA fragments separated by gel electrophoresis. *J. Mol. Biol.* **98,** 503–517.
3. Alwine, J. C., Kemp, D. J., Parker, B. A., Reiser, J., Renart, J., Stark, G. R., and Wahl, G. W. (1979) Detection of specific RNA's or specific fragments of DNA by fractionation in gels and transfer to diazobenzyloxymethyl-paper. *Methods Enzymol.* **68,** 220–242.
4. Thomas, P. S. (1980) Hybridization of denatured RNA and small DNA fragments transferred to nitrocellulose. *Proc. Natl. Acad. Sci. USA* **77,** 5201–5205.
5. Seed, B. (1982) Diazotizable arylamine cellulose papers for the coupling and hybridization of nucleic acids. *Nucleic Acids Res.* **10,** 1799–1810.
6. Renart, J., Reiser, J., and Stark, G. R. (1979) Transfer of proteins from gels to diazobenzyloxymethyl-paper and detection with antisera: A method for studying antibody specificity and antigen structure. *Proc. Natl. Acad. Sci. USA* **76,** 3116–3120.
7. Bowen, B., Steinberg, J., Laemmli, U. K., and Weintraub, H. (1980) Detection of DNA-binding proteins by protein blotting. *Nucleic Acids Res.* **8,** 1–20.
8. Burnette, W. N. (1981) Western blotting: Electrophoretic transfer of proteins from sodium dodecyl sulphate-polyacrylamide gels to unmodified nitrocellulose and radiographic detection with antibody and radioiodinated protein A. *Anal. Biochem.* **112,** 195–203.
9. Hames, B. D. and Rickwood, D., eds. (1981) *Gel Electrophoresis of Proteins: A Practical Approach.* IRL, London, UK.

10. Laemmli, U. K. (1970) Cleavage of structural proteins during the assembly of the head of bacteriophage T4. *Nature* **227,** 680–685.
11. Harlow, E. and Lane, D., eds. (1989) *Antibodies: A Laboratory Manual.* Cold Spring Harbor Laboratory, New York.
12. Gershoni, J. M. and Palade, G. E. (1983) Protein blotting: Principles and applications. *Anal. Biochem.* **131,** 1–15.
13. Towbin, H. and Gordon, J. (1984) Immunoblotting and dot immunoblotting—current status and outlook. *J. Immunol. Methods* **72,** 313–340.
14. Beisiegel, U. (1986) Protein blotting. *Electrophoresis* **7,** 1–18.
15. Howe, J. G. and Hershey, J. W. B. (1981) A sensitive immunoblotting method for measuring protein synthesis initiation factor levels in lysates of *Escherichia coli. J. Biol. Chem.* **256,** 12836–12839.
16. Torey, E. R, Ford, S. A., and Baldo, B. A. (1987) Protein blotting on nitrocellulose: Some important aspects of the resolution and detection of antigens in complex extracts. *J. Biochem. Biophys. Methods* **14,** 1–17.
17. Kaufmann, S. H., Ewing, C. M., and Shaper, J. H. (1987) The erasable Western blot. *Anal. Biochem.* **161,** 89–95.
18. Erickson, P. F., Minier, L. N., and Lasher, R. S. (1982) Quantitative electrophoretic transfer of polypeptides from SDS polyacrylamide gels to nitrocellulose sheets: A method for their re-use in immunoautoradiographic detection of antigens. *J. Immunol. Methods* **51,** 241–249.
19. Bolen, J., Garfinkle, J. A., and Consigli, R. A. (1982) Detection and quantitation of Newcastle disease virus proteins in infected chicken embryo cells. *Appl. Environ. Microbiol.* **43,** 193–199.
20. McLellan, T. and Ramshaw, J. A. M. (1981) Serial electrophoretic transfers: A technique for the identification of numerous enzymes from single polyacrylamide gels. *Biochem. Genet.* **19,** 647–654.
21. Bio-Rad Immun-blot™ assay manual, Bio-Rad Laboratories Ltd., Watford, UK.
22. Brower, M. S., Brakel, C. L., and Garry, K. (1985) Immunodetection with streptavidin-acid phosphatase complex on Western blots. *Anal. Biochem.* **147,** 382–386.
23. Hauber, R. and Geiger, R. (1987) A new, very sensitive bioluminescence-enhanced detection system for protein blotting *J. Clin. Chem. Clin. Biochem.* **25,** 511–514.

CHAPTER 25

Erasable Western Blots

Scott H. Kaufmann and Joel H. Shaper

1. Introduction

Western blotting (reviewed in *1–3; see also* this vol., Chapter 24) refers to formation and detection of an antibody–antigen complex between an antibody and a polypeptide that is immobilized on derivatized paper. Most commonly, polypeptides in a complex mixture are separated by electrophoresis through polyacrylamide gels in the presence of sodium dodecylsulfate (SDS), electrophoretically transferred to thin sheets of nitrocellulose or nylon, and reacted sequentially with one or more antibody-containing solutions. This sequence of manipulations can be utilized to determine whether a polypeptide recognized by a specific antiserum is present in a particular biological sample (cell type, subcellular fraction, or biological fluid), to follow the purification of the polypeptide, or to assess the location of epitopes within the polypeptide during chemical or enzymatic degradation. Alternatively, the same series of manipulations can be utilized to determine whether antibodies that recognize a particular polypeptide are detectable in a sample of biological fluid. Since Western blotting takes advantage of the power of electrophoresis for separating complex mixtures of polypeptides, it is possible to derive large amounts of information from this technique without necessarily purifying the antigen being studied.

There are certain circumstances in which it is convenient to be able to dissociate the antibodies from a Western blot after detection of antibody–antigen complexes. If, for example, the experiment gives an unexpected

From: *Methods in Molecular Biology, Vol. 10: Immunochemical Protocols*
Ed.: M. Manson ©1992 The Humana Press, Inc., Totowa, NJ

result regarding the subcellular distribution of a polypeptide, it is convenient to be able to reprobe the blot with an antibody that recognizes a second polypeptide in order to confirm that the samples have been properly prepared, loaded, and transferred. Likewise, if the polypeptides being analyzed are derived from a precious source (e.g., biological fluid, tissue, or organism that is not readily available), it is convenient to dissociate the antibody–antigen complexes and reutilize the blots.

Over the last decade, several methods for removing antibodies from Western blots have been described. In early experiments, proteins were covalently bound to diazotized paper. Antibodies that were subsequently (noncovalently) bound to the paper were removed by treating the paper at 60°C with 10*M* urea *(4)* or 2% (w/v) sodium dodecylsulfate *(5)* under reducing conditions. Because of several undesirable properties (reviewed in *1–3*), diazotized paper has been largely replaced by nitrocellulose, a solid support to which proteins are noncovalently bound. It has been reported that treatment of nitrocellulose blots with glycine at pH 2.2 *(6)* or with 8*M* urea at 60°C *(7)* will remove antibodies and permit reuse of blots. Whereas these techniques are effective at disrupting low-affinity interactions between antigens and antibodies, they have been found to be ineffective at disrupting interactions between immobilized antigens and high-affinity antibodies *(6,8)*.

Two subsequent observations have allowed the development of a more widely applicable technique for the removal of antibodies from Western blots. First of all, it was observed that treatment of nitrocellulose with acidic solutions of methanol would "fix" transferred polypeptides to the nitrocellulose (reviewed in *3; see also 9,10*). Polypeptides treated in this fashion remained bound to the nitrocellulose even during treatment with SDS at 70–100°C under reducing conditions *(8,9)*. Second, it was observed that removal of antibodies from nitrocellulose after Western blotting could be facilitated by reincubation of the blot with a large excess of irrelevant protein immediately prior to drying and autoradiography *(8)*. Based on these observations, we present a technique that allows the reutilization of Western blots after reaction with a wide variety of antibodies or with lectins. In brief, polypeptides immobilized on nitrocellulose are stained with dye dissolved in an acidic solution of methanol. After unoccupied binding sites have been saturated with protein, the nitrocellulose is treated sequentially with unlabeled primary antibodies and radiolabeled secondary antibodies. Prior to drying, the blot is briefly incubated in a protein-containing buffer. After subsequent drying and autoradiography, the antibodies are removed by treating the nitrocellulose with SDS at 70°C under reducing conditions.

2. Materials

1. Apparatus for transferring polypeptides from gel to solid support. (Design principles are reviewed in refs. *2,3*.)
   a. TE52 reservoir-type electrophoretic transfer apparatus (Hoefer Scientific, San Francisco, CA) or equivalent.
   b. Polyblot semidry blotter (ABN, Hayward, CA) or equivalent.
2. Paper support for binding transferred polypeptides.
   a. Nitrocellulose.
   b. Nylon (e.g., Genescreen from New England Nuclear, Boston, MA, or Nytran from Schleicher and Schuell, Keene, NH).
   c. Polyvinylidene difluoride PVDF (e.g., Immobilon; Millipore, Bedford, MA).
3. Fast green FCF, for staining polypeptides after transfer to solid support.
4. Penicillin 10,000 U/mL and streptomycin 10 mg/mL.
5. Reagents for electrophoresis (acrylamide, *bis*-acrylamide, 2-mercaptoethanol, sodium dodecylsulfate) should be electrophoresis grade.
6. All other reagents (Tris, glycine, urea, methanol) are reagent grade.
7. Transfer buffer: 0.02% (w/v) Sodium dodecylsulfate (SDS), 20% (v/v) methanol, 192 m*M* glycine-HCl, and 25 m*M* Tris base. Prepare enough buffer to fill the chamber of the transfer apparatus and a container for assembling the cassette.
8. TS buffer: 150 m*M* NaCl and 10 m*M* Tris-HCl (pH 7.4). This can be conveniently prepared as a 10X stock (1.5*M* NaCl containing 100 m*M* Tris-HCl, pH 7.4). The 10X stock can be stored indefinitely at 4°C and then used to prepare 1X TS buffer, TSM buffer, and the other buffers described below.
9. TSM buffer: TS buffer containing 5% (w/v) powdered milk, 100 U/mL penicillin, 100 µg/mL streptomycin, and 1 m*M* sodium azide. This buffer can be stored for several days at 4°C. Note that sodium azide is poisonous and can form explosive copper salts in drain pipes if not handled properly.
10. TS buffer containing 2*M* urea and 0.05% (w/v) Nonidet P–40. Prepare 300 mL per blot by combining 0.15 g of Nonidet P-40, 30 mL of 10X TS buffer, 75 mL of 8*M* urea (freshly deionized over Bio-Rad AG1X–8 mixed bed resin to remove traces of cyanate), and 195 mL of water.
11. TS buffer containing 0.05% (w/v) Nonidet P-40. Prepare 300 mL per blot.

12. Fast green stain: 0.1% (w/v) Fast green FCF in 20% (v/v) methanol-5% (v/v) acetic acid. This stain is reusable. Prepare 50–100 mL per blot.
13. Fast green destain: 20% (v/v) Methanol in 5% (v/v) acetic acid.
14. Blot erasure buffer: 2% (w/v) Sodium dodecylsulfate (SDS), 62.5 m*M* Tris-HCl (pH 6.8), and 100 m*M* 2-mercaptoethanol. The SDS/Tris-HCl solution is stable indefinitely at 4°C. Immediately prior to use, 2-mercaptoethanol is added to a final concentration of 6 µL/mL.
15. Primary antibody.
16. [^{125}I]-labeled secondary antibody. Secondary antibodies can be labeled as previously described *(8)* or purchased commercially. Radiolabeled antibodies should only be used by personnel trained to properly handle radioisotopes.

3. Methods

3.1. Transfer of Polypeptides to Nitrocellulose

The following description is appropriate for transfer in a transfer reservoir. If a semidry transfer apparatus is to be used, follow the manufacturer's instructions (*see* Note 1).

1. Perform SDS-polyacrylamide gel electrophoresis using standard techniques (*see* refs. *11,12* and vol. 1, Chapter 6 for a description of this method).
2. Wear disposable gloves while handling the gel and nitrocellulose at all steps. This avoids cytokeratin-containing fingerprints.
3. Cut nitrocellulose sheets to a size slightly larger than the polyacrylamide gel (*see* Note 2).
4. Fill the transfer apparatus with transfer buffer (*see* Note 3).
5. Fill a container large enough to accomodate the transfer casettes with transfer buffer. Assemble the cassette under the surface of the buffer in the following order:
   a. Back of the cassette.
   b. Two layers of filter paper.
   c. The gel.
   d. One piece of nitrocellulose—gently work bubbles out from between the nitrocellulose and gel by rubbing a gloved finger or glass stirring rod over the surface of the nitrocellulose.
   e. Two layers of filter paper—again, gently remove bubbles.
   f. Sponge or flexible absorbant pad.
   g. Front of the cassette.
6. Place the cassette in the transfer apparatus so that the front is oriented toward the POSITIVE pole.

7. Transfer at 4°C in a cold room with the transfer apparatus partially immersed in an ice-water bath. Power settings: 90 V for 5–6 h or 60 V overnight.
8. Place the fast green stain in a container with a surface area slightly larger than one piece of nitrocellulose (*see* Note 4). After the transfer is complete, place all the pieces of nitrocellulose in the stain and incubate for 2–3 min with gentle agitation. Decant the stain solution, which can be reused (*see* Note 5).
9. Destain the nitrocellulose by rinsing it for 3–5 min in fast green destain solution with gentle agitation. Decant the destain (which can also be reused). Rinse the nitrocellulose four times (5 min each) with TS buffer (200 mL/rinse).
10. Mark the locations of lanes, standards, and any other identifying features by writing on the blot with a standard ball point pen.
11. Coat the remaining protein-binding sites on the nitrocellulose by incubating the blot in TSM buffer (50–100 mL per blot) for 6–12 h at room temperature (*see* Note 6). Remove the blot from the TSM buffer. Wash the blot four times in quick succession with TS buffer (25–50 mL per wash) and dry the blot on fresh paper towels. Either before or after coating of the unoccupied protein binding sites, blots can be dried and stored indefinitely in an appropriate container, e.g., Ziplock™ disposable food storage bags (1–2 blots/bag).

3.2. Formation of Antibody–Antigen Complexes

1. Place the nitrocellulose blot in an appropriate container for reaction with the antibody. A 15- or 20-lane sheet can be reacted with 15–20 mL of antibody solution in a Ziplock™ bag. A 1- or 2-lane strip can be reacted with 2–5 mL of antibody solution in a disposable 15 mL conical test tube.
2. If the nitrocellulose has been dried, rehydrate it by incubation for a few minutes in an appropriate volume of TSM buffer (*see* previous step).
3. Add an appropriate dilution of antibody to the TSM buffer and incubate overnight (10–15 h) at room temperature with gentle agitation (*see* Notes 7 and 8).
4. Remove the antibody solution and save for reuse (*see* Note 9).
5. Wash the nitrocellulose with the following solutions (100 mL/wash for each large blot or 15–50 mL/wash for each individual strip): TS buffer containing 2*M* urea and 0.05% NP–40: three washes (15 min each); TS buffer: one wash (5 min) (*see* Note 10).
6. Add fresh TSM buffer to the nitrocellulose sheets or strips. For nitrocellulose sheets (or pooled strips) in Ziplock™ bags, it is convenient to use

50 mL of TSM buffer. Add 5–10 μCi [^{125}I]-labeled secondary antibody (*see* Note 11). Incubate for 90 min at room temperature with gentle agitation.

7. Remove the radiolabeled antibody and discard appropriately.
8. Wash the sheets as follows using 100 mL/wash for each large blot or for each group of pooled strips: TS buffer containing 0.05% NP-40: three washes (15 min each). Additional washes can be performed until no further radioactivity elutes in the wash as measured by a hand-held gamma counter. TS buffer: one wash (5 min).
9. Before drying blots, incubate them for 5 min with TSM buffer. This incubation step facilitates subsequent removal of antibody and reuse of the blot (*see* Notes 12–14 and Fig. 1).
10. After incubating the blot with TSM buffer, immediately dry it between several layers of paper towels. After 5 min, move the blot to fresh paper towels to prevent the nitrocellulose from sticking to the paper towels. Allow the blot to dry thoroughly.
11. Mount the dried blot on heavy paper, cover it with clear plastic wrap, and subject it to autoradiography (*see* refs. *13,14* for details).

3.3. Dissociation of Antibodies from Western Blots

1. After the blot has been subjected to autoradiography for the desired length of time, remove it from its mounting and place it in a Ziplock™ bag.
2. Add 50 mL of erasure buffer, seal the bag, and incubate in a water bath at 70°C for 30 min with gentle agitation every 5–10 min (*see* Notes 13 and 14).
3. Decant and discard the erasure buffer. Wash the blot twice (5 min each) with 50–100 mL of TS buffer to remove SDS.
4. To ensure that nonspecific binding sites on the blot are well coated, incubate with TSM buffer for 6–8 h at room temperature with gentle agitation.
5. The blot is ready to be stored (Step 3.1.11.) or to be incubated with a new antibody as described in Section 3.2.

4. Notes

4.1. Transfer of Polypeptides to Nitrocellulose

1. The method described is for transferring polypeptides after electrophoresis in SDS-containing polyacrylamide gels. Alternative methods have been described for transferring polypeptides after acid–urea gels and after isoelectric focusing (reviewed in *2,3*).

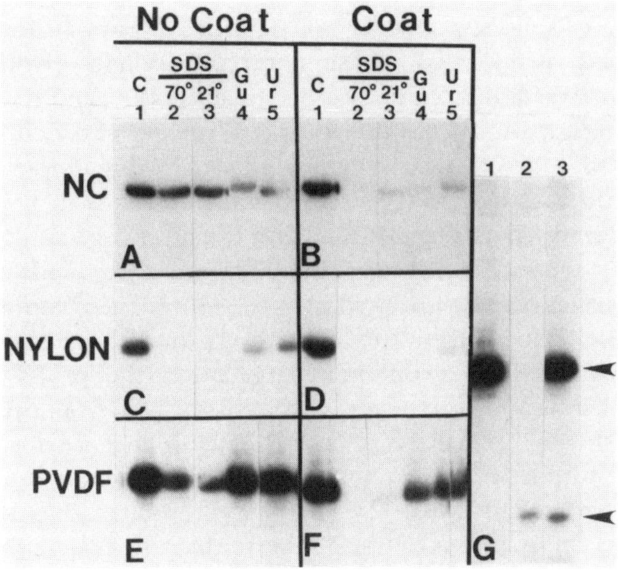

Fig. 1. Conditions for dissociating antibodies from Western blots after immobilization of polypeptides on various solid supports. Replicate samples containing 2×10^6 rat liver nuclei were subjected to polyacrylamide gel electrophoresis in the presence of SDS as, as previously described *(17)*. The separated polypeptides were transferred to nitrocellulose paper (panels A, B), Nytran nylon sheets (panels C, D), or Immobilon PVDF paper (panels E–G) (*see* Note 2). Unoccupied binding sites were blocked by incubation with milk-containing buffer (Step 3.1.11.). Blots were incubated with chicken polyclonal antiserum that reacts with the nuclear envelope polypeptide lamin B *(17)* followed by [^{125}I]-labeled rabbit antichicken IgG (Steps 3.2.1.–3.2.8.). One-half of each blot (panels B, D, F, and G) was coated with milk-containing buffer for 5 min prior to drying (Step 3.2.9.); the other half of each blot was dried without being recoated with protein (panels A, C, and E). Autoradiography (not shown) confirmed that the signal in all lanes of a given panel was identical prior to subsequent manipulation. Panels A–F: To investigate the efficacy of various conditions for dissociating antibodies, samples were incubated at 70°C for 30 min with SDS erasure buffer (lane 2), with 6*M* guanidine hydrochloride in 50 m*M* Tris-HCl (pH 6.8) containing 100 m*M* 2-mercaptoethanol (lane 4), or with 8*M* urea in 50 m*M* Tris-HCl (pH 6.8) containing 100 m*M* 2-mercaptoethanol (lane 5). Alternatively, samples were incubated at 21°C for 30 min with SDS erasure buffer (lane 3). Strips were then washed twice with TS buffer and dried for autoradiography. Untreated strips (lane l) served as controls. In each pair of panels, nonadjacent wells from a single autoradiograph have been juxtaposed to compose the figure. It is important to note that coating with milk prior to drying (Step 3.2.9.) does not affect the amount of radiolabeled antibody initially bound to the blots (cf lane 1 in panels A and B, C and D, E and F). The efficacy of various treatments at removing antibodies varies depending on the solid support. For nitrocellulose or PVDF, coating of the blots with protein prior to drying (panels B, F) greatly facilitates the dissociation of antibodies. In both cases, [continued on page 242]

2. Choice of solid support for polypeptides. Figure 1 shows the results obtained when various solid supports (nitrocellulose, nylon, PVDF) are used for Western blotting, stripped of antibodies, and reused. Nitrocellulose (Figs. 1A,1B) has the advantage of ease of use. It is compatible with a wide variety of staining procedures. With multiple cycles of blotting and erasing, however, nitrocellulose tends to become brittle. Derivatized nylon (Figs. 1C,1D) has the advantage of greater protein binding capacity and greater durability, but avidly binds many nonspecific protein stains (reviewed in *1–3*). The higher binding capacity of nylon is said to contribute to higher background binding despite the use of blocking solutions containing large amounts of protein (reviewed in ref. *3*). Antibodies can be more easily dissociated from nylon than from nitrocellulose (cf Figs. 1C, 1A). PVDF membranes are durable, are compatible with a variety of nonspecific protein stains, and are capable of being stripped of antibody (Fig. 1F) and reutilized (Fig. 1G).
3. Various compositions of transfer buffer have been described (reviewed in *1–3*). Methanol is said to facilitate the binding of polypeptides to nitrocellulose, but to retard the electrophoretic migration of polypeptides out of the gel. In the absence of SDS, polypeptides with molecular weights above 116 kDa do not transfer efficiently. Low concentrations of

---

SDS- containing buffer (lanes 2,3) is more effective than guanidine hydrochloride (lane 4) or urea (lane 5) at dissociating the antibodies. For Nylon, SDS-containing buffer is again more effective at dissociating the antibodies (cf lanes 2–5 in panel C). When SDS-containing erasure buffer is used, it is not necessary to recoat nylon with protein prior to drying for autoradiography (cf lanes 2 or 3 in panels C and D). On the other hand, when guanidine hydrochloride-containing buffer is used to dissociate antibodies, it is necessary to recoat the nylon (cf lane 4 in panels C and D). Panel G: Reutilization of blots after dissociation of antibodies. Nuclear polypeptides were immobilized on PVDF, reacted with antibodies, and treated with milk-containing buffer (Step 3.2.9.) prior to drying. Autoradiography (not shown) confirmed that all three lanes initially had indistinguishable signals for the 66 kDa lamin B polypeptide. After lane 2 was treated with SDS erasure buffer and recoated with milk (Steps 3.3.2.–3.3.5.), lanes 2 and 3 (a lane that was not erased) were reacted sequentially with chicken antiserum that recognizes the 38 kDa nucleolar polypeptide B23 *(17)* and [^{125}I]-labeled rabbit antichicken IgG. The signal for B23 (lower arrow) was readily detectable on the strip that had previously been erased (lane 2), as well as the strip that had not previously been erased (lane 3). Thus, treatment with SDS erasure buffer did not remove the nuclear polypeptides from the PVDF paper nor substantially alter their reactivity with polyclonal antibodies. The absence of a signal for lamin B after erasure (upper arrow, lane 2) indicates that the erasure buffer efficiently dissociated the antilamin B primary antibodies as well as the radiolabeled secondary antibodies from the PVDF-immobilized polypeptides.

SDS (0.01–0.1%) facilitate the transfer of larger polypeptides, but simultaneously increase the current generated during electrophoretic transfer, necessitating the use of vigorous cooling to prevent damage to the transfer apparatus.

4. Alternative staining procedures (reviewed in *1–3*) utilize Coomassie blue, Ponceau S, Amido black, India drawing ink, colloidal gold, or silver.
5. A washing step in acidified alcohol is probably essential to immobilize the polypeptides on the nitrocellulose *(3,9,10)*. The fast green staining procedure satisfies this requirement. Polypeptides are observed to elute from nitrocellulose under mild conditions if a wash in acidified alcohol is omitted *(10,15)*.
6. Various proteins have been utilized to block unoccupied binding sites on nitrocellulose (reviewed in *2,3*). These include 5% (w/v) powdered dry milk, 3% bovine serum albumin, 1% hemoglobin, and 0.1% gelatin. Blots of polypeptides immobilized on nitrocellulose have been successfully stripped of antibody and reutilized after coating of unoccupied binding sites with any of these protein solutions *(8)*, if the blot is recoated with the protein solution immediately prior to drying (Step 3.2.9.).

4.2. Formation of Antigen–Antibody Complexes

7. No guidelines can be provided regarding the appropriate dilution of antibody to use. Some antisera are useful for blotting at a dilution of greater than 1:20,000. Other antisera are useful at a dilution of 1:5 or 1:10. When attempting to blot with an antiserum for the first time, it is reasonable to try one or more arbitrary concentrations in the range of 1:10–1:500. If a strong signal is obtained at 1:500, further dilutions can be performed in subsequent experiments.
8. Different investigators incubate blots with primary antibodies for different amounts of time (reviewed in *3*). Preliminary studies with some of our antibodies have revealed that the signal intensity on Western blots is much greater when blots are incubated with antibody overnight rather than 1–2 h at room temperature (unpublished observations).
9. Diluted antibody solutions can be reused multiple times. They should be stored at 4°C after additional aliquots of penicillin/streptomycin and sodium azide have been added. Some workers believe that the amount of nonspecific (background) staining on Western blots diminishes as antibody solutions are reutilized. Antibody solutions are discarded when the intensity of the specific signal begins to diminish.
10. Choice of wash buffer after incubation with primary antibody: 2*M* urea is included in the suggested wash buffer to diminish nonspecific binding.

Alternatively, some investigators include a mixture of SDS and nonionic detergent (e.g., 0.1% (w/v) SDS and 1% (w/v) Triton X–100) in the wash buffers. For antibodies with low avidity (especially monoclonal antibodies), the inclusion of 2*M* urea or SDS might diminish the signal intensity. These agents are, therefore, optional depending on the properties of the primary antibody used for blotting.

11. [^{125}I]-labeled protein A can be substituted for radiolabeled secondary antibody. Protein A, however, can bind to the immunoglobulins present in milk (causing a high background on the blot). Hence, when [^{125}I]-labeled protein A is to be used, milk should not be utilized to block unoccupied binding sites (Step 3.1.11.), nor as a diluent for antibodies (Steps 3.2.2., 3.2.3., and 3.2.6.). Instead, bovine serum albumin, hemoglobin, or gelatin should be considered (*see* Note 6).

4.3. Dissociation of Antibodies from Western Blots

12. Reincubation of blots with protein-containing buffer prior to drying has been found to be essential for efficient dissociation of antibodies from nitrocellulose (cf Figs. 1B and 1A) or PVDF paper (cf Figs. 1F and 1E). Recoating the blots is not required in order to dissociate antibodies from Western blots performed on certain types of nylon Fig. 1C).
13. Choice of erasure buffer. Preliminary experiments have shown that the SDS/2-mercaptoethanol erasure buffer is more effective than urea, guanidine hydrochloride, or acidic glycine at dissociating polyclonal antibodies from Western blots on nitrocellulose (*8; see also* Fig. 1B) or PVDF (Fig. 1F). On the other hand, 6*M* guanidine hydrochloride is effective under certain conditions at removing antibodies from nylon (ref. *8* and Fig. 1D).
14. a. Temperature of incubation. When blotting is performed after immobilization of polypeptides on nitrocellulose, complete removal of antibodies requires heating of erasure buffer to ≥50°C for 30 min (*8; see also* Fig. 1B, lanes 2, 3). On the other hand, after immobilization of polypeptides on nylon, antibodies are efficiently dissociated by erasure buffer at room temperature (lane 3 in Figs. 1C,1D).
    b. Length of incubation. When blotting is performed on nitrocellulose, complete dissociation of antibodies at 70°C requires a minimum of 20 min incubation with erasure buffer *(8)*. Incubation times for removal of antibodies from nylon and PVDF have not been investigated.

4.4. General Notes

15. Epitopes recognized by some antibodies appear to be destroyed by the erasing process (*see* Fig. 3 in ref. *8;* for a similar effect of urea on some epitopes, *see* ref. *7*). Other epitopes on the same protein are preserved as confirmed by Western blotting with alternative antibodies against the same protein. The features that render certain epitopes particularly sensitive to this destruction are presently unknown.
16. The technique described above is not useful for removing colored peroxidase reaction products (e.g., diaminobenzidine oxidation products) from blots. Thus, we avoid the use of peroxidase-coupled secondary antibodies. On the other hand, recently described peroxidase-based luninescence assays *(16)* do not deposit a chemical reaction product on the blot and should be compatible with this erasure method.
17. The techniques described above can be applied to the detection of glycoproteins by radiolabeled lectins. For this application, blots would be coated with albumin or gelatin, reacted with radiolabeled lectin (Steps 3.2.6.–3.2.8.), and recoated with albumin or gelatin (Steps 3.2.9.,3.2.10.) prior to drying. After autoradiography, the radiolabeled lectin would be solubilized in warm SDS under reducing conditions (Steps 3.3.1.–3.3.5.).

References

1. Gershoni, J. M. and Palade, G. E. (1983) Protein blotting: Principles and applications. *Anal. Biochem.* **131,** 1–15.
2. Beisiegel, U. (1986) Protein blotting. *Electrophoresis* **7,** 1–18.
3. Stott, D. I. (1989) Immunoblotting and dot blotting. *J. Immunol. Methods* **119,** 153–187.
4. Renart, J., Reiser, J., and Stark, G. R. (1979) Transfer of proteins from gels to diazobenzyloxymethyl-paper and detection with antisera: A method for studying antibody specificity and antigen structure. *Proc. Natl. Acad. Sci. USA* **76,** 3116–3120.
5. Gullick, W. J. and Lindstrom, J. M. (1982) Structural similarities between acetylcholine receptors from fish electric organs and mammalian muscle. *Biochemistry* **21,** 4563–4569.
6. Legocki, R. P. and Verma, D. P. S. (1981) Multiple immunoreplica technique: Screening for specific proteins with a series of different antibodies using one polyacrylamide gel. *Anal. Biochem.* **111,** 385–392.
7. Erickson, P. F., Minier, L. N., and Lasher, R. S. (1982) Quantitative electrophoretic transfer of polypeptides from SDS polyacrylamide gels to nitrocellulose sheets: A method for their re-use in immunoautoradiographic detection of antigens. *J. Immunol. Methods* **51,** 241–249.
8. Kaufmann, S. H., Ewing, C. M., and Shaper, J. H. (1987) The erasable Western blot. *Anal. Biochem.* **161,** 81–95.

9. Parekh, B. S., Mehta, H. B., West, M. D., and Montelaro, R. C. (1985) Preparative elution of proteins from nitrocellulose membranes after separation by sodium dodecylsulfate-polyacrylamide gel electrophoresis. *Anal. Biochem.* **148,** 87–92.
10. Salinovich, O. and Montelaro, R. C. (1986) Reversible staining and peptide mapping of proteins transferred to nitrocellulose after separation by sodium dodecylsulfate-polyacrylamide gel electrophoresis. *Anal. Biochem.* **156,** 341–347.
11. Laemmli, U. K. (1970) Cleavage of structural proteins during the assembly of the head of bacteriophage T4. *Nature* **227,** 680–685.
12. Cooper, T. G. (1977) *The Tools of Biochemistry.* Wiley, New York, pp. 206–212
13. Laskey, R. A. and Mills, A. D. (1977) Enhanced autoradiographic detection of ^{32}P and ^{125}I using intensifying screens and hypersensitized film. *FEBS Lett.* **82,** 314–316.
14. Swanstrom, R. and Shank, P. R. (1978) X-ray intensifying screens greatly enhance the detection by autoradiography of the radioactive isotopes ^{32}P and ^{125}I. *Anal. Biochem.* **86,** 184–192.
15. Lin, W. and Kasamatsu, H. (1983) On the electrotransfer of polypeptides from gels to nitrocellulose membranes. *Anal. Biochem.* **128,** 302–311.
16. Leong, M. M. L., Fox, G. R., and Hayward, J. S. (1988) A photodetection devise for luminol-based immunodot and western blotting assays. *Anal. Biochem.* **168,** 107–114.
17. Kaufmann, S. H. (1989) Additional members of the rat liver lamin polypeptide family: Structural and immunological characterization. *J. Biol. Chem.* **264,** 13946–13955.

CHAPTER 26

Colloidal Gold Staining and Immunoprobing on the Same Western Blot

Denise Egger and Kurt Bienz

1. Introduction

Proteins, blotted from polyacrylamide gels onto nitrocellulose sheets (Western Blots) can be stained nonspecifically with a variety of dyes, or they can be identified individually by probing with appropriate antibodies. These procedures may be performed on duplicate blots, staining the total protein pattern on one blot and using the second blot for the immune reaction *(1,2)*. This chapter describes how to combine both methods on one blot, i.e., staining the blot first for total protein, followed by an indirect immune reaction *(3)*.

Possible applications of this method include:

1. Viewing the blot for artifacts, resolution, and so on, before probing with a (precious) antibody;
2. Cutting out the desired region for immunoprobing (e.g., in screening monoclonal antibodies);
3. Locating immunoreactive proteins or protein A-containing fusion proteins on the blot in relation to the total protein pattern of a given sample;
4. Testing the degree of purity obtained during purification of a protein. Examples are shown in Fig. 1.

The simultaneous demonstration of the whole protein pattern and the individual immunoreactive proteins on one single blot can be achieved in

From: *Methods in Molecular Biology, Vol. 10: Immunochemical Protocols*
Ed.: M. Manson

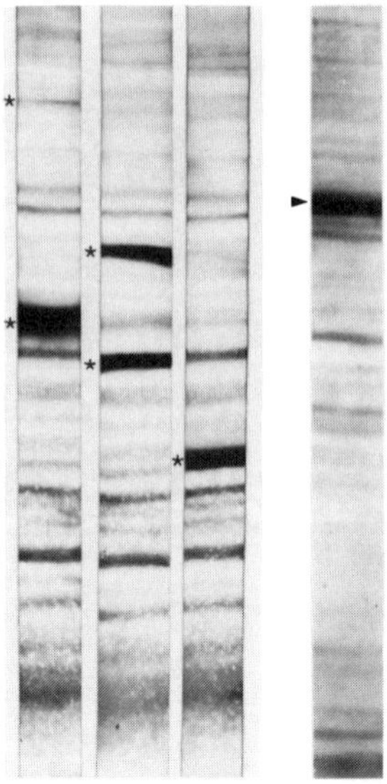

Fig. 1. Examples of Western blots stained with colloidal gold for total protein followed by immunostaining of individual antigens. Lanes 1–3: Proteins on a Western blot from a cytoplasmic extract of poliovirus-infected HEp-2 cells were stained with colloidal gold. The probing monoclonal antibodies, recognizing the viral proteins VP1 and precursor, VPO and VP2, and VP3, respectively, are detected by peroxidase-coupled rabbit-antimouse antibody (asterisks). Lane 4: Western blot of an *E. coli* lysate, containing a fusion protein composed of protein A and the poliovirus protein 2B. The fusion protein (arrowhead) is detected on the gold-stained blot by peroxidase-coupled IgG that binds to the protein A moiety.

two different ways: Either the immune reaction is performed first and the total protein stain is applied afterward. This method, using "AuroDye" (formerly obtainable from Janssen, Belgium, now from Amersham, UK, or Aurion, Wageningen, Holland) as a protein stain, is described in detail in vol. 3, Chapter 34 of this series. Or, as presented here, the blot is first stained for total protein, using colloidal gold (referred to as "citrate gold" because of the method of preparation) and then, after blocking free protein binding sites on the nitrocellulose as usual, the immune reaction is performed.

In deciding which method to use, the following differences should be considered: in contrast to the "AuroDye" method, in which only Tween-20 can be used to block free binding sites on the nitrocellulose, the citrate gold method allows for optimal blocking with one of a variety of proteins (gelatin, ovalbumin, milk powder, and so on) or Tween-20. This often results in a much lower background and preserves the highest possible immunoreactivity. Even more important is that the two gold preparations differ at least 30-fold in their sensitivity for protein staining. This should be kept in mind when the relative amount of immunoreactive antigen in the protein mixture under study is small and, therefore, a large amount of total protein has to be loaded on the gel to obtain enough antigen on the blot for immunological detection. In this case, the lower sensitivity of the citrate gold is preferable over "AuroDye" in order not to overstain the blot and to obtain clear staining of the protein bands. The immunoreactivity of the blotted proteins after staining with citrate gold is qualitatively and quantitatively the same as in unstained blots. Using peroxidase-labeled antibody for the immune reaction, a blue reaction product is obtained that contrasts well with the red stained protein pattern, and allows an easy documentation, even by black and white photography.

2. Materials

1. Colloidal gold: This can be prepared by the citrate method ("citrate gold") (*see also* this vol., Chapter 16) using gold chloride and sodium citrate. $H(AuCl_4)$: 2% stock solution in water, stable for 1–2 yr at 4°C. Dilute to 0.01% in water just before use. Sodium-citrate: 1% in water, freshly prepared. All solutions are made up in double-distilled water of highest purity. Glassware has to be extremely clean (*see* Note 1).
2. Filters: To filter the working solutions from above, use 0.45 µm nitrocellulose filters (e.g., Millex-HA, Millipore).
3. Siliconized Erlenmeyer flask: A 200 mL flask is siliconized by rinsing it with silicone solution (e.g., SERVA) and curing the silicone film for 1 h at 100–150°C.
4. Nitrocellulose for blotting: Nitrocellulose sheets with 0.45 µm pore size are available from several suppliers.
5. Blotting buffer: 25 m*M* Tris, 192 m*M* glycine, pH 8.4, 20% methanol, and 0.02% SDS.
6. Wash after blotting: 100 m*M* Tris-HCl, pH 7.4 or 100 m*M* Tris-HCl, pH 7.4 in 150 m*M* NaCl (*see* Table 1).
7. Blocking solutions (*see also* vol. 3, Chapter 33): The following blocking solutions have been successfully used (*see* Table 1): 0.25% gelatin and 3% ovalbumin; 5% skimmed milk (commercial milk powder, dissolved in the appropriate buffer); 0.3% Tween-20. Other solutions may work equally well, depending on the nature of the proteins under study.
8. The primary antibody is diluted according to the blocking solution employed (*see* Table 1). The dilution factor has to be found empirically. We use hybridoma supernatant diluted to 1:10.

Table 1
Immunoprobing with Different Blocking Agents

Blocking agent:	Ovalbumin/Gelatin	Skimmed milk	Tween-20
Wash after blotting	100 m*M* Tris, pH 7.4, room temperature (T_r), 5 min	100 m*M* Tris, pH 7.4, 150 m*M* NaCl, T_r, 5 min	100 m*M* Tris, pH 7.4, T_r, 5 min
Blocking solution	100 m*M* Tris, pH 7.4, 3% ovalbumin, 0.25% gelatin, 40°C, 60 min	100 m*M* Tris, pH 7.4, 150 m*M* NaCl, 5% (w/v) skimmed milk, T_r, 30 min	PBS, 0.3% Tween-20,37°C, 30 min
Wash	—	100 m*M* Tris, pH 7.4, 150 m*M* NaCl, 0.1% NP40, T_r, 3× 5 min, rinse in 100 m*M* Tris, 150 m*M* NaCl	—
Primary antibody	Diluted in blocking solution, T_r, overnight	Diluted in 100 m*M* Tris, pH 7.4, 150 m*M* NaCl, T_r, overnight	Diluted in PBS, 0.05% Tween-20, T_r, overnight
Wash	50 m*M* Tris, pH 7.4, 5 m*M* EDTA, 150 m*M* NaCl, 0.25% gelatin, 0.5% NP40, T_r, 5× 10 min	100 m*M* Tris, pH 7.4, 150 m*M* NaCl, 0.1% NP40, T_r, 3× 5 min, rinse in 100 m*M* Tris, 150 m*M* NaCl	PBS, 0.05% Tween-20, T_r, 3× 10 min
Labeled secondary antibody	Diluted in blocking solution, T_r, 2.5 h	Diluted in 100 m*M* Tris, pH 7.4, 150 m*M* NaCl, T_r, 20 min–2.5 h	Diluted in PBS, 0.05% Tween-20, T_r, 2.5 h
Wash	50 m*M* Tris, pH 7.4, 5 m*M* EDTA, 450 m*M* NaCl, 0.4% Sarkosyl, T_r, 5 × 10 min	As after primary antibody, T_r, 3× 5 min	PBS, 0.05% Tween-20, T_r, 3× 10 min

9. The secondary, peroxidase-coupled antispecies antibody is diluted 1:200–1:2000 in the buffer indicated in Table 1.
10. Washing solutions: After primary and secondary antibody stages, washes are done in the solutions given in Table 1. PBS: 8 g of NaCl, 0.2 g of KCl, 1.44 g of $Na_2HPO_4 \bullet 2H_2O$, 0.2 g of KH_2PO_4, 1000 mL of distilled water, final pH 7.4.
11. 4-chloro-1-naphthol: This substrate for the peroxidase is prepared as stock solution of 0.3% in methanol. Add 0.6 mL of the stock solution and 2 µL of H_2O_2 (30%) to 9.4 mL of distilled water just before use.
12. KODAK TP-Film (Technical Pan 2415).
13. Yellow or orange filter (e.g., OG1 barrier filter from I.F.-microscope).
14. KODAK D19-developer.

3. Methods

3.1. Preparation of Colloidal Gold (see *Note 1*)

1. Filter the 0.01% gold chloride and the 1% sodium citrate solutions through 0.45 µm nitrocellulose filters.
2. Bring 100 mL of the gold chloride solution to a vigorous boil in a siliconized 200 mL Erlenmeyer flask.
3. Add 4 mL of the sodium citrate solution at once while shaking the flask. Keep the mixture boiling constantly. Don't worry if some of the fluid evaporates; reflux cooling is not necessary. The solution will stay colorless for 2–3 min, then turn purple for 9 min, and change to bright red thereafter. Keep boiling for a total of 12–15 min until the *bright* red color is obtained. Let it cool. This is the final staining solution. It consists of 20 nm gold grains with an OD_{515} of 1–1.1 and a pH of 6. If kept at 4°C, the shelf-life is at least one year.

3.2. Blotting

1. Prepare SDS-polyacrylamide gels or 2-D-gels (*see* vol. 1, Chapter 6 and vol. 3, Chapters 16 and 17).
2. Blot the gels onto nitrocellulose by standard procedures (*see* this vol., Chapter 24). An additional nitrocellulose sheet on the cathodic side of the gel might be helpful to absorb impurities. We blot routinely at 48 V with cooling during 1–1.5 h.
3. Wash the blots briefly with 100 m*M* Tris-HCl pH 7.4, for 5 min.
4. Air-dry the blots and store at 4°C or process for gold staining.

3.3. Gold Staining of Protein Bands (see *Note 2*)

1. Wash the blot in double-distilled water before gold staining.
2. Stain the blots with the gold solution at room temperature on a shaker

for 10–45 min until the protein bands are visible. Use sufficient gold solution to cover the blot easily.

3. After staining, rinse the blot in double-distilled water.

3.4. Blocking and Immune Reaction

1. To block free binding sites on the nitrocellulose, use the blocking solution that gives the most intense immunostaining of the antigen and the lowest background for the other proteins. This has to be found empirically; Table 1 may be used as a guide.
2. After blocking, incubate the blots in a suitable dilution of antibody (Table 1) overnight in an airtight container at room temperature on a shaker.
3. Wash and incubate the blots in peroxidase-labeled antispecies antibody at room temperature, as indicated in Table 1.
4. Wash again, rinse in distilled water, and incubate in freshly prepared substrate for 5–30 min until the immunoreactive bands are clearly visible.
5. Rinse with distilled water and air-dry.

3.5. Photography (see *Note 3*)

1. Take black-and-white photographs of the blots on Technical Pan film 2415 (KODAK), setting 125/22°. If an OG1 filter is used to enhance contrast further, the setting is 75/19°.
2. Develop for 3–4 min in undiluted (or in a 1:2 dilution) of D19 developer. This developer yields especially high contrast.

4. Notes

1. Colloidal gold. For preparation of the citrate gold, highly purified water and cleanliness of the glassware is essential. Adjusting the pH of the citrate gold in the range between 3.7 and 8.0 does not change its staining properties for the blots, but may impair its stability. It is best to use it without any addition or pH adjustment.
2. Gold staining of blot. Blots that are to be stained with colloidal gold have to be handled with extreme care, since they are very sensitive to mechanical damage (scratches, impression marks) and dirt, such as grease from the seal of the glass plates in the gel apparatus or impurities from the blotting buffer. Handle the blots at the edge only with clean forceps.

   As outlined above, the gold stain should be matched in sensitivity to the amount of protein loaded on the gel. This can be estimated as follows: if a strip of the gel from which the blot is to be made can be adequately stained with Coomassie blue, citrate gold will yield a good stain of the blot, whereas "AuroDye" will heavily overstain it. If the gel contains so little protein that it has to be detected by silver staining, "AuroDye" and the corresponding method (*see* vol. 3, Chapter 34) should be used.

3. Photography. On black and white photographs, the contrast can be enhanced by using an orange or yellow filter, i.e., a filter similar in color to the gold stain and complementary to the (blue) immunoreactive bands. This renders the immune reaction bands on the final print intensely black and the red gold-stained proteins contrastingly gray.

   Faint immunoreactive bands tend to fade during drying. They are easier to recognize and photograph on wet blots laid on a clean glass plate.

References

1. Rohringer, R. and Holden, D. W. (1985) Protein blotting: Detection of protein with colloidal gold, and of glycoproteins and lectins with biotin-conjugated and enzyme probes. *Anal. Biochem.* **144,** 118–127.
2. Hancock, K. and Tsang, V. C. W. (1983) India ink staining of proteins on nitrocellulose paper. *Anal. Biochem.* **133,** 157–162.
3. Egger, D. and Bienz, K. (1987) Colloidal gold staining and immunoprobing of proteins on the same nitrocellulose blot. *Anal. Biochem.* **166,** 413–417.

CHAPTER 27

Colloidal Gold Staining and Immunodetection in 2-D Protein Mapping

Anthony H. V. Schapira

1. Introduction

This chapter extends the use of the technique described in the previous chapter to two-dimensional (2D) protein gels, as well as containing some alternatives and modifications to the method.

Two-dimensional sodium dodecyl sulfate-polyacrylamide gel electrophoresis (SDS-PAGE) provides a rapid and reproducible method for the separation and analysis of complex mixtures of proteins. Proteins may be separated in the first dimension either by isoelectric focusing (IEF) *(1)* or nonequilibrium pH gradient electrophoresis (NEPHGE) *(2)*, and by SDS-PAGE in the second dimension. Such gels have the capacity to resolve over 1000 individual polypeptides. The identification and characterization of individual polypeptides separated by these techniques is the natural extension of the study of proteins by electrophoresis. A protein may be identified directly from the gel by cutting out the specific gel segment in which it is contained, and then eluting and sequencing the protein. Alternatively, a protein may be identified by specific antibody binding and detected with enzyme-linked or radiolabeled second antibodies. Detection by these methods is generally performed after the proteins in the gel have been transferred to a solid matrix, such as nitrocellulose. The mapping of individual proteins is then dependent on the identification of the protein(s) of interest within the whole protein map. This

From: *Methods in Molecular Biology, Vol. 10: Immunochemical Protocols*
Ed.: M. Manson

chapter describes a colloidal gold method of staining all proteins transferred to nitrocellulose followed by antibody binding, which allows the precise mapping of an individual protein within the whole two-dimensional gel picture.

2. Materials

1. Tube gel and slab gel electrophoresis equipment.
2. Electroblotting equipment.
3. Nitrocellulose (*see* Note 1).
4. 1% Milk powder or 1% bovine serum albumin in phosphate buffered saline, pH 7.4.
5. 1% (w/v) Analar grade gold chloride. This can be made up in a solution of 100 mL and stored at 4°C in a dark bottle.
6. 20% (v/v) Tween-20 in distilled water.
7. Stannous chloride solution: Dissolve 250 mg of stannous chloride in 1.25 mL of 1 *M* HCl and make up to 25 mL with distilled water. Prepare fresh.
8. Citric acid solution: Dissolve 2.42 g of anhydrous citric acid in 250 mL of distilled water.
9. Photographic facilities for high resolution photography.
10. Specific primary antibodies for individual polypeptides.
11. Appropriate antispecies second antibodies linked to horseradish peroxidase (*see* Note 2).
12. 4-chloro-1-naphthol (*see* Chapter 28, Section 2.8.) (*see* Note 2).
13. Phosphate buffered saline (PBS), pH 7.4.

3. Method

The techniques of 1- and 2-D SDS-PAGE (vols. 1 and 3), electroblotting (vol. 3), and immunoblotting (this vol., Chapter 24) are described elsewhere.

First dimension gels may be run either by IEF or NEPHGE. Following the electrophoretic separation of proteins, the gel may be equilibrated for 10–15 min in Tris-glycine-methanol before electroblotting to nitrocellulose. This reduces any distortion from swelling or contraction of the gel relative to nitrocellulose sheet. Prolonged equilibration beyond this time-period may lead to loss of protein from the gel. Electroblotting is now most conveniently performed by the semidry method. This reduces transfer time considerably, utilizes very small amounts of buffer, and the uniform field strength produces consistent transfer of protein over the whole gel.

For consistent results, the gold stain should be prepared fresh for each batch of nitrocellulose filters. The gold stain is prepared essentially as described by Righetti et al. *(3),* with some modifications *(4)* (*see also* this vol., Chapter 16).

1. To a clean 2-L glass flask or beaker, add 750 mL of distilled water. Place on a magnetic stirrer and begin stirring—this will be continued throughout (*see* Note 3).
2. Add dropwise 10 mL of the 1% gold chloride solution. Leave to stir for 5 min.
3. Slowly add 100 mL of 20% (v/v) Tween-20 (*see* Note 4). Leave to stir for 15 min.
4. Add dropwise 4 mL of freshly prepared stannous chloride solution to the gold/Tween mixture (*see* Note 5). The color of the solution will change from gold to burgundy. Leave to stir for 5 min.
5. Slowly add 250 mL of citric acid solution to the gold solution. Leave to stir for 20–30 min. The color of the solution should develop into red (*see* Note 6).
6. Following transfer of proteins from the gel, quickly rinse the nitrocellulose in two changes of distilled water. Pour 150–200 mL of the gold stain into a clean glass or plastic container and lay the nitrocellulose face up on top of the gold stain. Shake gently, allowing the stain to cover the filter. Proteins will begin to stain within 15 min, often appearing as a "ghost" before fully developing. A good fresh gold stain will give a pink color to the proteins.
7. Filters should be photographed before blocking overnight with a solution of 1% milk powder or 1% BSA in PBS. Immunoblotting may then be performed as described in the previous Chapter (*see* Notes 7 and 8).

3.1. Example of Technique

Figure 1 shows the gold stain of a section of a 2-D gel of beef heart mitochondria proteins separated by isoelectric focusing and then by SDS-PAGE, and transferred to nitrocellulose. The filter was then probed with antibody specific to the 49-kDa iron sulfur protein of NADH CoQ reductase. The blot was photographed with (Fig. 2) and without (Fig. 3) a red filter following development with 4-chloro-1-naphthol. The position of the 49-kDa protein can then be determined by back reference to Fig. 1 (*see* arrow).

4. Notes

1. The nitrocellulose should always be handled with gloves. Otherwise, the gold stain will bring out fingerprints.
2. It is best to use horseradish peroxidase-linked compounds and to develop with 4-chloro-1-naphthol, as this blue-black stain provides a good contrast to the pink of the gold stain.
3. The volumes of solutions used can be adjusted to the number of filters to be stained. The method described provides sufficient stain for six filters of 14 × 14-cm nitrocellulose.

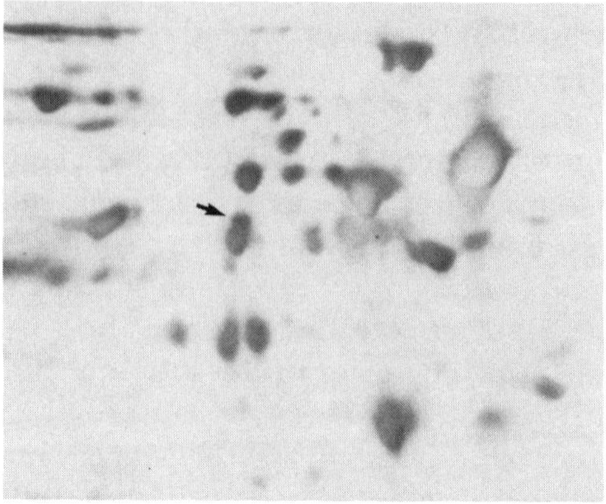

Fig. 1. Gold stain of 2-D separation of proteins from beef heart mitochondria. Arrow indicates the 49-kDa protein of NADH CoQ reductase (*see* Figs. 2 and 3).

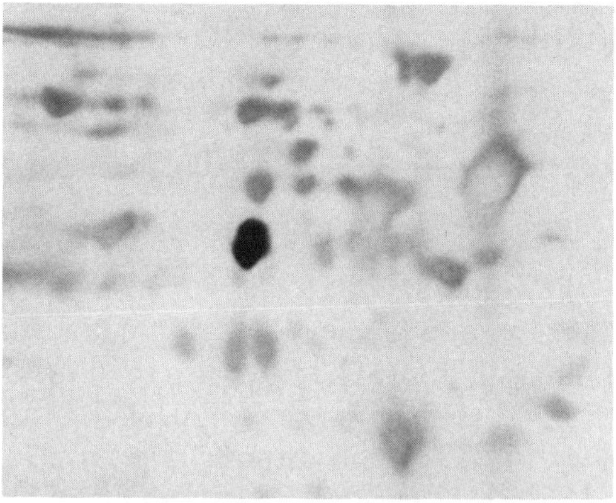

Fig. 2. Immunoblotting with antibody to the 49-kDa protein demonstrates its position on the 2-D separation. Photographed with red filter.

4. The addition of the Tween-20 before the stannous chloride enhances the stability of the gold colloid and leads to more reproducible results.
5. The slow, dropwise addition of the stannous chloride is also important in establishing a stable stain. The Tween-20 and citric acid solutions can be added at a slow pour. Constant stirring is mandatory.
6. The color of the final solution after 20–30 min of stirring is a good guide to stain quality: A beaujolais color is good; a darker purple hue indicates

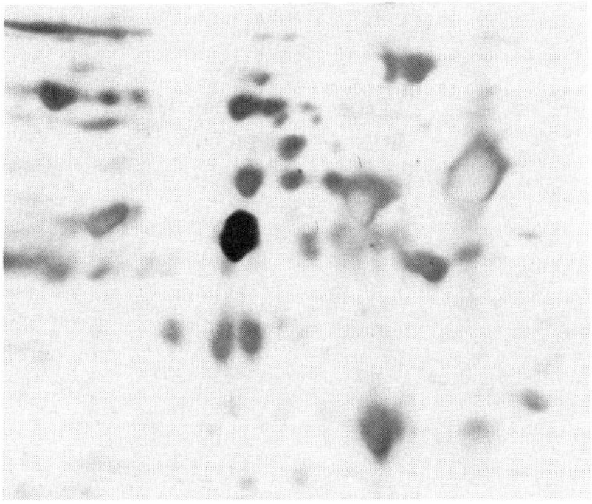

Fig. 3. As for Fig. 2, but photographed without a filter.

an unstable gel. The longer the stain is left before use, the more purple the stained proteins will be, and this will contrast less well with the 4-chloro-1-naphthol-developed immunoblot.

7. For greatest sensitivity, the biotin–streptavidin system should be used in antibody detection, especially since proteins separated in 2-D are often more difficult to detect than when separated in 1-D.
8. The best, most unequivocal results are obtained by immunoblotting a 2-D separation with an antibody to a single protein. This can then be developed and photographed. Further antibodies can then be used in sequence on the same filter, with developing and photography recording the position of each protein as the respective antibody is used. There is no need to requench filters between each blot. Washing for 30 min in three changes of PBS–0.1% Tween-20 is sufficient to clean the filter before the next overlay.

References

1. O'Farrell, P. H. (1975) High resolution two dimensional electrophoresis of proteins. *J. Biol. Chem.* **250,** 4007–4021.
2. O'Farrell, P. Z., Goodman, H. M., and O'Farrell, P. H. (1977) High resolution two dimensional electrophoresis of basic as well as acidic proteins. *Cell* **12,** 1133–1142.
3. Righetti, P. G., Casero, P., and Del Campo, G. B. (1986) Gold staining in cellulose acetate membranes. *Clin. Chem. Acta* **157,** 167–174 .
4. Schapira, A. H. V. and Keir, G. (1988) Two dimensional protein mapping by gold stain and immunoblotting. *Analyt. Biochem.* **169,** 167–171.

CHAPTER 28

Fluorescent Protein Staining on Nitrocellulose with Subsequent Immunodetection of Antigen

Boguslaw Szewczyk and Donald F. Summers

1. Introduction

The transfer of proteins from gels to nitrocellulose or other immobilizing matrices has become increasingly popular as a powerful tool for the subsequent analysis of proteins. Most frequently, the blotted proteins are analyzed for their antigenic properties by Western blotting.

A frequently encountered problem when performing immunoblotting is to make a direct correlation between the total electrophoretic pattern and the bands detected by reaction with antibodies. A number of stains, such as amido black, India ink, and colloidal gold, have been described for detection of proteins on nitrocellulose *(1–3)*. When these stains are used, a duplicate nitrocellulose blot has to be made, which may lead to discrepancies when comparing two patterns. The only stain that binds reversibly to proteins and does not give high background on nitrocellulose is Ponceau S *(4)*. The detection limit of proteins stained with Ponceau S is between 250–500 ng.

Here, we describe a staining method that is based on the coupling of a fluorescent reagent to proteins. The proteins stained in this way are colorless in visible light but can be detected by illuminating blots with long-range UV light. Dichlorotriazynylaminofluorescein (DTAF) is a reagent we have used

From: *Methods in Molecular Biology, Vol. 10: Immunochemical Protocols*
Ed.: M. Manson ©1992 The Humana Press, Inc., Totowa, NJ

for coupling fluorochrome to proteins on nitrocellulose. Most of the proteins are detectable at levels of about 50 ng. The coupling of the fluorochrome to proteins does not alter their antigenic properties, and the blots can be subsequently probed with antibodies using one of the many protocols for immunoblotting (*see* refs. *5–9* and this vol., Chapters 24–27).

2. Materials

1. Transfer buffer: Tris-glycine buffer (25 m*M* Tris/192 m*M* glycine), pH 8.3. It may be used with the addition of methanol to 20%.
2. TBS buffer: 0.47 g of Tris base, 2.54 g of Tris-HCl, 29.25 g of NaCl made up to 1 L with distilled water.
3. Borate-KCl buffer: 50 m*M* borate and 50 m*M* KCl, pH 9.3. It is prepared by mixing 50 mL of 0.1*M* of boric acid and 0.1*M* KCl with 29.3 mL of 0.1*M* NaOH and adding water to 100 mL.
4. Dichlorotriazynylaminofluorescein (DTAF) from Research Organics, Cleveland, OH. It should be stored desiccated at –20°C.
5. Blocking solution: 3% Nonfat dry milk in TBS.
6. Primary antibody.
7. Secondary antibody (e.g., goat antirabbit IgG–peroxidase conjugated if primary antiserum is from rabbit).
8. 4-Chloro-1-naphthol. The reagent should be discarded when crystals, originally white, turn to grey. It is stable for a few months at –20°C when stored desiccated. It can also be stored as a solution in ethanol at –20°C.
9. Color reagent solution: 60 mg of 4-Chloro-1-naphthol is dissolved in 20 mL of ice-cold 95% ethanol and added to 100 mL of TBS containing 60 µL of 30% H_2O_2. The reagent should be prepared just before using.
10. Nitrocellulose membrane filters (BA83 or BA85) from Schleicher and Schuell, Keene, NH.
11. Whatman 3MM filter paper.
12. Scotch-Brite pads (size depends on the size of the gel holder).
13. Glass vessels with flat bottom (e.g., Pyrex baking dishes). Their dimensions should be slightly greater than the size of the nitrocellulose sheet.
14. Rocker platform.
15. Transfer apparatus (e.g., Trans Blot Cell, Bio-Rad Laboratories, Richmond, CA).
16. Aluminum foil.
17. Long-range UV transilluminator (e.g., Spectroline TL302, Spectronics Corp., Westbury, NY).

3. Methods

3.1. Transfer of Proteins from an SDS-PAGE Gel to Nitrocellulose

1. Apply a sample containing antigen of interest to an SDS-PAGE gel and run the gel (*see* vol. 1, Chapter 6 and this vol., Chapter 24) (*see also* Note 2).
2. While running the gel, prepare transfer buffer (around 4 L for most of the transfer tanks) and four glass vessels, one of them large enough to accommodate the gel holder. Each of the vessels should contain 100–200 mL of transfer buffer.
3. Place the gel holder in one of the dishes. Put a wetted Scotch-Brite pad on one side of the holder and then three sheets of Whatman paper saturated with transfer buffer on the pad. Place the gel on the paper directly after taking it out from the electrophoresis apparatus. The upper stacking gel should be carefully removed because it sticks to nitrocellulose. Pour a few milliliters of transfer buffer on top of the gel and carefully place on it a sheet of prewetted nitrocellulose. Roll over the nitrocellulose with a glass rod to remove any air bubbles between the gel and the membrane. Next, place three Whatman 3MM sheets prewetted with transfer buffer on top of the nitrocellulose, and finally another wet Scotch-Brite pad.
4. Close the holder and place it in the transfer tank. Proteins have negative charges in an SDS-PAGE gel, so the nitrocellulose should face the anode.
5. Begin electroblotting. Apply 30 V for overnight uncooled runs or 60 V for 2 h for runs cooled with tap water, if the transfer buffer contains 20% methanol. Apply 20 V overnight when transfer buffer is without methanol.

3.2. Fluorescent Staining of Proteins

1. After transfer, wash the nitrocellulose in 50 mL of borate-KCl buffer for 5 min.
2. Immerse the membrane for 10 min in 50 mL of DTAF solution (0.5–2 µg/mL) in borate-KCl buffer. The dish should be wrapped in aluminium foil to avoid photodegradation of DTAF (*see* Notes 3–5).
3. Remove the excess reagent by two washings (100 mL each) in borate-KCl buffer for 5 min.
4. Place the nitrocellulose on a long-range UV transilluminator. The protein bands appear green on a light yellow background (*see* Note 7.)

5. Mark the bands of interest with a pencil or photograph the blot using a Polaroid™ camera equipped with an orange filter. The exposures range from 30–120 s (*see* Notes 8 and 9).

3.3. Immunodetection

1. Wash the nitrocellulose sheet that was previously subjected to staining with DTAF in 200 mL of TBS for 5 min.
2. Immerse the sheet in 200 mL of blocking solution and agitate for 1 h at room temperature.
3. Transfer the membrane to a second dish containing 50 mL of primary antibody solution (the antiserum is diluted 1:50 to 1:3000 with blocking solution). Incubate with shaking at room temperature for 1 h (*see* Note 10).
4. Wash the membrane with 200 mL of blocking solution, 10 min each wash.
5. Transfer the membrane to 50 mL of peroxidase-conjugated secondary antibody solution (diluted as suggested by the manufacturer in blocking solution) and incubate with agitation for 1 h at room temperature. Instead of the system with peroxidase-conjugated secondary antibody, a number of other conjugates described in refs. *5–9* can be also used (*see also* this vol., Chapters 10 and 15).
6. Wash the membrane once in 200 mL of blocking solution, and then twice with 200 mL of TBS alone, 10 min each wash.
7. Prepare the color reagent solution just before using.
8. Incubate the membrane for 5–10 min in color reagent solution. The bands (dark blue in color) should normally be visible within 1–2 min.
9. Transfer the membrane to distilled water for 10 min. Change the water and leave the membrane for an additional 10 min.
10. Dry the membrane in air. Keep it away from light, otherwise, the background will turn yellow.

4. Notes

1. It is advisable to run protein mol wt standards in one of the electrophoresis lanes. Their position after transfer to the membrane is helpful in the assessment of antigen mol wt and also, their intensity after staining provides information on the quality of transfer.
2. Fluorescein isothiocyanate (FITC) can be used instead of DTAF. The detection limit for staining of proteins with FITC is 2–3 times lower than for staining with DTAF.

3. After fluorescent staining, the nitrocellulose sheet can be stored practically indefinitely at 4°C, provided it is wrapped in aluminium foil or otherwise protected from light.
4. Most of the other fluorochromes for protein labeling (e.g., TRi TC-tetramethylrhodamine isothiocyanate) cannot be used for protein staining on nitrocellulose because these fluorescent reagents bind irreversibly to nitrocellulose.
5. Nylon membranes cannot be stained with DTAF because nylon binds the stain. However, Immobilon P membranes (Millipore), which are composed of polyvinylidine difluoride, are stained equally as well as nitrocellulose.
6. The limit of protein detection with DTAF will vary depending on the intensity of UV radiation. The power of a transilluminator should be at least 100 W. In case of transilluminators with more than one wavelength (e.g., 302 nm and 365 nm), the longer wavelength should be used (absorption maximum for DTAF is 489 nm).
7. Staining before immunodetection permits unambiguous cutting of the membrane into strips for incubation in more than one type of primary antibody (or more than one concentration of the same antibody).
8. When immunodetection is performed with peroxidase-conjugated secondary antibody, the fluorescent protein bands disappear during incubation with the color reagent solution (oxidation with H_2O_2). If immunodetection is done with radioiodinated protein A (*see* vol. 3, Chapter 31), the fluorescence of the proteins remains visible under UV light through all steps of immunoblotting. It is, however, advisable to take a photograph of the membrane immediately after finishing the fluorescent staining because long incubations during immunodetection lead to some quenching of fluorescence.
9. The sensitivity of antigen detection by immunoblotting is not affected by coupling DTAF to proteins if polyclonal antibodies are used. However, this may not always be the case for monoclonal antibodies because, depending on the epitope, DTAF binding may have a more or less drastic effect on the conformation of the antigen, and hence, on its reactivity with antibodies.

References

1. Gershoni, J. M. and Palade, G. E. (1982) Electrophoretic transfer of proteins from sodium dodecyl sulfate-polyacrylamide gels to a positively charged membrane filter. *Anal. Biochem.* **124,** 396–405.

2. Hancock, K. and Tsang, V. (1983) India ink staining of proteins on nitrocellulose paper. *Anal. Biochem.* **133,** 157–162.
3. Moeremans, M., Daneels, G. and de Mey, J. (1985) Sensitive colloidal metal (gold or silver) staining of protein blots on nitrocellulose membranes. *Anal. Biochem.* **145,** 315–321.
4. Salinovich, O. and Montelaro, R. C. (1986) Reversible staining and peptide mapping of proteins transferred to nitrocellulose after separation by sodium dodecyl sulfate-polyacrylamide gel electrophoresis. *Anal. Biochem.* **156,** 341–347.
5. Burnette, W. N. (1981) "Western blotting": Electrophoretic transfer of proteins from sodium dodecyl sulfate-polyacrylamide gels to unmodified nitrocellulose and radiographic detection with antibody and radioiodinated protein A. *Anal. Biochem.* **112,** 195–203.
6. Hawkes, R., Niday, E. and Gordon, J. (1982) A dot immunobinding assay for monoclonal and other antibodies. *Anal. Biochem.* **119,** 142–147.
7. Blake, M. S., Johnston, K. H., Russell-Hones, G. J., and Gotschlich, E. C. (1984) A rapid, sensitive method for detection of alkaline phosphatase anti-antibody on Western blots. *Anal. Biochem.* **136,** 175–179.
8. Brada, D. and Roth, J. (1984) "Golden blot"—detection of polyclonal and monoclonal antibodies bound to antigens on nitrocellulose by protein A–gold complexes. *Anal. Biochem.* **142,** 79–83.
9. Davis, L. G., Dibner, M. D., and Batley, J. F. (1986) *Basic Methods in Molecular Biology.* Elsevier, New York.

Chapter 29

Competitive ELISA

Kitti Makarananda and Gordon E. Neal

1. Introduction

Enzyme-linked immunosorbent assay (ELISA) is a very useful technique for the specific and sensitive assay of certain compounds, in which suitable antibodies, monoclonal or polyclonal, to the compounds are available. The technique has found particular application in the monitoring of environmental contaminants and toxins, either studying the primarily contaminated materials, e.g., foodstuffs, or body fluids of potentially exposed humans. The technique has been increasingly applied to monitoring the carcinogenic mycotoxins, the aflatoxins.

The principle of the direct ELISA system is a double-antibody sandwich technique, which is illustrated in Fig. 1. The binding of the primary antibody to the antigen, which is immobilized on the bottom of the wells in multiwell ELISA plates, is followed by the addition of the secondary antibody, which has an affinity for the primary antibody. The secondary antibody is linked to an enzyme that is then reacted with a chromagenic substrate. The color formed is proportional to the amounts of both of the antibodies, and hence, to the amount of antigen. The use of the secondary antibody can amplify the signal from the antigen-bound primary antibody. In the competitive ELISA, two antigens are involved; one, the sample antigen to be assayed and the other, a constant level of "binding" antigen immobilized in the multiwell plate. The primary antibody is allowed to react in solution with the sample antigen it is required to assay, before applying to the ELISA plates. The amount of binding antigen with which the wells are coated, necessary to give quantitative competition curves over a useful range, is determined in preliminary assays.

From: *Methods in Molecular Biology, Vol. 10: Immunochemical Protocols*
Ed.: M. Manson

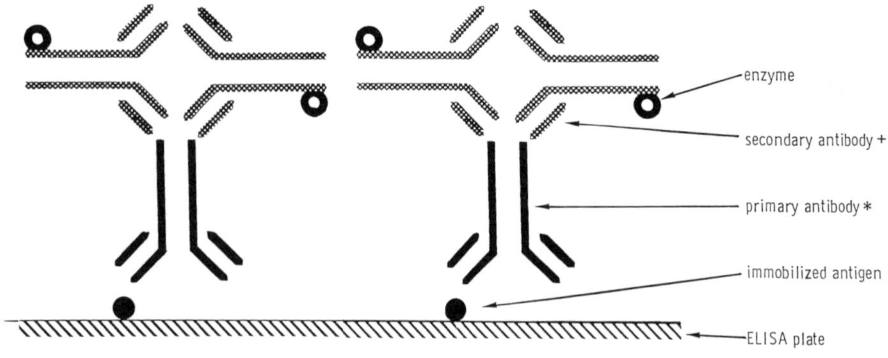

Fig. 1. Diagrammatic representation of the double antibody labeling technique used in the ELISA.

Since only unbound primary antibody, i.e., that which has not undergone reaction with the sample antigen, is now available to bind to the immobilized, constant level of antigen coating the wells, there is an inverse relationship between the level of enzymically formed color and the amount of antigen in the sample. Calibration curves are constructed using a range of concentrations of antigen in the competition reaction. This method of assaying antigens can be widely used. It is illustrated here as an assay for aflatoxins (*see* Note 1), in which case it is usually carried out using aflatoxin B_1 (AFB_1) as the standard. Polyclonal and monoclonal antibodies raised against aflatoxins usually detect a wide spectrum of aflatoxin metabolites, but with differing sensitivities. If the nature of the aflatoxin under investigation is known, a suitable calibration curve can be constructed. If it is not known, or if it is a mixture of aflatoxins, the results can be expressed as AFB_1 equivalents.

2. Materials

1. Rabbit anti-AFB_1 serum.
2. Antirabbit IgG–peroxidase conjugate.
3. Bovine serum albumin–AFB_1 conjugate (BSA–AFB_1).
4. Phosphate buffered saline, pH 7.3 (PBS): 8 g of NaCl, 0.2 g of KCl, 1.15 g of $Na_2HPO_4 \cdot 2H_2O$, and 0.2 g of KH_2PO_4 made up to 1 L with distilled water.
5. Standard concentrations of AFB_1 ranging from 0.01 to 100 ng/mL in PBS (*see* Note 1).
6. 0.25% Gelatin in PBS (PBS-gelatin).
7. 0.05% Tween-20 in PBS (PBS-Tween-20).
8. 3,3',5,5'-Tetramethylbenzidine (*see* Note 2).

9. Dimethyl sulfoxide (DMSO) (Spectrosol grade).
10. 0.1 *M* Sodium acetate buffer, pH 6.0.
11. Hydrogen peroxide (100 vol).
12. 2 *M* H_2SO_4.
13. Microtiter plates, model M129E (Dynatech Laboratories Inc., VA).
14. Plate sealing tapes.
15. Microtiter plate shaker.
16. Microtiter plate reader.

3. Methods

1. Determine the protein content of BSA–AFB_1 prepared as described by Sizaret et al. *(1)*, using Lowry's method *(2,3)*. The molar ratio of BSA:AFB_1 usually obtained is in the order 1:7.
2. Dilute the BSA–AFB_1 to 5 ng of protein/50 µL of PBS.
3. Coat the microtiter plates with the diluted BSA–AFB_1 by adding 50 µL to each well.
4. Leave the plates to dry overnight at 37°C (*see* Note 3).
5. The plates can be stored at –40°C until required.
6. Wash the plates four times with PBS-Tween-20, using the immersion technique. This is carried out by totally immersing each plate in approx 800 mL of the PBS-Tween-20 solution in a plastic box. Care must be taken to avoid air locks causing some wells not to be washed. This can be achieved by passing a small roller over the submersed plates (*see* Note 4).
7. Dry the plates by vigorously banging the inverted plate on several layers of absorbent paper (*see* Note 5).
8. Incubate each well with 100 µL of PBS-gelatin (to block the nonspecific sites) for 60 min at room temperature (*see* Note 6).
9. Prepare a concentration range of AFB_1 standards (0.01–100 ng/mL) and pipet 200 µL of each into small thoroughly clean glass test tubes (2-mL capacity). One tube is also prepared using PBS to serve as a noninhibited standard.
10. Dilute "unknown" samples appropriately (all assays are carried out using a 200 µL sample volume).
11. Dilute rabbit anti-AFB_1 serum 1:10,000 using 2 µL of antibody in 20 mL of PBS-gelatin. Optimal dilution of rabbit serum is determined in a preliminary experiment using a range of serum dilutions against a range of AFB_1 concentrations (*see* Notes 7 and 8).
12. Add the diluted antibody to the AFB_1 concentration standards or "unknown" samples, using a 1:1 ratio of antibody to sample. Then incubate

the tubes at 37°C with continuous shaking for 60 min. The tubes should be covered to prevent evaporation.

13. After the incubation to block nonspecific sites on the plate (Step 8), discard the solution and wash the plates twice with PBS-Tween-20, then dry the plates (as in Steps 6 and 7).
14. Load each well with 50 µL of the mixture obtained from Step 12, cover the plates with plate-sealing tape, and incubate, with continuous shaking using a microtiter plate shaker, for 90 min at room temperature.
15. Dilute rabbit anti-IgG-peroxidase conjugate 1:5000, using 4 µL of antibody in 20 mL of PBS-gelatin, approx 5 min before use and leave it in ice (*see* Note 9).
16. Stop the plate shaker after 90 min (Step 14), remove the sealing tape, and discard the solution into a waste bucket containing bleach solution to destroy the toxin. Wash the plates five times with PBS-Tween-20 and dry them as in Steps 6 and 7.
17. Add 50 µL of diluted antirabbit IgG–peroxidase conjugate (from Step 15) into each well and then seal the plates and incubate at room temperature for another 90 min with continuous shaking.
18. Prepare the substrate by warming (approx 40°C) 24.75 mL of 0.1*M* sodium acetate buffer, pH 6.0 for 30 min, then adding 250 µL of tetramethylbenzidine solution (10 mg in 1 mL of DMSO) and warming it for another 30 min (*see* Note 10).
19. Stop the plate shaker after the 90-min incubation (in Step 17), remove the tape, and discard the solution. Wash the plates five times with PBS-Tween-20, followed by a single distilled water wash, and dry the plates as in Step 7.
20. Add 10 µL of 100 vol hydrogen peroxide to the substrate solution (from Step 18) *immediately* before use and thoroughly mix by shaking (*see* Note 11).
21. Dispense 50 µL of substrate solution into each well. Incubate at room temperature for 30 min. A blue color will develop.
22. Stop the reaction by adding 50 µL of 2*M* H_2SO_4 to each well. The color will change to yellow. Leave the plates for 15 min.
23. Read the absorbance at 450 nm using a microtiter plate reader (*see* Notes 11 and 12).
24. Calculate the percentage inhibition using the non-AFB_1 containing PBS standard as the uninhibited control absorbance. Absorbance of the PBS standard is routinely in the range 1.4–1.6. Inhibition of 50% is usually achieved using concentrations in the order of 0.25 ng of AFB_1/mL.

4. Notes

1. The aflatoxins are extremely potent hepatotoxins and hepatocarcinogens, exerting their biological effect at the microgram level in animal model systems. Extreme caution is therefore necessary when using these materials to avoid contact with them. Decontamination of contaminated glassware using diluted bleach, followed by extensive rinsing in water, should be routinely carried out. All combustible contaminated materials should be sealed in plastic bags and incinerated. Solutions of aflatoxins to be discarded should also be exposed to diluted bleach for several hours, followed by extensive rinsing of the container in running tap water.
2. Tetramethylbenzidine is used because of its noncarcinogenic property.
3. When the plates are coated with BSA–AFB_1 and dried overnight, ensure that they are *completely* dry, otherwise, binding antigen will be lost from the plates during subsequent procedures.
4. *All* the plate washings are carried out using the immersion technique. Care must be taken to ensure that all wells are filled with the washing solution.
5. The plates are dried by inverting, followed by vigorous banging on absorbent paper to ensure that PBS-Tween-20 is completely removed, otherwise, it may interfere with the subsequent binding of the antibodies or the substrate.
6. In each incubation step, the plates should be properly sealed to prevent evaporation, otherwise varying concentration of the reactants will lead to lack of constant results between wells.
7. All the dilutions of the antibodies required should be prepared shortly before use and they should be kept cooled in ice.
8. The appropriate dilutions of the antibodies giving suitable curves in the competitive ELISA should be determined in preliminary experiments. Suitable antiserum for use in ELISA has been found to have working dilutions of between 1 in 10,000 and 1 in 20,000.
9. Peroxidase is routinely used because it is sufficiently sensitive to permit the detection of small quantities of antigen. Other enzymes linked to the second antibody, e.g., alkaline phosphatase, have been examined, but have generally been less satisfactory than peroxidase *(4)*.
10. Sodium acetate buffer has to be warmed to 40°C before the tetramethylbenzidine is added to avoid precipitation.
11. Hydrogen peroxide should be added to the substrate solution *immediately* before use.

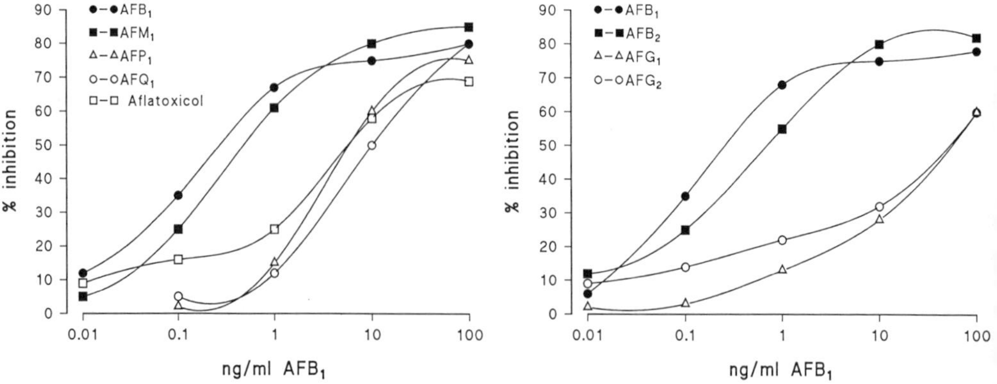

Fig. 2. Competitive ELISA curves obtained using primary aflatoxins (left), and some aflatoxin metabolites (right).

12. Standard curves should be included in every batch of assays. Figure 2 shows typical competition curves obtained using a range of aflatoxins.
13. Use replicate wells (usually 6) for each concentration assayed.

References

1. Sizaret, P., Malaveille, C., Montesano, R., and Frayssinet, C. (1982) Detection of aflatoxins and related metabolites by radioimmunoassay. *J. Natl. Cancer Inst.* **69,** 1375–1381.
2. Lowry, O. H., Rosebrough, N. J., Farr, A. L., and Randall, L. J. (1961) Protein measurement with the Folin phenol reagent. *J. Biol. Chem.* **193,** 265–275.
3. Waterborg, J. H. and Matthews, H. R. (1984) The Lowry method for protein quantitation, in *Methods in Molecular Biology,* vol. 1 (Walker, J., ed.), Humana, Clifton, NJ, pp. 1–3.
4. Martin, C. N., Garner, R. C., Tursi, F., Garner, J. V., Whittle, H. C., Ryder, R. W., Sizaret, P., and Montesano, R. (1984) An enzyme linked immunosorbent procedure for assaying aflatoxin B_1, in *Monitoring Human Exposure to Carcinogenic and Mutagenic Agents* (Berlin, A., Draper, M., Hemminki, K., and Vainio, H., eds.), IARC Scientific Publication No. 59, IARC, Lyon, France, pp. 313–321.

CHAPTER 30

Twin-Site ELISAs for *fos* and *myc* Oncoproteins Using the AMPAK System

John P. Moore and David L. Bates

1. Introduction

Twin-site ELISA is a simple technique for quantitation of specific proteins in cell or tissue extracts. The application of this method to *fos* and *myc* oncoproteins is described. There are two basic procedures:

1. Extraction of *fos* and *myc* proteins from biological material in a form suitable for immunoassay.
2. Determination of the amount of *fos* or *myc* protein in the extract by ELISA using specific antibodies, one of which is conjugated to alkaline phosphatase (AP) and is detected by the AMPAK™ amplifier system.

The *fos* and *myc* proteins are located in the nuclei of eukaryotic cells, where they are components of the machinery by which cell growth and differentiation are regulated *(1,2)*. The *fos* protein contributes to the early alteration of the pattern of gene expression that occurs when resting cells are activated by external stimuli *(2,3)*: the *myc* protein functions later in the cell cycle *(4)*. Both proteins bind to DNA and their extraction from tissues into solution requires conditions that are vigorous, yet compatible with immunoassay.

From: *Methods in Molecular Biology, Vol. 10: Immunochemical Protocols*
Ed.: M. Manson

We extract *fos* and *myc* proteins from cells and tissues by boiling the biological material in a solution of 1% sodium dodecyl sulfate (SDS) containing a disulfide-reducing agent and protease inhibitors. The DNA liberated from the nuclei is sheared by passage through a needle or by sonication, disulfide bonds are alkylated by iodoacetamide, the sample is diluted tenfold into a solution containing 1% Nonidet-P40 (NP40), and insoluble material is removed by centrifugation *(5,6)*.

The extract is added to microtiter plate wells containing adsorbed anti-*fos* or anti-*myc* peptide antibodies. Captured *fos* or *myc* proteins are then recognized by a second anti-*fos* or anti-*myc* antibody raised against a different peptide sequence, and which has been directly coupled to alkaline phosphatase. Bound alkaline phosphatase is assayed using the AMPAK amplifier system *(7,8)*. The amplifier uses NADPH as the primary substrate, which is dephosphorylated by the bound label to produce NADH. The NADH subsequently acts catalytically in a redox cycle consisting of the two enzymes alcohol dehydrogenase and diaphorase, and the end product is a red formazan dye that is monitored in a plate-reading spectrophotometer. Coupling of two catalytic steps in this way produces a substantial amplification factor (about 100-fold over *p*-nitrophenyl phosphate) that allows high sensitivity in this type of immunoassay *(8)*. The assay is illustrated schematically in Fig. 1.

This method has been used to estimate the concentration of c-*myc* protein in a number of normal and transformed cell lines *(5,9)*; to monitor the increases in c-*myc* and c-*fos* expression that occur during mitogenic stimulation of eukaryotic cells *(5,6)*; to distinguish mitogenic pathways activated by different mitogens *(10,11)*; and to analyze levels of c-*myc* expression in solid tumors of the head and neck *(12)* and breast *(13)*.

2. Materials

1. Extraction buffer: 1% SDS, 1% aprotinin, 0.5 m*M* phenylmethylsulfonylfluoride (PMSF), and 50 m*M* dithiothreitol in Tris-buffered saline (TBS; 144 m*M* NaCl, 25 m*M* Tris, pH 7.6). Aprotinin is a protease inhibitor obtainable in solution from Sigma: the stock solution is diluted 1:100. PMSF, a protease inhibitor, is toxic and can be omitted in many instances. It is added from a 100 m*M* solution in isoamyl alcohol. Dithiothreitol is added to the extraction buffer immediately before use. For some purposes, the assay buffer is prepared at 2X normal strength by doubling the concentrations of SDS, aprotinin, and PMSF in TBS.
2. Dilution buffer: 1% NP40, 1% aprotinin, and 0.5 m*M* PMSF in TBS. Aprotinin and PMSF can probably be omitted without serious consequences.
3. Iodoacetamide: a saturated solution (approx 1*M*) in water.

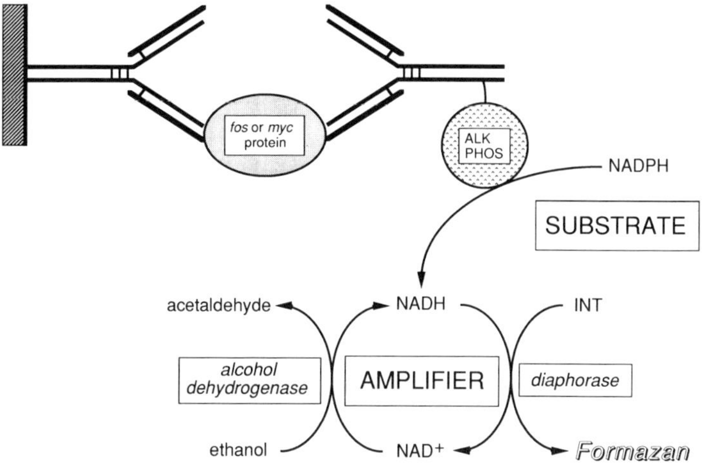

Fig. 1. A schematic representation of the twin-site ELISA for *fos* and *myc* proteins. The signal from the alkaline phosphatase label is amplified via the AMPAK™ enzyme cycle to generate the red formazan dye.

4. Needles, 19–26 gage with 0.5–1 mL syringe; or probe sonicator.
5. Immunoassay plates. Immulon II (M129B) from Dynatech Ltd. (Billingshurst, UK) and Immunoplate (F8) from Nunc Ltd. (Roskilde, Denmark) are suitable.
6. Antibodies and alkaline-phosphatase-conjugated antibodies (described in refs. *5* and *6*). The same antibodies and conjugates are available from Novo Nordisk Diagnostics Ltd. (Cambridge, UK).
7. Multichannel pipet (8 barrel).
8. Coupling buffer: 100 m*M* sodium hydrogen carbonate, pH 9.6.
9. 2% Marvel: skimmed milk powder (Cadbury's Marvel) in TBS at 2% w/v.
10. TMT: 2% Marvel, 0.5% Tween-20 in TBS.
11. AMPAK wash buffer, substrate, and amplifier reagents are available in kit form from Novo Nordisk Diagnostics.
12. Reagents for the protein assay.
    a. Stock Solutions: 0.8*M* NaOH; 10% SDS; 0.15% sodium deoxycholate; 72% trichloroacetic acid; 20% Na_2CO_3. Cu solution: 0.5% $CuSO_4.5H_2O$, 1% sodium citrate.$2H_2O$ (store dark, refrigerated); Folin-Ciocalteu reagent (Merck Art 9001).
    b. Working Solutions: CTC Solution: 2 parts Cu solution; 5 parts 20% Na_2CO_3; 3 parts water. Solution A: 1 part 0.8*M* NaOH; 1 part 10% SDS; 1 part CTC solution (as above); 1 part water. Solution B: 1 part Folin-Ciocalteu; 5 parts water.

3. Methods

3.1. Extraction

1. Remove adherent cells rapidly from the substratum with a scraper. This is best done at 4°C to slow cellular metabolism. Spin the cells rapidly into a pellet (e.g., 3 min, 250*g*), remove the medium and dry the tube with, for example, tissue wrapped around a Pasteur pipet. Resuspend the cell pellet in extraction buffer. Suspension-grown cells are pelleted directly from solution and treated in the same way (*see* Notes 1 and 2).
2. For solid tissue such as that obtained from tumors, rapidly grind to a powder while frozen (liquid nitrogen or –70°C), then lyophilize on a freeze-dryer. Redissolve the residue in extraction buffer (*see* Notes 1 and 2).
3. Transfer the extract into an Eppendorf tube, puncture the lid with a needle, and boil the extract for 5 min.
4. Shear the DNA. This can be done by passage through a fine-gage needle (26-gage) although preliminary passage through a 23-gage needle may be necessary if the solution is very viscous. Alternatively, 30-s sonication with a probe will shear the DNA adequately.
5. Add one-tenth vol of saturated iodoacetamide (gives approx 100 m*M*), pass the solution through the same syringe and needle, and incubate the extracts for about 30 min on ice (*see* Note 3).
6. Add 9 vol of dilution buffer and rinse the syringe and needle with several passages of diluted solution (*see* Note 4). Remove any insoluble material by centrifugation (e.g., 30 s, 14,000*g*). Transfer the supernatant solution into a fresh Eppendorf tube and store at –70°C (–20°C will suffice if necessary). This sample is then used for both the ELISA (Section 3.2.) and protein assay protocols (Section 3.3.).

3.2. ELISA

1. Dilute the first antibody to 2–5 µg/mL in coupling buffer, add 100 µL to each well of the microtiter plate (*see* Note 5). Cover the plate with cling film and incubate overnight at room temperature.
2. Shake out unbound antibody and wash the wells twice with 200 µL of TBS. Add 200 µL of 2% Marvel and incubate for approx 30 min at room temperature (*see* Notes 6 and 7). Wash the wells twice with 200 µL of TBS.
3. Add 100 µL of dilution buffer to each well, followed by 100 µL of sample. Cover with cling film and incubate for 2–3 h at room temperature. During this period, *fos* or *myc* proteins are captured from solution by the adsorbed first antibody (*see* Note 7).

4. Shake out the wells and wash twice with 200 µL of TBS to remove unbound *fos* or *myc* proteins.
5. Add 100 µL of second antibody–AP conjugate diluted to approx 1 µg/mL in TMT. Incubate for 1 h at room temperature, to bind the second antibody to captured *fos/myc* protein (*see* Notes 8 and 9).
6. Shake out the conjugate and wash the wells five times with 200 µL of AMPAK wash buffer (1:20 dilution of concentrate in water; *see* Note 10).
7. Add 100 µL of AMPAK substrate solution and incubate for 1 h at room temperature (*see* Note 11).
8. Add 100 µL of AMPAK amplifier solution and incubate for a timed 5 min. Terminate the reaction with 100 µL of AMPAK stopping solution (0.3*M* H_2SO_4; *see* Note 11).
9. Read the plate at 492 nm on a standard ELISA plate reader within 1–2 h of termination (*see* Notes 12–14).

3.3. Protein Assay

1. Dilute samples to 1 mL in water as necessary to give a concentration of <100 µg/mL.
2. Make a series of suitable standards (1 mL). Use BSA and dilute in water to give standards in the range of 1–100 µg/mL.
3. To all samples and standards, add 100 µL of sodium deoxycholate and shake.
4. Stand for 10 min at room temperature, then add 100 µL of 72% trichloroacetic acid and mix.
5. Centrifuge at 3000*g* for 10 min, discard the supernatant, and drain the last drop by absorbing on paper.
6. Add 250 µL of water, then 250 µL of Solution A, vortex until fully mixed, then stand at room temperature for 10 min.
7. Add 125 µL of Solution B and mix. Stand at room temperature for 30 min.
8. Read the absorbance between 600–750 nm, either in a conventional spectrophotometer, or pipet 3 × 200 µL aliquots into the wells of a microtiter plate and use a plate reader (*see* Note 15). The color is stable for 1–2 h. The standard curve should be linear if plotted on log–log paper.

4. Notes

1. The amount of cells or tissue to extract depends on the amount of *fos* or *myc* protein in the tissue. In general, the more the better, but the practical limit is set by the amount of DNA released from the tissue by SDS. Too much DNA and the solution becomes so viscous that complete resuspension and extraction becomes impossible, as does pipeting. Thus,

the upper limit is about 5–10 × 10^6 cells/100 µL of extraction buffer, or 5 mg of tumor tissue (wet wt)/100 µL of extraction buffer.

2. It is important to resuspend the cells efficiently in the extraction buffer; this is complicated by the released DNA, but it is usually relatively straightforward to pipet the extract into an Eppendorf tube. This shears some of the DNA. Transfer all of the solution, including the foam. In many instances, it helps extraction if the cells are resuspended in (e.g., 50 µL) culture medium, then an equal volume of double-strength extraction buffer added as soon as possible.
3. DNA is most easily sheared while hot, i.e., immediately after boiling. If the extract has been inadequately resuspended in extraction buffer, fragments of coagulated tissue will block the needle. Discard these. Keep the needle and syringe on ice during the 30-min incubation with iodoacetamide, since much of the extract will still be in the barrel. It is essential for the *fos* ELISA not to add iodoacetamide while the sample is still hot. At elevated temperatures, iodoacetamide alkylates a lysine residue in a *fos* antibody binding site and destroys the epitope.
4. The function of the dilution buffer is to lower the SDS concentration to a level that will not interfere with the ELISA. The final solution should be homogeneous and not contain visible lumps of DNA. If these are present, shear again through a 26-gage needle and/or discard the lumps.
5. Use a polypropylene tube for dilution of antibody into coupling buffer and vortex vigorously (typical dilution of antibody stock is 1:1000). If antibodies (and conjugates) are in short supply, carefully plan how many wells are needed and make only the necessary volume of solution. Avoid the peripheral wells on the microplate as these give greatest variation. Maximal antibody adsorption takes only a few hours, but the plates are conveniently prepared the night before use.
6. The purpose of the milk powder is to block protein binding sites on the plates. This is often unnecessary. However, if significant first-antibody-independent absorption of *fos* or *myc* protein does occur, this can be prevented by the inclusion of 1% Marvel or BSA in the buffer during the capture stage. Detergent-containing solutions are best pipeted by taking up more than 100 µL into the tip, then ejecting a fixed 100 µL—use the twin actions of the pipet in the opposite way to the conventional method, otherwise much of the solution remains in the tip and is lost.
7. Capture of *fos* or *myc* proteins is 80–90% complete after 2 h and maximal after 3 h. For very low concentrations of *fos* or *myc* proteins, a prolonged incubation period *may* increase sensitivity. Adding 200 µL of extract instead of 100 µL plus 100 µL of dilution buffer increases sensitivity, but

not by twofold, owing to the increased SDS concentration. There are some indications that PMSF and aprotinin slightly reduce the sensitivity of the assay: they may be omitted from the dilution buffer at this stage.

8. The quality of the conjugate is an important consideration in any immunoassay, but especially so in an assay employing a highly sensitive detection system such as the AMPAK. The best results (i.e., high activity with low nonspecific binding) are obtained with a directly linked antibody–enzyme conjugate prepared from highly purified components, using a well-characterized coupling chemistry. Additionally, the conjugate should be fractionated according to size and a working reagent prepared from fractions containing only small oligomers (dimers to tetramers). In some circumstances, better performance can be achieved using an antibody fragment (Fab, Fab', or $[Fab']_2$) rather than a whole antibody. Protocols for all these methods are given in *(14)*. If nonspecific binding occurs (*see also* Note 9), it can often be reduced by including 10% sheep or calf serum in the conjugate buffer.
9. As an alternative to a directly conjugated antibody, an unconjugated antibody may be used, followed by a subsequent incubation with an anti-IgG–AP conjugate appropriate to the species in which the second antibody was raised (e.g., antimouse IgG–AP). Obviously, this must not crossreact with the first antibody, and can only be applied if the first two antibodies are from different species. Again, the inclusion of an appropriate serum may help to minimize nonspecific interactions. This approach generally results in higher assay backgrounds than a direct conjugate, but may be acceptable if the primary antibody is scarce, or as a screening process to select an antibody for direct conjugation.
10. The washing stage after binding of AP-conjugate is crucial. Nonspecifically adsorbed conjugate is detected as readily as that specifically bound to *fos* or *myc* protein. Thus, the washing must be thorough. If using a multipipet, change the tips after every 1–2 washes to avoid carryover; avoid sucking wash-buffer into the pipet barrel; use a separate receptacle for wash buffer and add only enough for 1–2 washes at a time; empty the wells thoroughly—shake onto a pile of tissues after the initial emptying to remove the last few microliters.
11. Reconstitute AMPAK solutions as recommended by the manufacturer, and store in aliquots (5 mL) at –20°C. Increasing the period of incubation with substrate solution increases the subsequent rate of color generation by the amplifier solution. The length of the amplifier stage determines the final extent of color generation. For this reason, it is very important that all wells to be compared have amplifier added for the

same time, which can be 2–10 min depending on the sensitivity required. This is most conveniently done by adding amplifier to each row of wells with a multichannel pipet at timed 5–10 s intervals, then terminating each row of reactions with acid at the same 5–10 s intervals (e.g.) 5 min later. The sensitivity of the assay can, of course, be altered by changing the length of the substrate and amplifier incubations. In general, it is not advisable to use amplifier (second) incubations of longer than about 20 min, but additional sensitivity can be achieved by using a longer substrate (first) incubation, e.g., up to 3 h *(8)*.

12. If the optical density of a well is off-scale (i.e., >1.5 or 2.0, depending on the plate reader), dilute half the well's contents with an equal volume of water. The dynamic range of the assay can also be extended considerably by monitoring the color development in the amplification step kinetically rather than as an end point *(8)*. This, however, requires a more sophisticated plate reader than is often available. A suitable instrument for kinetic reading is the Molecular Devices (Palo Alto, CA) Vmax.
13. The absorbance of assay blanks (no added extract) should normally be in the range of 0.100–0.200. The c-*myc* ELISA can be calibrated with recombinant c-*myc* protein *(5,15)*, but this is not likely to be available routinely. The assays are therefore best calibrated for determination of *relative fos* and *myc* values by preparation of a standard curve using different amounts of that extract known or suspected to contain the most *fos* or *myc* protein. Add appropriate buffer (extraction/dilution 1:9) to make all wells in the standard curve up to the same volume. Note that the relationship between A_{492} and the amount of *fos*/*myc* in the added extract is nonlinear.
14. Samples from the same extract assayed in duplicate or triplicate should give similar A_{492} values (+/– 10%). Duplicate extracts will produce rather larger variations, the magnitude of the variations depending on the care taken in the preparation of the extracts. The occasional well or row of wells might give an anomalously high value, probably because of bad washing. These samples should be reassayed.
15. The function of the protein assay is to enable data from different extracts to be normalized as amount of *fos*/*myc*/µg protein. This controls for variation in the number of cells extracted and the efficiency of extraction and sample recovery. It is unimportant where cell suspensions or monolayers are to be compared under similar conditions, but rather important for comparison of, for example, *myc* levels in different tumor samples *(12,13)*.

Acknowledgments

This version of the protein assay was developed by Anders Wennborg, Karolinska Institute, Stockholm, to overcome inhibitory effects of detergents on protein determination. We are grateful to him for allowing us to detail the method here.

References

1. Alitalo, K., Koskinein, T. P., Saksela, K., Sistonen, L., and Winqvist, W. (1987) *Myc* oncogenes: Activation and amplification. *Biochimica et Biophysica Acta* **907,** 1–32.
2. Distel, R. J., Ro, R.-S., Rosen, B. S., Groves, D. L., and Spiegelman, B. M. (1987) Nucleoprotein complexes that regulate gene expression in adipocyte differentiation: Direct participation of c-*fos*. *Cell* **49,** 835–844.
3. Sassone-Corsi, P., Sisson, J. C., and Verma, I. M. (1988) Transcriptional autoregulation of the proto-oncogene *fos*. *Nature* **334,** 314–319.
4. Heikkila, R., Schwab, G., Wikstrom, E., Loke, S. L., Pluznik, D. H., Watt, R., and Neckers, L. M. (1987) A c-*myc* antisense oligodeoxynucleotide inhibits entry into S phase but not progress from G_0 to G_1. *Nature* **328,** 445–449.
5. Moore, J. P., Hancock, D. C., Littlewood, T. D., and Evan, G. I. (1987) A sensitive and quantitative enzyme-linked immunosorbence assay for the c-*myc* and N-*myc* oncoproteins. *Oncogene Res.* **2,** 65–80.
6. Moore, J. P., Littlewood, T. D., Hancock, D. C., and Evan, G. I. (1988) A sensitive enzyme-linked immunosorbence assay for c-*fos* and v-*fos* oncoproteins. *Biochimica et Biophysica Acta* **964,** 60–67.
7. Bates, D. L. (1987) Enzyme amplification in diagnostics. *Trends Biotechnol.* **5,** 204–209.
8. Johannsson, A. and Bates, D. L. (1988) Amplification by second enzymes, in *ELISA and Other Solid Phase Immunoassays* (Kemeny, D. M. and Challacombe, S. J., eds.), John Wiley, NY, pp. 85–106.
9. Erisman, M. D., Scott, J. K., Watt, R. A., and Astrin, S. M. (1988) The c-*myc* protein is constitutively expressed at elevated levels in colorectal carcinoma cell lines. *Oncogene* **2,** 367–378.
10. Mehmet, H., Sinnett-Smith, J. W., Moore, J. P., Evan, G. I., and Rosengurt, G. (1988) Differential induction of c-*fos* and c-*myc* by cyclic AMP in Swiss 3T3 cells: Significance for the mitogenic response. *Oncogene Res.* **3,** 281–286.
11. Mehmet, H., Moore, J. P., Sinnett-Smith, J. W., Evan, G. I., and Rosengurt, G. (1989) Dissociation of c-fos induction from protein kinase C-independent mitogenesis in Swiss 3T3 cells. *Oncogene Res.* **4,** 215–227.
12. Field J. K., Spandidos, D. A., Stell, P. M., Vaughan, E. D., Evan, G. I., and Moore, J. P. (1989) Elevated expression of the c-*myc* oncoprotein correlates with poor prognosis in head and neck squamous cell carcinoma. *Oncogene Res.* **4,** 1463–1468.
13. Spandidos D. A., Field, J. K., Agnatis, N. J., Evan, G. I., and Moore, J. P. (1989) High levels of c-myc protein in human breast tumours determined by a sensitive ELISA technique. *Anticancer Res.* **9,** 821-826.
14. Ishikawa, E., Imagawa, M., Hashida, S., Yoshitake, S., Hamaguchi, Y., and Ueno, T. (1983) Enzyme-labelling of antibodies and their fragments for enzyme immunoassay and histochemical staining. *J. Immunoassay* **4,** 209–327.
15. Watt, R., Shatzman, A. R., and Rosenberg, M. (1985) Expression and characterization of the human c-*myc* DNA binding protein. *Mol. Cell. Biol.* **5,** 448–456.

CHAPTER 31

Preparation of Cytotoxic Antibody–Toxin Conjugates

Alan J. Cumber and Edward J. Wawrzynczak

1. Introduction

Conjugates of antibodies with plant toxins, such as ricin and abrin, are potent cytotoxic agents that selectively eliminate target cells from mixed cell cultures in vitro, and have great promise as antitumor agents in cancer therapy *(1)*. Ricin and abrin are protein toxins consisting of two different polypeptide subunits, the A and B chains, which are of similar size (between 30 and 34 kDa) and are joined by a single disulfide bond. The A chain is a ribosome-inactivating protein (RIP) that inactivates eukaryotic ribosomes by a specific irreversible covalent modification of the ribosomal RNA *(2)*. The B chain binds to cell surface galactose-containing oligosaccharide residues. Following receptor-mediated endocytosis of toxin bound to the cell surface, the A chain gains access to the cytosol and destroys the ability of the cell to make protein *(3)*.

Antibody–toxin conjugates prepared with intact toxins are invariably cytotoxic, and their preparation and use requires special care owing to the hazardous nature of the toxins *(4)*. An alternative way to construct conjugates is to couple the isolated A chain directly to the antibody. This is conveniently achieved by using the single free sulfhydryl group of the A chain, which is revealed when the B chain is removed. Such conjugates, also referred to as immunotoxins, retain the catalytic activity of the A chain and possess a target specificity conferred solely by the antibody component. Many constructs of this type are highly potent and specific cytotoxins *(5)*.

From: *Methods in Molecular Biology, Vol. 10: Immunochemical Protocols*
Ed.: M. Manson ©1992 The Humana Press, Inc., Totowa, NJ

The preparation of cytotoxic antibody-A chain conjugates using chemical crosslinking methods has two main requirements. First, the number of crosslinkers introduced per antibody molecule should be low. This minimizes the risk that the antigen-binding capability of the antibody will be compromised by modification. Secondly, the subsequent reaction of the derivatized antibody with the A chain should lead to the formation of a disulfide bond between the two components. This disulfide linkage is necessary for maximal expression of cytotoxic activity by the conjugate *(6)*.

The most commonly used crosslinking reagent is *N*-succinimidyl 3-(2-pyridyldithio)propionate (SPDP). This heterobifunctional agent attaches covalently to the antibody at lysyl ε-amino groups introducing an *S*-pyridyl group that is attached to the linker by a disulfide bond *(7)*. In the conjugation step, the derivatized antibody is mixed with an excess of freshly reduced A chain. The sulfhydryl group of A chain molecules displaces *S*-pyridyl groups with concomitant formation of a disulfide bond between A chain and antibody. A useful feature of the SPDP reagent is the formation of pyridine-2-thione, which is a chromophore, on release of the *S*-pyridyl group. Therefore, the level of derivatization and the degree of reaction can be measured spectrophotometrically.

RIPs, which act in a fashion identical to the toxin A chains and are similar in size (about 30 kDa), also occur naturally in plants as single chain polypeptides *(8)*. A variety of RIPs of this type, including gelonin and momordin, have been used to make cytotoxic conjugates *(9,10)*. The single chain RIPs differ from the toxin A chains and from one another in primary structure, isoelectric point, and glycosylation. They also lack cysteinyl residues because they are not associated with the equivalent of a cell-binding toxin B chain. For conjugation to antibody, such RIPs must first be reacted with SPDP. Subsequently, free sulfhydryl groups are revealed by treating the derivatized RIP with a reducing agent to detach the *S*-pyridyl group.

The methods described below for the preparation of antibody–toxin conjugates containing A chains isolated from toxins or single-chain RIPs are generally applicable to the synthesis of conjugates using any type of antibody and all known RIPs of plant origin.

2. Materials

1. Antibody solution in PBSE *(see* Step 6 *below)* containing between 10 and 15 mg, at a concentration of 5–10 mg/mL (*see* Notes 6–8).
2. Solution of toxin A chain or RIP in PBSE, containing between 5 and 7.5 mg, at a concentration of about 1.5 mg/mL (*see* Notes 1–5).
3. *N*-Succinimidyl 3-(2-pyridyldithio)propionate (SPDP) (Pharmacia).
4. Dimethylformamide (DMF), (Sequanal grade, Pierce).

5. Dithiothreitol (DTT), (Sigma).
6. Phosphate/saline buffer (PBSE): 28.4 g of Na_2HPO_4, 11.7 g of NaCl, 0.744 g of EDTA disodium salt. Adjust to pH 7.5 with 1*M* HCl and make up to a final vol of 2 L in distilled water.
7. Acetate/saline buffer: 16.4 g of CH_3COONa, 11.7 g of NaCl, 0.744 g EDTA disodium salt. Adjust to pH 4.5 with 1*M* HCl and make up to a final vol of 2 L in distilled water.
8. Column chromatography apparatus, including a suitable low-pressure pump, on-line UV monitoring at 280 nm and a fraction collector.
9. Chromatography media: Sephacryl S-200 (SF), Sephadex G-25 (SF) (Pharmacia).
10. Chromatography columns:
    a. Sephacryl S-200 (SF); dimensions, 80 cm × 1.6 cm internal diameter (id); bed vol, 160 mL; equilibrated with PBSE (for purification of antibody and antibody–toxin conjugate).
    b. Sephadex G-25 (SF); dimensions, 30 cm × 1.6 cm (id); vol, 60 mL; equilibrated with PBSE (for purification of derivatized antibody).
    c. Sephadex G-25 (SF); dimensions, 30 cm × 1.6 cm (id); vol, 60 mL; equilibrated with acetate/saline buffer (for purification of derivatized RIP).
    d. Sephadex G-25 (SF); dimensions, 45 cm × 1.6 cm (id); vol, 90 mL; equilibrated with PBSE (for purification of reduced RIPs).
11. N_2 cylinder.
12. Ultrafiltration cell (10 mL vol) containing a suitable membrane with a cut-off of M_r >10,000 for globular proteins (PM10 membrane, Amicon).
13. Low protein binding 0.22 µm filtration units (Millex GV, Millipore).

3. Methods

3.1. Derivatization of Antibody

1. Apply the antibody solution at a concentration between 5 and 10 mg/mL to the S-200 column (a) and elute with PBSE at a flow rate of between 10 and 20 mL/h as a preliminary purification step (*see* Notes 6–8).
2. Pool all the fractions contained within the main antibody peak, i.e., corresponding to an M_r of 150,000 relative to protein standards. Discard any flanking fractions that may contain aggregated protein or contaminants of lower mol wt.
3. Concentrate the antibody solution to about 10 mg/mL in the ultrafiltration cell under N_2 pressure and transfer the solution to a suitable reaction vessel holding 10 mL of solution.

4. Make up a fresh solution of SPDP at 2 mg/mL in DMF in a glass vessel (*see* Note 9).
5. To 1 mL of the antibody solution at 10 mg/mL (a total of 67 nmoles), slowly add 42 µL of the SPDP solution (a total of 0.27 µmoles, i.e., a fourfold molar excess of reagent added) while stirring rapidly. Once the addition is complete, stir the mixture gently for 30 min at room temperature (*see* Note 10).
6. Apply the reaction mixture to the G-25 column (b) and elute with PBSE at a flow rate of 30 mL/h to remove low mol wt byproducts of the reaction. Collect the protein peak that elutes at the void vol of the column (total vol, about 7 mL).
7. Remove 0.5 mL of the pooled protein solution to measure the loading of *S*-pyridyl groups on the antibody (*see* Notes 11 and 12).
8. Concentrate the remainder of the derivatized antibody solution to a final vol of about 1 mL by ultrafiltration. Store in the ultrafiltration cell at 4°C in preparation for the conjugation procedure.

3.2. Derivatization of Single-Chain RIPs

1. Make up a fresh solution of SPDP at 2 mg/mL in DMF.
2. To a solution (about 3 to 5 mL) of the single-chain RIP in PBSE containing 7 mg (a total of 0.22 µmol), slowly add 67 µL of the SPDP solution (a total of 0.43 µmol, i.e., a twofold molar excess of reagent added) while stirring rapidly. Once the addition is complete, stir gently for 30 min at room temperature.
3. Apply the reaction mixture to the G-25 column (c) and elute with acetate/saline buffer at a flow rate of 30 mL/h to remove the low mol wt byproducts of the reaction (*see* Note 13). Collect the protein peak that elutes at the void vol of the column (total vol, about 5 mL).
4. Remove 0.25 mL of the pooled RIP solution to measure the extent of modification with SPDP (*see* Note 14). Use the remainder of the derivatized RIP preparation for conjugation to the antibody.

3.3. Conjugation

1. Prepare a fresh solution of 1*M* DTT in PBSE.
2. To the solution of toxin A chain or derivatized single-chain RIP, add DTT solution to a final concentration of 50 m*M*. Leave the solution for 30 min at room temperature.
3. Apply the solution of reduced RIP to the G-25 column (d) and elute with PBSE at a flow rate of 30 mL/h to remove the excess DTT and low mol wt byproducts of the reaction (*see* Notes 15 and 16). Collect the protein that elutes at the void vol of the column (total vol, about 7 mL).

4. Determine the protein concentration of the RIP solution by measuring the optical density of the solution at 280 nm.
5. Add the appropriate vol of the freshly reduced RIP solution that contains a total of 5 mg of RIP to the solution of derivatized antibody in the ultrafiltration cell. Stir gently to mix and then leave overnight at room temperature, without stirring, under a gentle stream of N_2.
6. Concentrate the reaction mixture, with gentle stirring, to about 3.5 mL and leave for between 4 and 6 h at room temperature under a gentle stream of N_2 (*see* Note 17).
7. Pass the reaction mixture through a 0.22 µm filtration unit.
8. Apply the filtered reaction mixture to the S-200 column (a) and elute with PBSE at a flow rate of between 10 and 20 mL/h to separate the antibody–toxin conjugate from the excess of the RIP and the pyridine-2-thione released during the reaction. Collect fractions of approx 2 mL in vol (*see* Notes 18 and 19).
9. Remove samples from the column fractions making up the broad band containing conjugate and unconjugated antibody, and subject to SDS-PAGE in the absence of reducing agent to determine their content (*see* Note 20).
10. Pool fractions according to the determined content of conjugate and unconjugated antibody (*see* Note 21).
11. Pass the final conjugate preparation through a 0.22 µm filtration unit in a sterile hood and store in suitable containers at 4°C for short-term use, or at –70°C for long-term storage following rapid freezing.

4. Notes

4.1. Toxin A-Chains and Single-Chain RIPs

1. Toxin A chains are isolated from ricin and abrin by reductive cleavage of the toxin, followed by separation of the chains. These procedures are hazardous and should not be undertaken without the proper safeguards *(4)*.
2. Commercially obtained preparations of ricin A chain and abrin A chain may require further purification to eliminate traces of contaminating toxin B chains. The simplest procedure is to pass the A chain preparation over a column of Sepharose-linked asialofetuin, to which the B chains bind avidly *(4,11)*.
3. RIPs should generally appear as a single band on SDS-PAGE. The presence of additional bands is a clear indication of impurity, and may complicate the interpretation of electrophoretic analysis of conjugate products. The exception is ricin A chain that has been isolated from the native toxin. Native ricin A chain exists as a mixture of two differently

glycosylated forms giving the appearance of a doublet on SDS-PAGE: bands with apparent M_rs of about 32,000 and 34,000 occur in the approximate ratio of 2:1.

4. Caution must be exercised with the handling of ricin A chain which has a tendency to form "stringy" precipitates. The concentration of ricin A chain in solution should not exceed about 1.5 mg/mL. At higher concentration, even gentle agitation will cause aggregation. Frothing of the solution should also be avoided. Ricin A chain is much less prone to precipitation following conjugation to antibody.
5. The single-chain RIPs are generally more stable to a wider range of experimental conditions than ricin A chain. However, loss of material can occur in the course of procedures based on selective passage through membranes. Ultrafiltration may be performed using a membrane with a lower M_r cut-off.

4.2. Derivatization of Antibody

6. The method described has been used to prepare antibody–toxin conjugates containing mouse and rat monoclonal antibodies of various IgG subtypes and polyclonal antibody from several animal species. Modifications of this method have also been used to prepare conjugates with antibodies of other classes.
7. The starting antibody preparation should be as highly purified as possible, e.g., by using ion-exchange chromatography, affinity chromatography, or protein A chromatography, as appropriate. The gel filtration of the antibody on Sephacryl S-200 before derivatization removes protein aggregates that are frequently present even in the most carefully prepared antibody samples, and any low mol wt contaminants that may interfere with the subsequent derivatization reaction. Precise knowledge of the elution position of the antibody at this step also assists in the analysis of the chromatographic profile of the conjugation reaction mixture *(see below)*.
8. The method is readily adapted for the preparation of conjugates starting with between 20 and 50 mg of antibody by using columns of the same length but increased id (2.6 cm) and an ultrafiltration cell with 50 mL capacity. For preparations using less than 5 mg of antibody, HPLC gel filtration apparatus is more suitable for purification purposes.
9. SPDP is very stable when stored dry. As a solution in DMF, it is hydrolyzed only slowly, provided that moisture is rigorously excluded from the solution. It is advisable to prepare this solution freshly before each conjugation.

10. The *S*-pyridyl group on the derivatized antibody is also relatively stable. In practice, the conjugation procedure can be completed comfortably within a day. If this proves to be impossible, the antibody solution may be stored under nonreducing conditions for a few days at 4°C with little loss of *S*-pyridyl groups.
11. The level of substitution of the antibody after reaction with SPDP is determined spectrophotometrically. A sample of the derivatized antibody is treated with DTT at a final concentration of 5 m*M*, and the optical density measured at 280 nm and 343 nm. The pyridine-2-thione released by reduction has a molar extinction coefficient at 343 nm of $8.08 \times 10^3 M^{-1}$ cm^{-1} *(12)*. This product also absorbs at 280 nm with a molar extinction coefficient of $5.1 \times 10^3 M^{-1}$ cm^{-1}. The true protein absorbance is determined from the formula:

$$A_{280}(\text{protein}) = A_{280}(\text{observed}) - (\text{C} \times 5.1 \times 10^3)$$

where C is the molar concentration calculated for the pyridine-2-thione from the absorbance at 343 nm. The molar extinction coefficient for antibody at 280 nm is $2.1 \times 10^5 M^{-1}$ cm^{-1}.

The sample used for this analysis must be discarded and not returned to the bulk of the derivatized antibody because of the presence of the added DTT.

12. Using the conditions described, i.e., a fourfold molar excess of SPDP over antibody, the level of derivatization should be between 1.5 and 2.0 *S*-pyridyl groups per antibody molecule on average. At higher levels of modification, there is an increased risk of inactivating or precipitating the antibody.

4.3. Derivatization of Single-Chain RIPs

13. The acetate/saline buffer was originally used for the column chromatography of the derivatized RIP because the *S*-pyridyl group is easily displaced by sulfhydryl reagents under the conditions of low pH at which protein disulfide bonds are relatively stable. Most single-chain RIPs are stable to these conditions.
14. The level of substitution of the single-chain RIPs obtained using the method described should be close to 1.0 *S*-pyridyl group per RIP molecule on average. Slight impairment of the enzymic function of several RIPs has been reported as a result of derivatization with SPDP. Alternative procedures for the introduction of sulfhydryl groups using different crosslinkers (e.g., 2-iminothiolane) may be used to circumvent this problem *(8,13)*.

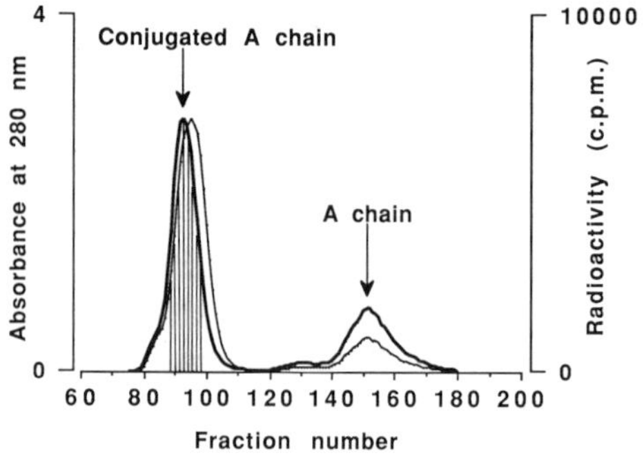

Fig. 1. Gel permeation chromatography of an antibody–toxin conjugate reaction mixture. The reaction mixture obtained following the conjugation of a mouse monoclonal antibody (50 mg) and [^{125}I]-labeled abrin A chain was chromatographed on a column of Sephacryl S-200 (SF), dimensions: 80 cm × 2.6 cm (id). Fractions eluting from the column were monitored spectrophotometrically at 280 nm (—) to measure total protein, and by gamma counting (▬) to measure the A chain in its free or conjugated form. The hatched area indicates a typical pooled conjugate preparation.

4.4. Conjugation

15. It is essential that all traces of DTT should be removed from the RIP after the reduction step. The presence of even very low concentrations of DTT at this stage will release *S*-pyridyl groups from the antibody in preference to the formation of conjugate. For this reason, the G-25 column should be rigorously cleaned before use. EDTA in the column buffer inhibits disulfide bond formation, catalyzed by trace amounts of metal ions.
16. The number of reactive sulfhydryl groups following reduction and chromatography can be measured by mixing a sample of the RIP with Ellman's reagent. Reaction of sulfhydryl groups with Ellman's reagent leads to the quantitative release of the 3-carboxylato-4-nitrothiophenolate anion, a chromophore with a molar extinction coefficient of $1.36 \times 10^4 M^{-1}$ cm^{-1} at 412 nm *(14)*.

    The sample used for this analysis should be discarded. The remaining RIP solution should then be added to the derivatized antibody solution without delay.
17. The final reaction mixture should not be stirred overnight to avoid the risk of precipitating the RIP. The course of the reaction between the derivatized antibody and the RIP can be followed by monitoring the release of pyridine-2-thione spectrophotometrically. The reaction mixture can be stored at 4°C for several days without deleterious effect.

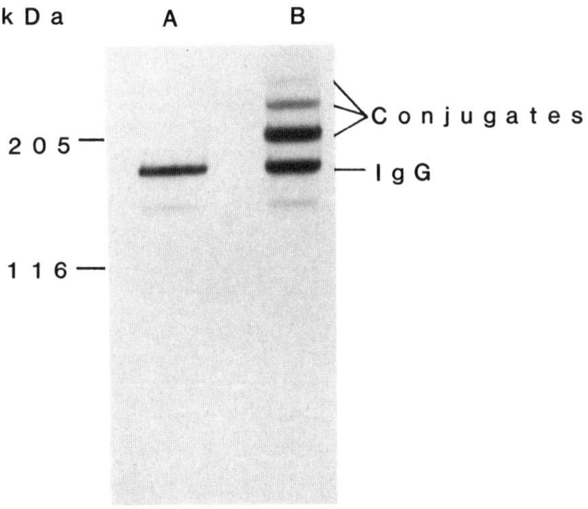

Fig. 2. SDS-PAGE of an antibody–toxin conjugate preparation. **A.** Antibody (starting material). **B.** Abrin A chain conjugate (pooled fractions shown in Fig. 1). Samples were prepared under nonreducing conditions and run on a 2–27% gradient polyacrylamide gel.

18. The conjugate product consists of a mixture of antibody molecules crosslinked to one or more RIP molecules. Using the conditions described, the 1:1 conjugate is obtained in the highest yield. On S-200 column chromatography, the 1:1 conjugate elutes at a position well separated from unreacted RIP and released pyridine-2-thione. Any aggregated protein and the multiple-substituted conjugates elute near the void volume of the column. This procedure does not efficiently separate the 1:1 conjugate from unconjugated antibody. (*See* Fig. 1 and also Chapter 32).
19. In the case of ricin A chain, a proportion of unreacted RIP binds irreversibly to the S-200 column during the final purification. The column can be cleaned by washing with 0.1*M* NaOH before further chromatography.
20. The pooled conjugate preparation contains the 1:1 conjugate as the major product and unconjugated antibody as the major contaminant. Smaller amounts of multiple-substituted conjugates are also present. The composition of the conjugate preparation can be determined by densitometric analysis of the pattern of bands on SDS-PAGE before and after reduction of the sample (Figs. 2 and 3). An alternative procedure involves trace radiolabeling of the RIP component before conjugation to allow a precise measurement of the RIP content of the conjugate *(15)*.

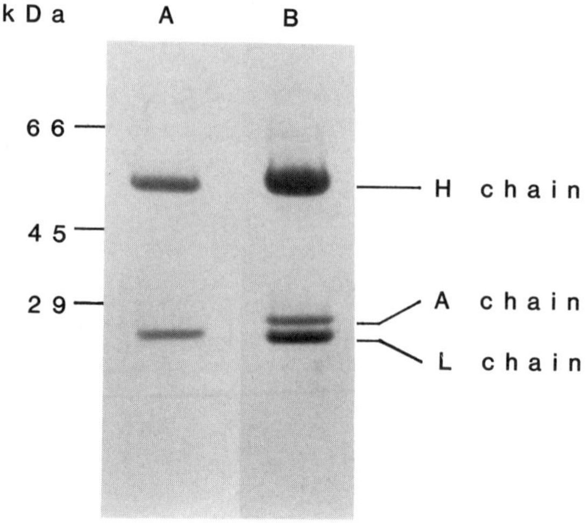

Fig. 3. SDS-PAGE of reduced antibody–toxin conjugate. **A.** Antibody (starting material). **B.** Abrin A chain conjugate (as in Fig. 2). Samples were prepared in the presence of 5% w/v DTT and run on a 2–27% gradient polyacrylamide gel.

21. Antibody–toxin conjugates made with ricin A chain, abrin A chain, gelonin, and momordin can be stored for at least 4 yr at –70°C without detectable loss of activity. The bond between the antibody and the RIP breaks down very slowly at 4°C in PBSE but, provided that care is taken to ensure the sterility of the solution, conjugates can be stored under these conditions for up to one year with little deterioration in quality.

References

1. Blakey, D. C., Wawrzynczak, E. J., Wallace, P. M., and Thorpe, P. E. (1988) Antibody toxin conjugates: A perspective, in *Monoclonal Antibody Therapy* (Waldmann, H., ed.), Karger, Basel, pp. 50–90.
2. Endo, Y. (1988) Mechanism of action of ricin and related toxins on the inactivation of eukaryotic ribosomes, in *Immunotoxins* (Frankel, A. E., ed.), Kluwer Academic Publishers, Boston, pp. 75–89.
3. Olsnes, S. and Sandvig, K. (1988) How protein toxins enter and kill cells, in *Immunotoxins* (Frankel, A. E., ed.), Kluwer Academic Publishers, Boston, pp. 39–73.
4. Cumber, A. J., Forrester, J. A., Foxwell, B. M. J., Ross, W. C. J., and Thorpe, P. E. (1985) Preparation of antibody-toxin conjugates. *Methods Enzymol.* **112,** 207–225.
5. Thorpe, P. E. (1985) Antibody carriers of cytotoxic agents in cancer therapy: A review, in *Monoclonal Antibodies '84: Biological and Clinical Applications* (Pinchera, A., Doria, G., Dammacco, F., and Bargellesi, E., eds.), Editrice Kurtis s. r. l., Milan, pp. 475–506.
6. Wawrzynczak, E. J. and Thorpe, P. E. (1988) Effect of chemical linkage upon the

stability and cytotoxic activity of A chain immunotoxins, in *Immunotoxins* (Frankel, A. E., ed.), Kluwer Academic Publishers, Boston, pp. 239–251.
7. Carlsson, J., Drevin, H., and Axen, R. (1978) Protein thiolation and reversible protein-protein conjugation. *N*-succinimidyl 3-(2-pyridyldithio)propionate, a new heterobifunctional crosslinking reagent. *Biochem. J.* **173,** 723–737.
8. Lambert, J. M., Blattler, W. A., McIntyre, G. D., Goldmacher, V. S., and Scott, C. F., Jr. (1988) Immunotoxins containing single chain ribosome-inactivating proteins, in *Immunotoxins* (Frankel, A. E., ed.), Kluwer Academic Publishers, Boston, pp. 175–209.
9. Thorpe, P. E., Brown, A. N. F., Ross, W. C. J., Cumber, A. J., Detre, S. I., Edwards, D. C., Davies, A. J. S., and Stirpe, F. (1981) Cytotoxicity acquired by conjugation of an anti-Thy1.1 monoclonal antibody and the ribosome-inactivating protein, gelonin. *Eur. J. Biochem.* **116,** 447–454.
10. Stirpe, F., Wawrzynczak, E. J., Brown, A. N. F., Knyba, R. E., Watson, G. J., Barbieri, L., and Thorpe, P. E. (1988) Selective cytotoxic activity of immunotoxins composed of a monoclonal anti-Thy1.1 antibody and the ribosome-inactivating proteins bryodin and momordin. *Br. J. Cancer* **58,** 558–561.
11. Fulton, R. J., Blakey, D. C., Knowles, P. P., Uhr, J. W., Thorpe, P. E., and Vitetta, E. S. (1986) Purification of ricin A_1, A_2, and B chains and characterization of their toxicity. *J. Biol. Chem.* **261,** 5314–5319.
12. Stuchbury, T., Shipton, M., Norris, R., Malthouse, J. P. G., Brocklehurst, K., Herbert, J. A. L., and Suschitsky, H. (1975) A reporter group delivery system with both absolute and selective specificity for thiol groups and an improved fluorescent probe containing the 7-nitrobenzene–2-oxa–1,3-diazole moiety. *Biochem. J.* **151,** 417–432.
13. Wawrzynczak, E. J. and Thorpe, P. E. (1987) Methods for preparing immunotoxins: Effect of the linkage on activity and stability, in *Immunoconjugates: Antibody Conjugates in Radioimaging and Therapy of Cancer* (Vogel, C.-W., ed.), Oxford University Press, New York and Oxford, pp. 28–55.
14. Ellman, G. L. (1959) Tissue sulphydryl groups. *Arch. Biochem. Biophys.* **82,** 70–77.
15. Thorpe, P. E. and Ross, W. C. J. (1982) The preparation and cytotoxic properties of antibody-toxin conjugates. *Immunol. Rev.* **62,** 119–158.

CHAPTER 32

Immunoaffinity Purification and Quantification of Antibody–Toxin Conjugates

Edward J. Wawrzynczak and Alan J. Cumber

1. Introduction

Cytotoxic antibody–toxin conjugates made using antibodies and ribosome-inactivating proteins (RIPs) are prepared using chemical crosslinking methods (*1,2* and this vol., Chapter 31). Gel permeation chromatography is used as a first step to purify conjugate molecules from the reaction mixture. This procedure removes protein aggregates, the excess of RIP employed in the conjugation reaction, and low molecular weight byproducts. However, a significant fraction of the resulting conjugate preparation consists of unconjugated antibody that cannot be completely separated from the conjugate on the basis of size discrimination alone (*see* Chapter 31).

The efficient separation of some antibody–toxin conjugates from unconjugated antibody can be achieved using methods exploiting the physicochemical properties of the RIP. Separation by ion-exchange chromatography is possible in cases where the RIP and the antibody have sufficiently distinct isoelectric points *(3)*. Affinity chromatography on Blue Sepharose CL-6B can be used to purify conjugates containing ricin A chain or abrin A chain *(4)*. However, these methods are not certain to succeed with different antibody-RIP combinations because the chromatographic properties of RIPs may be altered following their attachment to antibody *(5)*.

From: *Methods in Molecular Biology, Vol. 10: Immunochemical Protocols*
Ed.: M. Manson ©1992 The Humana Press, Inc., Totowa, NJ

An alternative approach to purification is immunoaffinity chromatography using RIP-specific antibody immobilized on a column matrix. Affinity chromatography using soft gel matrices leads to poor recoveries and large sample dilution. An effective high performance immunoaffinity chromatography procedure that purifies antibody–toxin conjugates free from contaminating antibody and overcomes the problems associated with soft gel chromatography is described below. This generally applicable method is based on the selective binding of antibody-conjugated RIP molecules to affinity-purified anti-RIP antibody immobilized on a silica matrix.

The concentration of antibody–toxin conjugates made with mouse monoclonal IgG antibodies and a number of different RIPs can be accurately measured, even in the presence of free antibody and serum proteins by means of a highly sensitive enzyme-linked immunosorbent assay (ELISA). The noncompetitive indirect ELISA procedure presented below uses immobilized affinity-purified rabbit anti-RIP antibody to capture conjugate molecules by binding to the RIP component. The conjugate bound to the anti-RIP antibody is then detected using an enzyme-linked reagent that specifically recognizes the mouse antibody component of the conjugate.

2. Materials

2.1. Production of RIP-Specific Antiserum

1. New Zealand white rabbits.
2. Filtration units: 0.45 μm (Millex HA, Millipore); 0.22 μm (Millex GV, Millipore).
3. Dulbecco's phosphate-buffered saline, solution A (PBSA): 0.2 g of KCl, 0.2 g of KH_2PO_4, 8.0 g of NaCl, 2.16 g of $Na_2HPO_4 \cdot 7\ H_2O$ in 1 L of distilled, deionized water. Sterile.
4. Freund's complete adjuvant.
5. RIP solution: 1 mg in PBSA.
6. Sodium azide. CARE: TOXIC.

2.2. Affinity Purification of Anti-RIP Antibody

7. CNBr-activated Sepharose 4B, (Pharmacia).
8. Low protein binding 0.22 μm filtration units (Millex GV, Millipore).
9. Glass sinter (No. 1).
10. 1*M* ethanolamine solution in distilled water, 50 mL.
11. 1 m*M* HCl (1 L).
12. Bicarbonate buffer: 8.4 g of $NaHCO_3$, 29.2 g of NaCl. Make up to a final vol of 1 L in distilled water. The pH should be 8.3.

13. Acetate/saline buffer: 8.2 g of CH_3COONa, 29.2 g of NaCl. Adjust to pH 4 with CH_3COOH and make up to a final vol of 1 L in distilled water.
14. Borate/saline buffer: 6.18 g of H_3BO_3, 29.2 g of NaCl. Adjust to pH 8.0 with 1*M* NaOH and make up to a final vol of 1 L in distilled water.
15. Phosphate/saline buffer: 1.42 g of Na_2HPO_4, 8.53 g of NaCl. Adjust to pH 7.4 with 1*M* HCl and make up to a final vol of 1 L in distilled water.
16. Saline solution: 8.53 g of NaCl in 1 L of water.
17. Eluting solution: 71.2 g of $MgCl_2 \cdot 6H_2O$ in 100 mL of distilled water.
18. RIP solution: 10–20 mg in 10 mL of bicarbonate buffer.

2.3. High Performance Immunoaffinity Chromatography

19. High performance liquid chromatography (HPLC) apparatus (isocratic system) fitted with a 2-mL sample loop.
20. Ultraffinity-EP HPLC column; dimensions, 50 mm × 4.6 mm id, (Beckman).
21. Loading buffer: 78.0 g of $NaH_2PO_4.2H_2O$. Adjust to pH 6.8 with 1*M* NaOH and make up to a final vol of 1 L in distilled water.
22. Running buffer: 2.84 g of Na_2HPO_4, 11.7 g of NaCl. Adjust to pH 6.8 with 1*M* HCl and make up to a final vol of 1 L in distilled water. This buffer should be degassed before use.
23. Eluting buffer: 0.751 g of glycine. Adjust to pH 2.5 with 1*M* HCl and make up to a final vol of 100 mL in distilled water.
24. Neutralizing buffer: 1*M* Tris-HCl, pH 7.5 (100 mL).
25. Affinity-purified rabbit anti-RIP antibody solution: 6 mg in 6 mL of loading buffer.
26. RIP solution: 1 mg in 1 mL of running buffer.

2.4. Enzyme-Linked Immunosorbent Assay

27. Flat-bottomed microtiter plates (96-well) treated to enhance protein binding (Immulon2, Dynatech).
28. Humidified chamber, e.g., a sandwich box containing a wet tissue.
29. Sheep antimouse immunoglobulin–horseradish peroxidase (SAMIg-HRP), (Amersham).
30. Casein (hammersten grade), (BDH).
31. *O*-phenylenediamine (OPD). MUTAGENIC: HANDLE WITH CARE.
32. Thimerosal. HIGHLY TOXIC: HANDLE WITH CARE.
33. Conc.HCl.
34. 12.5% v/v H_2SO_4 solution in distilled water. Make up by adding conc.H_2SO_4 slowly to excess water.

35. 100 vol H_2O_2 solution.
36. PBSA (*see* Section 2.1.3.).
37. Carbonate/bicarbonate buffer: 0.159 g of Na_2CO_3, 0.294 g of $NaHCO_3$. Adjust the pH to 9.6 and make up to 100 mL in distilled water. Make up the solution freshly before each assay.
38. Casein buffer: 25 g of casein, 6.05 g of Tris base, 45 g of NaCl, 1 g of thimerosal in 5 L of distilled water. Warm the buffer briefly to 60°C with gentle stirring and leave stirring overnight at room temperature. Add conc.HCl to adjust the pH of the buffer to 7.6 and keep stirring until any precipitates that appear on addition of the acid have redissolved. Store at 4°C and use within one month. Warm to room temperature before use.
39. Substrate buffer: 12 mL of 0.1*M* citric acid (stock solution stored at 4°C), 13 mL of 0.2*M* Na_2HPO_4 (stock solution) and 25 mL of distilled water. Check that the pH of the solution is 5.0. Add 20 mg of OPD and shake to dissolve. Then add 10 µL of 100 vol H_2O_2 solution and mix well. Prepare this solution shortly before use and cover to protect from light.
40. Affinity-purified rabbit anti-RIP antibody solution in carbonate/bicarbonate buffer, 20–40 µg per microtiter plate.

3. Methods

3.1. Production of RIP-Specific Antiserum

1. Prepare 4 mL of a solution of RIP at a concentration of 0.2 mg/mL in PBSA. Sterilize by passing the solution through a 0.22 µm filtration unit in a sterile hood.
2. Mix 2 mL of the RIP solution with 2 mL of Freund's complete adjuvant and emulsify, using a high-speed mixer.
3. Inject the rabbit intramuscularly in the two hind-legs with 0.5 mL of the RIP emulsion per site, i.e., 50 µg of RIP per site (*see* Notes 1–3).
4. Five weeks later, administer four subcutaneous injections of 0.25 mL of the RIP solution in PBSA prepared in Step 1, i.e., 50 µg of RIP per site, to boost the antibody response of the animal.
5. One week later, bleed out the rabbit (*see* Note 4).
6. Allow the blood to clot for 1 h at room temperature in a glass container and then leave to stand overnight at 4°C to allow the clot to shrink in size.
7. Dislodge the clot from the sides of the tube and transfer the supernatant to a centrifuge tube.
8. Centrifuge the clot at 2500*g* for 30 min at 4°C. Remove any liquid carefully and combine with the supernatant from Step 7.

9. Centrifuge the pooled supernatants at 1500*g* for 15 min at 4°C to remove intact blood cells.
10. Treat the serum for 30 min at 56°C to inactivate complement and pass through 0.45 µm and 0.22 µm filtration units in succession. Dispense 5-mL aliquots into suitable containers, freeze rapidly, and store at –70°C in the presence of 0.02% w/v NaN_3.

3.2. Affinity Purification of Anti-RIP Antibody

1. Prepare a solution containing between 10 and 20 mg of RIP in about 10 mL of bicarbonate buffer. Pass the solution through a 0.22 µm filtration unit to ensure that it is free of particulate matter.
2. Swell 2 g of CNBr-activated Sepharose 4B (equivalent to about 6–7 mL of swollen gel) in 1 m*M* HCl and wash with a further 500 mL of 1 m*M* HCl on a glass sinter, removing the liquid under vacuum from a water pump.
3. Mix the acid-washed gel directly with the RIP solution prepared in Step 1 and incubate for 2 h at room temperature with gentle inversion (*see* Notes 5 and 6).
4. Remove the solution from the gel by careful filtration on a glass sinter, followed by washing of the gel with a small volume of bicarbonate buffer. Measure the optical density of the pooled filtrate at 280 nm to determine the amount of protein that has not coupled to the gel (*see* Note 7).
5. Mix the gel immediately with ethanolamine solution to block remaining active groups on the gel and incubate for 2 h at room temperature with gentle inversion.
6. Remove the ethanolamine solution by careful filtration on the glass sinter.
7. Wash the gel alternately with at least three cycles of the acetate/saline and borate/saline buffers to remove noncovalently adsorbed protein.
8. Pour the gel into a glass chromatography column and equilibrate with phosphate/saline buffer.
9. Prewash the column by pumping through one column volume of saline solution, followed by the eluting solution to remove any noncovalently adsorbed protein.
10. Reequilibrate the column with phosphate/saline buffer.
11. Apply the antiserum (*see* Note 8) to the column and elute with phosphate/saline buffer.
12. Monitor fractions from the column by measuring the optical density at 280 nm.
13. Continue eluting until the nonbinding material has passed through the column, i.e., when the optical density readings have returned to the baseline value.
14. Apply one column volume of saline solution, followed by 10 mL of eluting solution to remove the adsorbed antibody. Then reequilibrate the column with phosphate/saline buffer (*see* Note 9).

15. Dialyze the affinity-purified antibody against two changes of 2 L of phosphate/saline buffer, pass through a 0.22 μm filter, and store at 4°C in the presence of 0.02% w/v NaN_3.

3.3. High Performance Immunoaffinity Chromatography

1. Prepare a solution containing affinity purified rabbit anti-RIP antibody at a concentration of approx 1 mg/mL of loading buffer and pass through a 0.22 μm filtration unit.
2. Load the antibody solution onto the HPLC column at a flow rate of 0.5 mL/min for 5 min. Then recycle the solution through the column at a flow rate of 0.2 mL/min for 20 h at room temperature (*see* Notes 10–12).
3. Wash the column with 5 mL of the loading buffer and measure the optical density of the pooled cycling and washing solutions at 280 nm to determine the amount of antibody that has not coupled to the column.
4. Equilibrate the column with the running buffer.
5. Expose the column to several cycles of the low pH elution procedure (*see* Step 10 *below*) to remove any noncovalently adsorbed antibody.
6. Before using the column to purify antibody–toxin conjugate, apply a solution of the RIP at a concentration of 1 mg/mL to block any high affinity binding sites. Remove the RIP using the elution procedure (*see* Note 13).
7. Load the sample (1 mL) of the antibody–toxin conjugate preparation (at a concentration up to 1 mg/mL in running buffer) and elute with the running buffer at a flow rate of 0.5 mL/min.
8. Monitor fractions from the column by measuring the optical density at 280 nm.
9. Continue eluting with the running buffer until the nonbinding material has passed through the column, i.e., when the optical density readings have returned to the baseline value.
10. Apply a pulse of eluting buffer (2.0 mL) by means of the sample loop at a reduced flow rate of 0.2 mL/min to prevent trailing of the eluted conjugate peak. Collect fractions into tubes containing neutralizing buffer to minimize the exposure of the conjugate to the low pH elution conditions (*see* Note 14).
11. Dialyze the affinity-purified antibody–toxin conjugate into PBSA, filter-sterilize in a sterile hood, and store at 4°C or, freeze rapidly and store at –70°C (*see* Notes 15 and 16).

3.4. Enzyme-Linked Immunosorbent Assay

1. Prepare a solution of affinity-purified anti-RIP antibody at a concentration between 2 and 4 μg antibody/mL in carbonate/bicarbonate buffer.

Add 100 μL to each well of a microtiter plate. This step requires at least 10 mL of the solution per plate (*see* Note 17).

2. Incubate the plate overnight at 4°C in a humidified chamber to minimize evaporation.
3. Perform all subsequent steps at room temperature. Keep plates in the humidified box during incubation steps.
4. Prepare dilutions of the antibody–toxin conjugate samples with casein buffer to give an approximate concentration of between 0.5 and 5 ng of conjugated RIP/mL of buffer in each case. At least 300 μL of each dilution is required for Step 7 (*see* Note 18).
5. Wash the plate four times with casein buffer (*see* Note 19).
6. Fill the wells with casein buffer and incubate the plate for 30 min.
7. Remove the casein buffer. Add 100 μL of each diluted antibody–toxin conjugate sample (prepared in Step 4) to the plates in triplicate.
8. Incubate the plates for 2 h. Wash the plates four times with casein buffer.
9. Add 100 μL of SAMIg–HRP solution, freshly prepared by diluting the stock solution by 1 in 4000 to 1 in 10,000 with casein buffer. This step requires at least 10 mL of the solution per plate (*see* Note 20).
10. Incubate the plate for 1 h. Wash the plate four times with casein buffer, then twice with PBSA (*see* Note 21).
11. Add 100 μL of the substrate solution to each well. This step requires at least 10 mL of the solution per plate.
12. Incubate the plate for 10 min and stop the reaction by adding 50 μL of H_2SO_4 solution. This step requires at least 5 mL of the acid solution per plate (*see* Note 22).
13. Read the optical density of the dark orange/brown solution in the wells at 492 nm (*see* Note 23).
14. Calculate the concentration of antibody–toxin conjugate in sample wells by comparison with a standard curve (*see* Note 23).

4. Notes

4.1. Production of RIP-Specific Antiserum

1. Animals must only be handled by properly trained personnel in accordance with the pertinent regulations.
2. RIPs of plant origin elicit a strong antibody response in the rabbit. It is advisable to immunize two rabbits at the same time, because the antibody responses of individual animals can differ.
3. The titer of anti-RIP antibody can be determined before bleeding out the rabbit, by comparison with a sample of serum taken from the animal before immunization.

4. A single rabbit can be expected to yield about 50 mL of antiserum, which should contain at least 10 mg of RIP-specific antibody.

4.2. Affinity Purification of Anti-RIP Antibody

5. Ricin A chain in solution is prone to precipitate with even mild agitation. It is recommended that the gel-coupling procedure be performed without constant inversion.
6. For the purification of antibody raised against ricin and abrin A chains, a useful alternative to affinity chromatography on columns made with the isolated A chains is the use of immobilized *Ricinus* and *Abrus* agglutinins, which crossreact immunologically with the corresponding toxins but are much less toxic.
7. The amount of RIP coupled to CNBr-activated Sepharose 4B, calculated as the difference between the amount applied and the measured amount of RIP that has failed to couple to the gel, is generally in excess of 95%.
8. The immobilized RIP columns bind at least 2 mg of RIP-specific antibody. The precise volume of antiserum that can be applied without exceeding the binding capacity of the column must be determined by experiment in each case.
9. RIP affinity columns can be stored at 4°C in the presence of 0.02% w/v NaN_3 for several years with little loss of performance.

4.3. High Performance Immunoaffinity Chromatography

10. Using the coupling procedure described, 3–4 mg of affinity-purified anti-RIP antibody can be coupled to the HPLC column matrix. The procedure requires a minimum of 6 mL of anti-RIP antibody solution to allow cycling of the solution via the liquid reservoir, prepump solvent filter, pump, precolumn filter, the column itself, and all the connecting tubing.
11. The entire loading and pulsing cycle as described takes approx 30 min. Using an increased flow rate of 1 mL/min and a single elution pulse, at least 20 mg of antibody–toxin conjugate can be purified within a day on this size of HPLC column.
12. The capacity of immunoaffinity HPLC columns prepared using affinity-purified antibody against ricin A chain, abrin A chain, gelonin, and momordin varies between 400 and 600 µg/mL of bed vol for the appropriate RIP. In the case of the antibody–toxin conjugates made with these RIPs, the capacity is at least 2 mg/mL of bed vol.

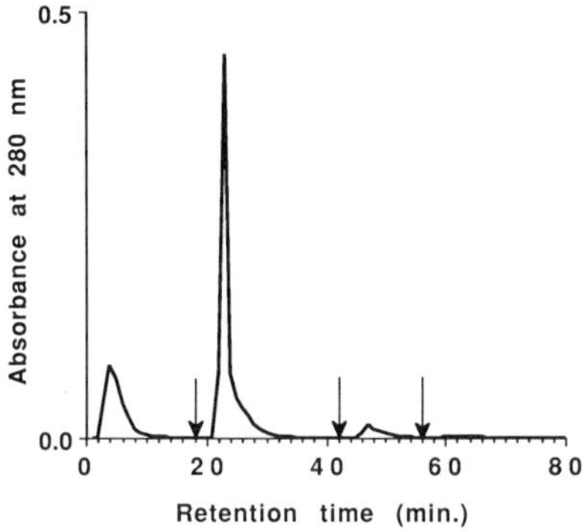

Fig. 1. High performance liquid immunoaffinity chromatography of an antibody–toxin conjugate preparation. An abrin A chain conjugate preparation obtained following Sephacryl S–200 (SF) chromatography was applied to an immunoaffinity HPLC column bearing immobilized affinity-purified rabbit anti-abrin A chain. Each arrow indicates the injection of a pulse of low pH eluting buffer. The first peak that emerges from the column in running buffer consists of unconjugated antibody only. The first elution pulse removes the majority of the bound conjugate from the column. The remainder of the bound conjugate is removed by the second pulse.

13. The ricin and abrin A chains used to block high affinity binding sites on the column should be alkylated to remove the free sulfhydryl group that could interfere with conjugate purification.
14. The elution procedure using the low pH buffer removes 85–95% of the affinity-bound conjugate. The remaining bound conjugate can be removed by a second pulse with the elution buffer (*see* Fig. 1).
15. All the antibody–toxin conjugates bind in their entirety to the appropriate immunoaffinity HPLC column and are completely separated from free antibody. There is no adverse affect on the integrity of the conjugates as judged by gel electrophoresis, size exclusion HPLC, and assays of cytotoxic potency *(5)*.
16. The anti-RIP antibody HPLC columns can be stored at 4°C in the presence of 0.02% w/v NaN_3 for several years with little loss of performance.

4.4. Enzyme-Linked Immunosorbent Assay

17. The quality of microtiter plates varies according to the supplier and the batch. It is advisable to check individual batches for their suitability and consistency in use.

18. Casein buffer is used during Steps 4–10 as a blocking agent to prevent the nonspecific adsorption of reagents to the plate, which otherwise gives rise to high background values of optical density *(6)*.
19. During washing (Steps 5 ,8, and 10), the best results are obtained by completely filling the wells of the microtiter plate with casein buffer. A plate-washing machine is recommended. It is important to flush the tubing of the machine immediately after each batch of washes with casein buffer to prevent the liquid dispensing system from becoming blocked. At the end of the washing steps, the plates should be inverted and slapped vigorously onto a pad of tissue to remove any remaining liquid from the wells.
20. The use of SAMIg–HRP that has been adsorbed against rabbit Ig to prevent crossreactivity with the rabbit anti-RIP antibody immobilized on the plate is recommended to prevent high background absorbance.
21. The final washes with PBSA (Step 10) are important to remove traces of casein that can cause turbidity on the addition of H_2SO_4 to the substrate solution.
22. The time for color development should be carefully controlled. Plates should be read soon after the color has developed. If this is impossible, store the plates in the dark to minimize the fading of color.
23. The ELISA method for measuring the concentration of antibody–toxin conjugates works successfully with conjugates containing several different types of ricin A chain, with abrin A chain, gelonin, and momordin. A sample of conjugate at a concentration of 5 ng/mL gives an optical density of about 0.5 at 492 nm using between 2 and 4 µg/mL of affinity-purified anti-RIP antibody and dilutions of SAMIg–HRP of 1 in 4000 to1 in 10,000. The optimal concentrations of these reagents must be established experimentally in each case.
24. The concentration of antibody–toxin conjugate in samples is determined from the optical density values at 492 nm, given by a standard curve. The standard curve is obtained by the same procedure using dilutions of the stock antibody–toxin solution having final concentrations between 0.125 and 20 ng/mL in casein buffer.

References

1. Cumber, A. J., Forrester, J. A., Foxwell, B. M. J., Ross, W. C. J., and Thorpe, P. E. (1985) Preparation of antibody-toxin conjugates. *Methods Enzymol.* **112,** 207–225.
2. Wawrzynczak, E. J. and Thorpe, P. E. (1987) Methods for preparing immunotoxins: Effect of the linkage on activity and stability, in *Immunoconjugates: Antibody Conjugates in Radioimaging and Therapy of Cancer* (Vogel, C.-W., ed.), Oxford University Press, New York and Oxford, pp. 28–55.

3. Lambert, J. M. and Blattler, W. A. (1988) Purification and biochemical characterization of immunotoxins, in *Immunotoxins* (Frankel, A. E., ed.), Kluwer Academic Publishers, Boston, pp. 323–348.
4. Knowles, P. P. and Thorpe, P. E. (1987) Purification of immunotoxins containing ricin A chain and abrin A chain using Blue Sepharose CL-6B. *Anal. Biochem.* **160**, 440–443.
5. Cumber, A. J., Henry, R. V., Parnell, G. D., and Wawrzynczak, E. J. (1990) Purification of immunotoxins containing the ribosome-inactivating proteins gelonin and momordin using high performance liquid immunoaffinity chromatography compared with Blue Sepharose CL-6B affinity chromatography. *J. Immunol. Methods* **135**, 15–24.
6. Kenna, J. G., Major, G. N., and Williams, R. S. (1985) Methods for reducing nonspecific antibody binding in enzyme-linked immunosorbent assays. *J. Immunol. Methods* **85**, 409–419.

CHAPTER 33

An Immuno-Slot-Blot Assay for Detection and Quantitation of Alkyldeoxyguanosines in DNA

Barbara I. Ludeke

1. Introduction

The detection and quantitation of DNA adducts plays a central role in the determination of dose-response relationships for chemical carcinogens, mutagens, and chemotherapeutic agents in experimental laboratory investigations. Furthermore, it serves in molecular dosimetry and risk estimation of humans chronically exposed to environmental carcinogens, and of cancer patients undergoing treatment with alkylating cytostatic drugs *(1,2)*. These investigations require sensitive methods that do not necessitate the administration of radioactively labeled compounds.

The immuno-slot-blot assay, a noncompetitive immunosorbent assay first described by Rajewsky and coworkers *(3)* and further developed in our laboratory *(4,5)*, offers the following advantages for the assessment of DNA adducts:

1. High sensitivity. Detection limits are in the range of 0.05–5 µmol/mol of unmodified parent base. This compares well to the sensitivities of commonly used methods, such as high pressure liquid chromatography with fluorescence detection *(6,7)*, radiochromatographic analysis *(8)*, radioimmunoassays, and enzyme-linked immunosorbent assays *(9,10)*.

From: *Methods in Molecular Biology, Vol. 10: Immunochemical Protocols*
Ed.: M. Manson ©1992 The Humana Press, Inc., Totowa, NJ

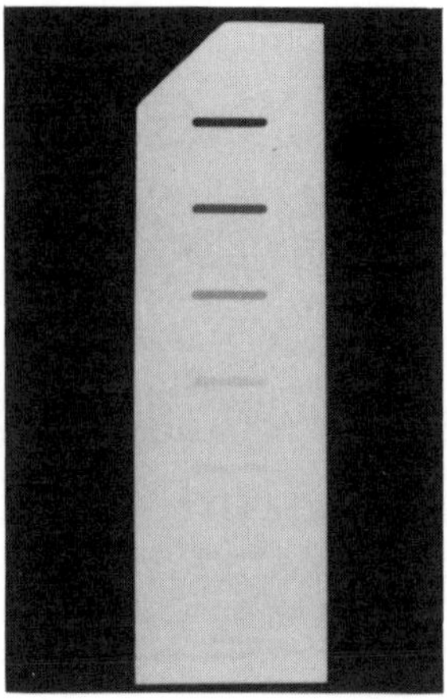

Fig. 1. An immuno-slot blot for the detection of O^6-hydroxyethyldeoxyguanosine. Calf thymus DNA was hydroxyethylated in vitro with 6.6 m*M* of hydroxyethylnitrosourea. Aliquots containing 300, 100, 33.3, 11.1, 3.7, 1.2, and 0 fmol of O^6-hydroxyethyldeoxyguanosine in 3 μg of heat-denatured DNA (corresponding to 217–0.9 μmol/mol of deoxyguanosine) were blotted onto nitrocellulose and incubated with a rabbit anti-O^6-hydroxyethyldeoxyguanosine serum (NPZ-146-2, 1:15,000; see ref. *4*). Bound antibodies were reacted with a goat–antirabbit IgG horseradish peroxidase conjugate and detected with hydrogen peroxide and 4-chloro-1-naphthol.

2. No radioactive isotopes. Since the assay is noncompetitive, DNA adducts are detected without the use of highly labeled tracer molecules, and immunocomplexes can be detected enzymatically using a colorimetric assay.
3. Small sample size. Only 3 μg of DNA are necessary for a single determination. Thus, this technique is readily applicable to small (50–100 mg) human biopsy samples, chromatin fractions, or DNA restriction fragments.
4. Minimal processing of DNA. Enzymatic digestion, hydrolysis, or chemical modifications are not required.

The immuno-slot-blot assay can be used to determine any heat or alkali stable DNA adduct for which a specific antibody has been raised. In contrast to nonimmunological methods (HPLC, radiochromatography, ^{32}P-postlabeling), the immuno-slot-blot requires that the DNA adduct of interest be known beforehand. A specific antibody must be available that does not show significant crossreactivity with either normal or other modified DNA bases.

This contribution will describe the use of the immuno-slot-blot assay for the quantitative assessment of promutagenic O^6-alkyldeoxyguanosines and 7-alkyldeoxyguanosines in the imidazole ring-opened form, as it is currently carried out in our laboratory. Briefly, alkylated DNA is denatured by heat or

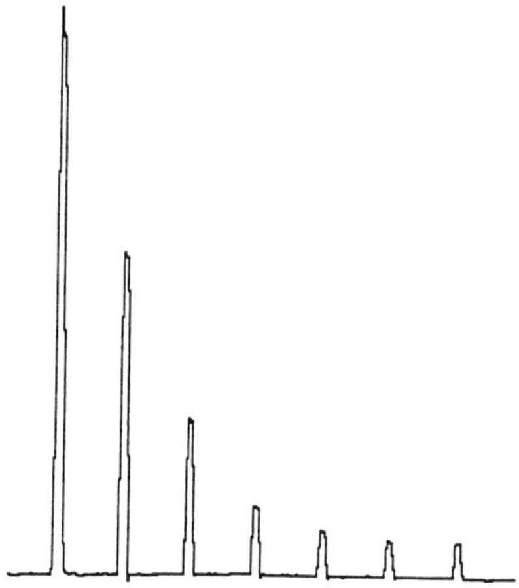

Fig. 2. Typical densitometric scan. Calf thymus DNA, methylated in vitro with 5 m*M* of methylnitrosourea, was denatured in 50 m*M* sodium hydroxide and blotted onto nitrocellulose. After incubation with a rabbit antiserum specific for O^6-methyldeoxyguanosine (NPZ-193-1, 1:8,000; see ref. *4*) and alkaline phosphatase conjugated goat–antirabbit IgG, the blot was developed with 5-bromo-4-chloro-3-indolylphosphate toluidine salt and nitroblue tetrazolium chloride in diethanolamine buffer, as described in the Methods section, and scanned by reflectance densitometry at 530 nm. Slots contain 283, 94.7, 31.3, 10.4, 3.4, 1.1, and 0 fmol of O^6-methyldeoxyguanosine, corresponding to 181–0.7 µmol/mol of deoxyguanosine.

at alkaline pH and immobilized on nitrocellulose filters using a manifold with slot-shaped openings—hence, the name of this technique. Antibodies raised against protein conjugates of the respective alkylguanosines *(11,12)* are used to stain the carcinogen adducts directly in the single-stranded DNA. After extensive washing, an enzyme-linked second antibody, which is directed against the first antibody, is added. Bound antibodies are subsequently detected by adding a chromogenic substrate that yields an insoluble colored product (Fig. 1). Densitometric evaluation of the spots, as shown in Fig. 2, can then be used for quantitative determinations of adduct levels (Figs. 3 and 4). Alternatively, radiolabeled, e.g., radioiodinated, second antibodies with detection by autoradiography can be used *(3)*. Both methods are equally sensitive. However, enzyme conjugates are nonhazardous, more stable, and thus, more convenient to use.

Antigens for generating antibodies to alkyldeoxyguanosines can be prepared by coupling the respective ribonucleoside haptens to a carrier protein, such as keyhole limpet hemocyanin *(13,14)*. The vicinal hydroxyl groups of the ribose group are oxidized with periodate to aldehyde moieties and reacted at high pH with terminal amino groups of the carrier protein to Schiff's bases. These are subsequently reduced with borohydride *(13)* or cyanogen-

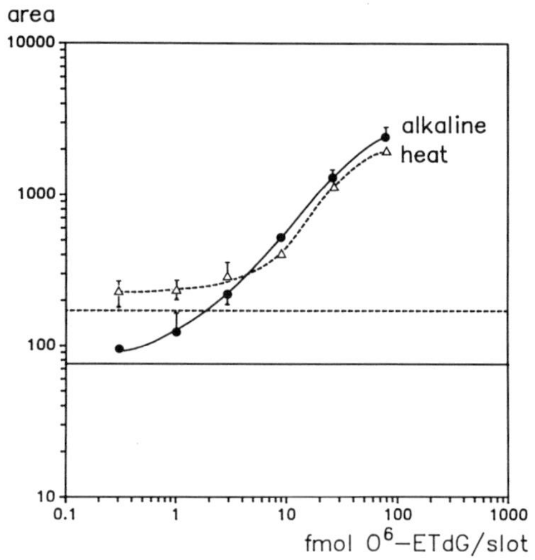

Fig. 3. Typical standard curves for O^6-ethyldeoxyguanosine (O^6-ETdG). Calf thymus DNA was ethylated in vitro with 5 m*M* of ethylnitrosourea. Aliquots containing 50.8, 16.7, 5.6, 1.9, 0.6, 0.2, and 0 µmol of O^6-ethyldeoxyguanosine/mol of dG in 3 µg of DNA were denatured by heat (dashed line) or alkali (solid line) and blotted onto nitrocellulose filters. Following incubation with rabbit antiserum 105-1 (1:6000; see ref. *5*) against O^6-ETdG, filters were developed and evaluated as described for Fig. 2. Peak areas (relative units) of the densitometric evaluation are plotted against concentrations of O^6-ETdG in a double-logarithmic plot. Nonspecific binding of the antiserum to unmodified DNA is indicated by the horizontal lines.

borohydride *(14)* to form stable protein conjugates. Both monoclonal and polyclonal antibodies can be used for the immuno-slot-blot assay *(3)*. In our experience, polyclonal rabbit antisera have been more useful for the detection of alkylated bases in immobilized DNA, since they have consistently exhibited lower crossreactivity with unmodified DNA. We will therefore describe the generation of rabbit antisera only. Binding to unmodified single-stranded DNA may pose a problem, and occasionally precludes the use of a certain antibody. Interestingly, this crossreactivity is not related to that of the same antibody to unmodified nucleosides in radioimmunoassays *(3)*.

2. Materials

2.1. Preparation of Nucleoside–Protein Conjugates

1. Alkylribonucleoside.
2. Sodium periodate: 0.1 *M* in distilled water. This solution should be freshly made.
3. Sodium carbonate: 2 *M* in distilled water.
4. Keyhole limpet hemocyanin.
5. Sodium borohydride: 30 mg/mL in distilled water, or sodium

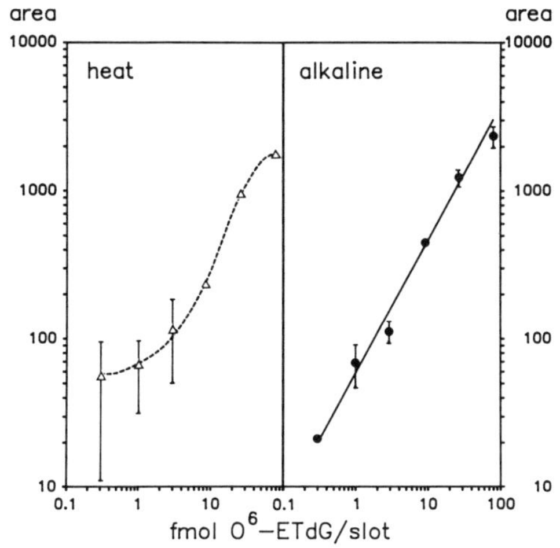

Fig. 4. Data of Fig. 2 were corrected for nonspecific background binding to unmodified DNA and plotted against O^6-ETdG concentrations in a double-logarithmic plot. The standard curve obtained with heat-denatured DNA (left) was fitted by two quadratic regression analyses, whereas the standard curve for alkaline denatured DNA (right) was calculated by linear regression analysis.

cyanogenborohydride: 2.5 mg/mL in distilled water. These solutions should be freshly made.

6. Acetic acid: 1*M* in distilled water.
7. Octanol.
8. Sephadex G-50 and gel filtration equipment.
9. Ethylene glycol (optional).
10. Dialysis tubing (optional).
11. Phosphate-buffered saline (PBS): 8.1 m*M* disodium hydrogenphosphate, 15 m*M* potassium dihydrogenphosphate, 140 m*M* sodium chloride, 2.7 m*M* potassium chloride, pH 7.2 (optional).

2.2. Immunization

12. Freund's complete and incomplete adjuvants.
13. Glass syringes and double-hub needle or plastic 3-way stopcock.
14. Sodium azide or thimerosal (sodium ethylmercurithiosalicylate). CARE—TOXIC.

2.3. Generation of Alkylated DNA Standards

15. Unmodified calf thymus DNA: 2 mg/mL in 80 m*M* Tris-HCl, pH 8.0.
16. The appropriate alkylnitrosourea. The nitrosourea can be used either as a solid or a stock solution can be prepared in 3 m*M* sodium citrate, pH

6.0, or in ethanol. Use care: alkylnitrosoureas are highly carcinogenic! For safety precautions and decontamination procedures, *see* ref. *15*.
17. 2-Ethoxyethanol.
18. Absolute ethanol.
19. Ether.
20. 0.1*M* HCl in distilled water.
21. HPLC or low pressure chromatography equipment.

2.4. Slot-Blot

22. TE buffer: 10 m*M* Tris-HCl, 1 m*M* EDTA, pH 7.8
23. 5X SSC (fivefold concentrated standard saline citrate): 75 m*M* NaCl, 75 m*M* trisodium citrate.
24. a. Ammonium acetate: 2*M* in distilled water, filtered (0.45 µm), or
    b. Ammonium acetate: 4*M* in distilled water, filtered (0.45 µm). Sodium hydroxide: 0.1*M* in distilled water. Acetic acid: 15% (v/v) in distilled water.
25. Autoclaved Eppendorf (1.5 mL) or sterile polypropylene tubes.
26. Nitrocellulose filters, pore size 0.45 µm (BA 85), blotting papers (GB002-SB), and Minifold II Slot blot apparatus, all from Schleicher and Schuell, or equivalent.
27. Blow drier.
28. Oven or vacuum oven set at 80°C.
29. Toluidine blue solution: 0.1% (w/v) toluidine blue in 1% (w/v) sodium tetraborate.

2.5. Immunodetection

30. PBS: *See* 11 above.
31. Casein buffer: PBS containing 0.5% (w/v) casein. Heat at 50°C to dissolve, cool, and then adjust the pH to 7.2.
32. PBS-HS: PBS containing an additional 160 m*M* sodium chloride.
33. PBS-HS-NP40: PBS-HS containing 0.05% (v/v) Nonidet P40.
34. TBS-HS: 300 m*M* sodium chloride in 10 m*M* Tris-HCl, pH 7.5.
35. Antialkyldeoxyguanosine antiserum or antibody.
36. Alkaline phosphatase conjugated second antibody.
37. Diethanolamine buffer: 100 m*M* Sodium chloride, 2 m*M* magnesium chloride, and 1 µ*M* zinc chloride in 100 m*M* Tris-HCl and 25 m*M* diethanolamine, pH 9.55.
38. BCIP stock solution: 40 mg/mL of 5-Bromo-4-chloro-3-indolylphosphate toluidine salt in dimethylformamide. This solution can be stored at 4°C and away from light for up to 3 wk.

39. Substrate solution *(16)*: Dissolve nitroblue tetrazolium chloride at a final concentration of 0.3 mg/mL in diethanolamine buffer. Immediately before use, add 3.4 µL/mL of the BCIP solution and mix well. Keep away from direct light.

2.6. Quantitation

40. Densitometry equipment (e.g., Shimadzu CS-930 Dual Wavelength TLC Scanner), fitted, if possible, with a microprocessor-based data acquisition system.

3. Methods

3.1. Preparation of Nucleoside-Protein Conjugates

1. Add 20 mg of ribonucleoside to 2 mL of the periodate solution and stir at room temperature for 20 min (*see* Note 1).
2. Dissolve 10 mg of keyhole limpet hemocyanin in 5 mL of distilled water, and adjust the pH to approx 9.5 with sodium carbonate (*see* Note 2).
3. Add the oxidized nucleoside to the keyhole limpet hemocyanin solution dropwise while stirring at room temperature. Maintain the pH of the reaction mixture at 9.5–9.7 with sodium carbonate (*see* Note 3).
4. Stir at room temperature for 1 h.
5. Add 20 mg of sodium borohydride or 25 µg of sodium cyanogenborohydride from the respective stock solutions. Stir in an ice/water bath for 3 h.
6. Optional: Add one drop of octanol to reduce foaming.
7. Neutralize to pH 6.5–7.5 with 1*M* acetic acid.
8. Separate the hemocyanin conjugate from low molecular weight reactants by gel filtration on Sephadex G-50 using PBS as the mobile phase. The protein conjugate elutes with the solvent front.
9. Determine the extent of coupling by UV spectroscopy vs unmodified keyhole limpet hemocyanin (*see* Note 4).
10. Dialyze against distilled water and lyophilize for long-term storage.

3.2. Immunization

1. Prepare a stable 1:1 (v/v) emulsion containing 0.5–1 mg of the hemocyanin–nucleoside conjugate in PBS and Freund's complete adjuvant by passing the mixture through the two glass syringes connected with a double-hub needle or a three-way plastic stopcock (*see* Notes 5–7 and also this vol., Chapter 1).
2. Place one drop of the emulsion in a beaker of cold water. The emulsion is stable if the drop remains on the surface as a discrete globule.

3. Immunize rabbits by a single or by multiple intramuscular or subcutaneous injection(s).
4. Administer similar booster injections in Freund's incomplete adjuvant after 8 and 16 wk.
5. Bleed animals two weeks after each booster. After storing the blood for 24 h at 4°C, decant the serum, and add sodium azide or thimerosal to a final concentration of 0.2% (w/v) to inhibit bacterial growth. Aliquot and store at –20°C.

3.3. Generation of Alkylated DNA Standards

1. Add the respective alkylnitrosourea to the DNA solution, either as a solid or from a stock solution, to a final concentration of 0.5–5 m*M* (*see* Note 8).
2. Incubate at 37°C for 1 h.
3. Cool the reaction vessel on ice and precipitate the DNA with 2 vol of ice-cold 2-ethoxyethanol.
4. Wash three times with absolute ethanol, once with ether, dry at room temperature, and store at –80°C.
5. Hydrolyze an aliquot in 0.1*M* HCl for 20 h at 37°C and calibrate by strong cation exchange HPLC on a Whatman Partisil 10 SCX column or by column chromatography on Sephasorb, using appropriate optical or radioactive standards. Detailed descriptions of these methods can be found elsewhere *(6,7)*.

3.4. Slot-Blot

1. Dissolve in vitro alkylated standard DNA and unmodified DNA in TE buffer (0.3 mg/mL). Prepare a standard curve by making serial dilutions. The DNA solutions can be stored at 4°C for several days. In order to minimize depurination of alkylated DNA, the dissolved standards should be aliquoted and frozen at –20°C for long-term storage.
2. Denature the DNA samples, which include the standard curve, samples, and unmodified DNA as a blank (*see* Note 9). This can be done in one of two ways (*see* Note 10):
   a. Dilute DNA samples (12 µg) to 30 µg/mL in TE buffer (final vol., 400 µL) in autoclaved Eppendorf tubes. Heat-denature in a boiling water bath for 10 min. Flash cool in ice/water for 5 min. Spin any droplets down in a tabletop centrifuge. Add 400 µL of 2*M* ammonium acetate.
   b. Dilute DNA samples (12 µg) to 60 µg/mL in TE buffer (final vol., 200 µL) in autoclaved Eppendorf tubes. Add 200 µL of 0.1*M* sodium hydroxide, mix well, and leave at room temperature for 10 min. Neutralize with 200 µL of 15% acetic acid solution and add 200 µL of 4*M* ammonium acetate.

3. Soak one nitrocellulose filter plus two sheets of blotting paper in 1*M* ammonium acetate for at least 15 min.
4. Assemble the filter apparatus and adjust the vacuum to approx 320 mm Hg. Rinse each slot (*see* Note 11) with 200 µL of 1*M* ammonium acetate.
5. Apply 200 µL of each sample (corresponding to 3 µg of DNA) to three slots. We have found results to be most consistent if triplicates are positioned on the filter in random order, but samples are applied row by row.
6. Rinse each slot with 200 µL of 1*M* ammonium acetate.
7. Remove the filter from the support and soak in 5X SSC for 5 min.
8. Carefully blow-dry and bake at 80°C for 2.5 h. At this stage, the filters can be stored in a desiccator at room temperature for several months. Handle carefully, as the baked filters are very brittle.
9. The suitability of a DNA solution for immobilization can be checked by postblot staining with toluidine blue *(17)*. However, the stained filter cannot be used for subsequent immunodetection. Cut off a portion of the filter, immerse in toluidine blue solution, and incubate at room temperature until the slots are dark blue. Destain the background in several changes of distilled water.

3.5. Immunodetection

1. Incubate the nitrocellulose filter in casein buffer for 30–45 min at 37°C to saturate ("block") nonspecific protein binding sites (*see* Note 12).
2. Incubate with the first antibody (diluted in casein buffer) overnight at 4°C with gentle agitation. The appropriate antibody concentration must be determined empirically.
3. Wash the filter extensively as follows, allowing 10–15 min for each buffer: 3 X casein buffer; 1X PBS-HS; 1X PBS-HS-NP40; 1X PBS-HS; 1X PBS-HS-NP40; 1X PBS.
4. Incubate with alkaline–phosphatase conjugated second antibody (diluted in casein buffer) for 2 h at room temperature or overnight at 4°C (*see* Note 13). Antibody concentration is usually indicated by the supplier.
5. Wash extensively as above, but substitute TBS-HS for PBS in the final step.
6. Add substrate solution and develop for 20–35 min at 37°C. Spots should be light to dark purple, depending on the concentration of alkylated base in the immobilized DNA, on an almost white background (*see* Note 14).
7. Rinse extensively with distilled water and store in distilled water at 4°C in the dark until densitometric evaluation.

3.6. Quantitation

1. Perform a densitometric scan. If possible, the wavelength should be set to 530 nm, i.e., the absorption maximum of the colored product.
2. Integrate peak areas. Alternatively, histograms of the light intensities per slot area can be printed out and peak heights determined manually (*see* Note 17).
3. Correct the peak areas or heights of the standard curve for binding to unmodified DNA and plot against the respective adduct concentrations in a double-logarithmic plot (Fig. 3).
4. Calculate by linear regression analysis. Reliable results can also be obtained if the standard curve is linear over only a narrow range of adduct concentrations by using nonlinear regression analysis (Fig. 4).

4. Notes

1. Optional: Excess periodate may subsequently be destroyed by adding 20 µL of ethylene glycol and stirring for an additional 5 min as suggested by Erlanger and Beiser *(13)*.
2. The procedure for coupling ribonucleosides to proteins can easily be scaled up or down. Also, the coupling ratio of ribonucleoside to carrier protein may be varied according to the amount of ribonucleoside available. However, the relative yield of coupled nucleoside increases with both the concentration of nucleoside and of protein during the reaction.
3. The coupling reaction can also be carried out at pH 8.5 when using alkali-sensitive haptens *(14)*.
4. The coupling ratio is determined most easily with a dual beam spectrophotometer with solutions containing equal concentrations of unmodified and ribonucleoside conjugated keyhole limpet hemocyanin in the reference and sample cuvettes, respectively. Protein concentrations can be determined by standard colorimetric assay (e.g., *18*), and the concentration of coupled ribonucleoside can be calculated using published extinction coefficients (e.g., *19*). An alternate method, based on the extinction coefficients of the carrier protein and of the coupled ribonucleoside, is described in ref. *20*.
5. Caution! Accidental eye contact or inoculations with Freund's adjuvant can lead to serious injury.
6. Double-hub needles can be made by gluing an 18-g needle to a 20-g needle with epoxy.
7. Alternatively, the antigen can be adsorbed to an equal vol of a 12.5% (w/v) aluminum hydroxide slurry before emulsification, and rabbits can

be immunized sc at multiple sites, as suggested by Müller and Rajewsky *(11)*. We have found both methods equally efficient at eliciting an antigenic response.

8. Conduct all alkylation reactions in a properly functioning chemical fume hood. For safety precautions and decontamination procedures, *see* ref. *13*.
9. As the antisera to modified DNA bases often exhibit significant crossreactivity to unmodified DNA, all slots must contain the same amount of DNA, e.g., 3 µg. Highly modified DNA should be diluted with unmodified DNA.
10. The method used for denaturing DNA depends both on the stability of the DNA adduct of interest and on the characteristics of the antiserum. For example, with the antisera developed in our laboratory, the sensitivity of the immuno-slot-blot assay for O^6-ethyldeoxyguanosine (Fig. 2) and O^6-methyldeoxyguanosine (not shown) is somewhat greater with alkaline denaturation than with heat denaturation, whereas the mode of denaturation has no significant effect on the detection of O^6-hydroxyethyldeoxyguanosine (not shown). DNA must be denatured under alkaline conditions when assaying for 7-alkyldeoxyguanosines as the base is readily lost on heating at neutral pH *(21)*. The high pH of the coupling reaction used in preparing the antigen and during DNA denaturation will cause the imidazole ring to open spontaneously. Antibodies are, therefore, directed against the ring-opened product rather than the parent alkyl purine *(12)*.
11. Although the principle of a dot-blot apparatus is the same, we recommend the use of a slot-shaped manifold for quantitative determinations, because this shape enables more uniform immobilization of DNA.
12. We use rectangular plastic containers and sufficient solution to cover the filters evenly for all incubations. To save on reagent, Nehls et al. *(3)* suggested the use of heat-sealed plastic bags.
13. Horseradish peroxidase conjugated second antibodies may also be used (ref. *4* and Fig. 1). The substrate solution is prepared immediately before use by mixing 1 vol of a solution of 4-chloro-1-naphthol in methanol (3 mg/mL) with 5 vol of TBS-HS containing 0.6 µL/mL of 30% hydrogen peroxide *(22)*. 3,3-Diaminobenzidine is not recommended since it leads to uniformly high background staining of the entire nitrocellulose filter. The filters are developed at 35°C and scanned at 590 nm. However, alkaline phosphatase conjugates used with the diethanolamine buffer described here produce darker signals that are significantly easier to evaluate. Also, we found the colored product more stable than that obtained with horseradish peroxidase.

14. Keep blot membranes away from direct light during and after development, to minimize background.
15. Expose blots with radioiodinated second antibodies to Kodak X-Omat AR film *(3)*, typically overnight to several days. Multiple exposures of the same filter may be required.
16. Depending on the crossreactivity of the antibody used, the sensitivity of the immuno-slot-blot assay may be increased by using sandwich-techniques for signal amplification, as used in immunohistochemistry, e.g., alkaline phosphatase–antialkaline phosphatase or peroxidase–antiperoxidase (*see* this vol., Chapter 10).
17. Since all slots are of equal width (Fig. 1), peak heights of histograms of the light intensities per slot area (Fig. 2) can be substituted for integrated peak areas for quantitative determinations. Peak heights can easily be measured without sophisticated equipment on printouts of the histograms.
18. On autoradiographs, the width of a spot increases with the amount of radioactivity, which necessitates integration of signal areas for quantitative determinations.

References

1. Harris, C. C. (1985) Future directions in the use of DNA adducts as internal dosimeters for monitoring human exposure to environmental mutagens and carcinogens. *Env. Health Pers.* **62,** 185–191.
2. Bartsch, H., Hemminki, K., and O Neill, I. K., eds. (1988), *Methods for detecting DNA damaging agents in humans: Applications in cancer epidemiology and prevention.* International Agency for Research on Cancer, Lyon.
3. Nehls, P., Adamkiewicz, J., and Rajewsky, M. F. (1984) Immuno-slot-blot: a highly sensitive immunoassay for the quantitation of carcinogen-modified nucleosides in DNA. *J. Cancer Res. Clin. Oncol.* **108,** 23–29.
4. Ludeke, B. I. and Kleihues, P. (1988) Formation and persistence of O^6-(2-hydroxyethyl)-2'-deoxyguanosine in DNA of various rat tissues following a single dose of *N*-nitroso-*N*-(2-hydroxyethyl)urea. An immumo-slot-blot study. *Carcinogenesis* **9,** 147–151.
5. Ludeke, B., Meier, T., and Kleihues, P. (1991) Bioactivation of asymmetric *N*-dialkylnitrosamines in rat tissues derived from the ventral entoderm, in *Relevance to Human Cancer of Nitroso Compounds, Tobacco, and Mycotoxins* (O'Neill, I. K., Chen, J. S., and Bartsch, H., eds.), IARC Scientific Publications No. 105. International Agency for Research on Cancer, Lyon, pp. 286–293.
6. Swenberg, J. A. and Bedell, M. A. (1982) Cell-specific DNA alkylation and repair: application of new fluorometric techniques to detect adducts, in *Banbury Report 13: Indicators of Genotoxic Exposure* (Magee, P. N., ed.), Cold Spring Harbor Laboratory, Cold Spring Harbor, NY, pp. 205–220.

7. von Hofe, E. and Kleihues, P. (1986) Comparative studies on hepatic DNA alkylation in rats by *N*-nitrosomethylethylamine and *N*-nitrosodimethylamine plus *N*-nitrosodiethylamine. *J. Cancer Res. Clin. Onc.* **112,** 205–209.
8. Hodgson, R. M., Schweinsberg, F., Wiessler, M., and Kleihues, P: (1982) Mechanism of esophageal tumor induction in rats by *N*-nitrosomethylbenzylamine and its ring-methylated analog *N*-nitrosomethyl(4-methylbenzyl)amine. *Cancer Res.* **42,** 2836–2840.
9. Adamkiewicz, J., Eberle, G., Huh, N., Nehls, P., and Rajewsky, M. F. (1985) Quantitation and visualization of alkyl deoxynucleosides in the DNA of mammalian cells by monoclonal antibodies. *Env. Health Persp.* **62,** 49–55.
10. Strickland, P. T. and Boyle, J. M. (1984) Immunoassay of carcinogen-modified DNA. *Progr. Nucleic Acid Res.* **31,** 1–58.
11. Müller, R. and Rajewsky, M. F. (1980) Immunological quantification by high affinity antibodies of O^6-ethyldeoxyguanosine in DNA exposed to *N*-ethyl-*N*-nitrosourea. *Cancer Res.* **40,** 887–896.
12. Degan, P., Montesano, R., and Wild, C. P. (1988) Antibodies against 7-methyldeoxyguanosine: Its detection in rat peripheral blood lymphocyte DNA and potential applications to molecular epidemiology. *Cancer Res.* **48,** 5065–5070.
13. Erlanger, B. F. and Beiser, S. M. (1964) Antibodies specific for ribonucleosides and ribonucleotides and their reaction with DNA. *Proc. Natl. Acad. Sci. USA* **52,** 68–74.
14. Meredith, R. D. and Erlanger, B. F. (1979) Isolation and characterization of rabbit anti-m7G-5'P antibodies of high apparent affinity. *Nucleic Acids Res.* **6,** 2179–2191.
15. Lunn, G., Sanson, E. B., Andrews, A. W., and Keefer, L. K. (1988) Decontamination and disposal of nitrosoureas and related *N*-nitroso compounds. *Cancer Res.* **48,** 522–526.
16. Ey, P. L. and Ashman, L. (1986) The use of alkaline phosphatase-conjugated anti-immunoglobulin with immunoblots for determining the specificity of monoclonal antibodies to protein mixtures. *Methods in Enzymology* **121,** 497–509.
17. Leary, J. J., Brigate, D. J., and Ward, D. C. (1983) Rapid and sensitive colorimetric method for visualizing biotin-labeled DNA probes hybridized to DNA or RNA immobilized on nitrocellulose: Bio blots. *Proc. Natl. Acad. Sci. USA* **80,** 4045–4049.
18. Bradford, M. (1976) A rapid and sensitive method for the quantitation of microgram quantities of protein utilizing the principle of protein-dye binding. *Anal. Biochem.* **72,** 248–254
19. Fasman, G. D., ed. (1975) *Handbook of Biochemistry and Molecular Biology: Nucleic Acids,* Vol. 1 CRC Press, Boca Raton, FL.
20. Inouye, H., Fuchs, S., Sela, M., and Littauer, U. Z. (1971) Anti-inosine antibodies. *Biochim. Biophys. Acta* **240,** 594–603.
21. Beranek, D. T., Wies, C. C., and Swenson, D. H. (1980) A comprehensive quantitative analysis of methylated and ethylated DNA using high pressure liquid chromatography. *Carcinogenesis* **1,** 595–606.
22. Hawkes, R., Niday, E., and Gordon, J. (1982) A dot-immunobinding assay for monoclonal and other antibodies. *Anal. Biochem.* **119,** 142–147.

CHAPTER 34

Production of Monoclonal Antibodies for the Detection of Chemically Modified DNA

Michael J. Tilby

1. Introduction

Monoclonal antibodies that are specific for sites of base modification in DNA have a number of applications in, for example, studies of carcinogenesis, drug action, or DNA repair *(1)*. The production of antibodies of the appropriate specificity entails practical considerations that are specific to this type of antigen.

Appropriate antibodies can be elicited by immunization either with modified polymeric DNA, the subject of this chapter, or with modified nucleosides conjugated to protein (*see* this vol., Chapter 33). Preparation of the latter type of immunogen is dependent on a knowledge of the chemistry of drug–nucleotide reactions, such that one can produce a purified reaction product and couple it to a carrier protein. The antibodies produced will tend to be specific for a known type of base modification and suitable for analysis of nucleic acid hydrolysates, e.g., HPLC fractions, but it is possible that they will not efficiently recognize modified bases in polymeric DNA (e.g., ref *2*). By contrast, preparation of drug-modified polymeric DNA is not dependent on detailed knowledge of the chemistry of the drug–nucleic acid reaction products. However, if a number of products are formed on the DNA, the exact specificity of the antibodies will not be immediately clear. The antibodies resulting from the use of this type of immunogen are less likely to recognize, with high affinity, modified nucleosides/nucleotides in DNA hydrolysates and

From: *Methods in Molecular Biology, Vol. 10: Immunochemical Protocols*
Ed.: M. Manson ©1992 The Humana Press, Inc., Totowa, NJ

are therefore less likely to provide a sensitive assay for such substances. However, this method of immunization may be the best for obtaining antibodies that recognize modified bases in polymeric DNA and, in addition, it offers the possibility of raising antibodies that recognize conformational changes induced in DNA by a drug. Therefore, this method may well yield the best antibodies for histological procedures or for affinity isolation of DNA fragments bearing drug-induced alterations.

In biological samples, the frequency of drug-induced base modification is generally very low. When the immunoassay uses polymeric DNA, it is particularly important that cross-reaction with unmodified DNA is minimal, because it is not possible to chromatographically separate modified from unmodified nucleotides before the analysis. This can be done when working with DNA hydrolysates.

In order to elicit antibodies against drug-modified regions, the DNA used for immunization should have a high frequency of base modification (0.1–0.02 modified bases per base). A potential problem is that the antibodies produced will preferentially recognize modified bases present at high frequencies, but will not provide a sensitive detection of modified bases present at low frequencies *(3,4)*. Furthermore, the inevitable presence of unmodified bases in the immunogen means that antibodies could be raised that bind to normal DNA. Fortunately, unmodified DNA regions appear to be less immunogenic than at least certain types of modified regions (*5,6* and as shown by the fact that several DNA modification-specific antisera have been produced by this method of immunization). These two types of undesirable antibody can only be eliminated by careful screening of hybridomas using various DNA preparations.

The methods are described for the preparation of DNA modified with the anticancer drug melphalan (phenylalanine mustard), and for the production of antibodies that specifically recognize the drug modified sites.

2. Materials

1. DNA is from calf thymus, highly purified.
2. Acid ethanol: 49 parts absolute ethanol plus one part 5*M* HCl.
3. Melphalan (from Wellcome, Beckenham, UK) should be dissolved in acid ethanol just before use and at a concentration 50 times its final concentration in the reaction mixture with DNA.
4. Radioactive Melphalan (^3H or ^{14}C): Obtained from Amersham (Slough, UK).
5. Buffer A: 50 m*M* NaCl, 50 m*M* sodium phosphate, pH 7.0.
6. Buffer B: 50 m*M* sodium phosphate buffer, 0.02% NaN_3 (sodium azide: TOXIC!), pH 7.0.

7. Methylated bovine serum albumin (mBSA) (from Sigma): Dissolve in water (1 mg/mL) and store at –20°C.
8. Animals: The volumes of immunogen described relate to immunization of rats. With mice, smaller volumes would suffice, and also, Section 3.2.2. on Peyer's patch immunization would be inappropriate. Of the rats that we have tested, the best strain for the present purposes seems to be F344 (*see* Note 1).
9. Phosphate buffered saline (PBS): 10 m*M* Na+K phosphates, 140 m*M* NaCl, pH 7.4.
10. PBSTw: PBS containing Tween-20 (0.1% v/v) and NaN_3 (0.02%).
11. Second antibody reagent (*see* Note 2): β-galactosidase conjugated sheep antirat antiserum (from Amersham) is diluted 1 in 1000 in PBSTw containing 1% BSA, 1 m*M* β-mercaptoethanol, and 10 m*M* $MgCl_2$.
12. Substrate solution: PBS containing 4-methyl-umbelliferyl β-D-galactoside (0.2 mg/mL), 100 m*M* β-mercaptoethanol, 10 m*M* $MgCl_2$ and 0.02% NaN_3.
13. Freunds complete adjuvant (FCA).
14. Freunds incomplete adjuvant (FIA).
15. Multiway pipets (5–50 and 50–250 μL ranges).
16. 96-well plates: Ordinary flat or V-bottomed for diluting solutions and flat-bottomed immunoabsorbent assay plates for coating with DNA.
17. Plate reader. If the fluorescence assay is to be used, this must be capable of reading fluorescence with excitation and emission wavelengths of 380 and 450 nm, respectively.

3. Methods

3.1. Preparation of Modified DNA

DNA preparations should be made that carry the appropriate base modifications at high and low frequencies (*see* Note 3). The following method, given as an example, is for melphalan.

1. Prepare a solution of DNA of about 1 mg/mL in buffer A (*see* Note 4), determine its concentration from the OD_{260} (*see* Note 5), and dilute part of it to give 3 × 10 mL aliquots containing DNA at 0.5 mg/mL in buffer A.
2. Add melphalan to these aliquots such that nonradioactive melphalan is at 500, 50, and 0 μg/mL and radioactive melphalan at about 10^4 Bq/mL in each. After incubation at 37°C for 1 h, store the reaction mixtures at –20°C.
3. Separate DNA from low molecular weight reaction products, preferably by gel-filtration chromatography (see Note 6) using Sephadex G–25 or G–75 and Buffer A.

4. Determine the DNA concentration from the OD_{260} (*see* Note 5) and the amount of melphalan bound from its specific radioactivity.

3.2. Immunogen and Immunization

The method for preparation of the immunogen is determined by the method of immunization chosen. (*see* Note 7).

3.2.1. Conventional Immunization

1. Dilute the highly modified DNA preparation with buffer A to 150 µg DNA/mL, allowing 0.6 mL per rat per immunization.
2. For every 1 mL of this solution, add 15 µL of mBSA solution, keeping the solution agitated during the addition. The solution immediately becomes cloudy as the result of the formation of an electrostatic complex between the mBSA and DNA (*see* Note 8).
3. Add to this suspension, 1.1 vol of FCA and vortex mix until an emulsion is formed (*see* Note 9).
4. Immunize about six rats, 10–15 wk old, each with 1 mL of immunogen, distributed evenly among four subcutaneous and one intraperitoneal sites. It is useful to take a prebleed of each animal before the first immunization (*see* Note 10) and store the sera at –20°C until later bleeds are tested.
5. After 2 wk, prepare a second batch of immunogen, as in Steps 1–3, except that FIA should be used. Immunize as in Step 4 and repeat immunization using incomplete adjuvant at 2-wk intervals. At the third and subsequent immunizations, take test bleeds (see Note 10) and test in order to determine which (if any) animals have responded most strongly and are, therefore, most suitable for fusion experiments.
6. Three or four days before carrying out a spleen cell fusion give the chosen animal an iv injection (0.5 mL) of a suspension of DNA/mBSA complex prepared as in Step 2.

3.2.2. Peyer's Patch Immunization

1. The DNA solution should be at a concentration of about 1 mg/mL (*see* Note 11). Allow about 80 µL of DNA solution per rat per immunization. Add mBSA solution, such that the weights of mBSA and DNA are equal. Then add 1.1 vol of adjuvant and emulsify as described in Step 3 of the previous section, taking care not to make the emulsion too viscous, as it must be injected via a very fine needle.
2. The details of immunization via Peyer's patches are described in Chapter 54 of vol. 5. With this procedure, lymph node cell fusions are often performed before a good serum antibody response develops. In those cases, analysis of test bleeds is inappropriate.

3.3. Testing Sera

1. In order to test the sera, coat 96-well immunoassay plates with control DNA and DNA carrying a high level of base modification. Place the appropiate DNA (50 µL of a 0.2 µg/mL solution in Buffer B) in each well of the plates using a multiway pipet and, after tapping the plates or placing them on a vortex plate shaker to ensure that the solutions completely cover the bottom of each well, leave the plates to dry, without lids, overnight in a 37°C incubator, the door of which is left slightly open. The plates can be stored dry until needed (*see* Note 12).
2. Before use, minimize the amount of insecurely attached DNA that might elute off the plastic during the assay by filling the wells of the plates with PBSTw, then leave them at room temperature for about 15 min before washing twice with PBSTw.
3. Dilute the sera to be tested initially 1 in 20, and then serially 1 in 2 in PBSTw. Then transfer 50 µL aliquots of each dilution to duplicate wells of plates coated (in Step 1) with (i) no DNA, (ii) control DNA, or (iii) highly modified DNA.
4. After a 1-h incubation (*see* Note 13), wash the plates five times and add 30 µL of second antibody reagent to each well (*see* Note 2). After incubation for 1–2 h, again wash the plates five times and then add 50 µL of substrate solution.
5. Incubate the plates and then read at intervals (starting at 5 min if the sensitive fluorescent ELISA system is used), until readings that are at least 10 times those of background are obtained. At high concentrations of serum, there is usually a small signal on the uncoated plates owing to nonspecific binding of serum components. This becomes insignificant at low to moderate serum dilutions. There are usually antibodies in the sera that bind to the plates coated with control DNA. These should be present at low concentrations or titers, as shown by the diminution of the signal on these plates at moderate serum dilutions. For animals that respond to immunization in the desired way, the assay signals on plates coated with modified DNA are high and remain high at greater dilutions than for the control DNA plates (1 in 100 up to perhaps 1 in 1000).

3.4. Production and Testing of Hybridomas

Hybridomas are produced and grown as described in vol. 5, Chapter 54 of this series, and in this vol., Chapter 6.

Hybridoma supernatants are screened using the same methods as those for testing sera, except that only undiluted culture supernatants (50 µL) are placed in the DNA coated wells. Diluted positive antiserum should be tested in one set of wells as a positive control. The initial screens should be carried out in wells coated with highly alkylated DNA and control DNA. Subsequent screening of selected hybridomas should make use of DNA preparations alkylated at lower frequencies. Although lower signals will be generated on such DNA preparations, antibodies that recognize structures only present in highly modified DNA will give disproportionately lower signals. The definitive characterization of selected antibodies must rely on other assay methods, such as competitive ELISA (*see* Chapter 35).

4. Notes

1. Nonimmunized rats contain various levels of antibodies against normal DNA. Such antibodies appear to be undesirable for the present purposes. We have observed that, of several strains tested, strain F344 contains the lowest levels of such antibodies (Johnson, Tilby, and Dean, unpublished observations) and we now use these animals for raising antibodies specific to modified polymeric DNA.
2. Any detection system can be used (e.g., ^{125}I-labeled or alkaline phosphatase-labeled antirat Ig). We find it convenient to use the β-galactosidase label plus a substrate, 4-methylumbelliferyl β-galactoside, which is converted to a highly fluorescent product. However, fluorescence detection is only important for high sensitivity assays as described in Chapter 35.
3. The method of preparation will be highly dependent on the DNA modification in question. In general, a base modification frequency in the range 0.1–0.02 should be used for immunization. Further DNA preparations with base modification frequencies down to 10^{-3} are necessary for screening the hybridomas. In the case of alkylation at the *N*-7 position of guanine, consideration should be given to the question of whether it might be preferable to produce antibodies that recognize the stable ring-opened structure formed on exposure of the alkylated bases to alkaline pH.
4. The starting DNA should be of high purity, and should be properly dissolved by first stirring gently for a few days in 5 m*M* NaCl at 4°C in a stoppered flask containing a few drops of toluene to prevent microbial growth. Then add concentrated Buffer A solution, remove the toluene by bubbling air through the solution and store at –20°C.
5. The OD_{260} of a 50 μg/mL solution of native DNA in a 10 mm pathlength cuvette is 1.0.
6. With gel filtration, one avoids any risk of delayed intramolecular second arm reactions, or difficulty in redissolving the DNA, problems that can

be encountered when the DNA has been reacted with crosslinking drugs. Ethanol precipitation would be the alternative method, but for this, a buffering agent other than phosphate would be necessary for the drug–DNA reaction because phosphate would be precipitated by the ethanol. When stable products are formed on the DNA, low molecular weight substances can alternatively be removed by dialysis.

7. Immunizations can be either conventional or via the Peyer's patches. Conventional immunizations involve the injection of immunogen at subcutaneous and intraperitoneal sites, and require 1 mL of immunogen per rat per immunization. Peyer's patch immunizations involve injection of small volumes of concentrated immunogen directly into these patches during a surgical procedure.
8. This procedure, which appears to be important for stimulating anti-DNA antibody formation, was described by Plescia *(7)*.
9. This should appear as a white liquid of a creamy consistency that does not separate into two phases on standing for at least 30 min. If difficulty is experienced in producing such an emulsion, another 0.1 or 0.2 vol of adjuvant should be added and the mixing continued.
10. Test bleeds (for rats, usually about 0.5 mL) are allowed to coagulate at room temperature for about 1 h, and the serum is harvested after centrifugation (about 3000*g* × 10 min).
11. The DNA can be conveniently concentrated using a centrifugational ultrafiltration device (10,000 MW cut-off), such as a "Centricon" (from Amicon).
12. We take the precaution of sealing the DNA-coated plates in polythene bags and storing them at –20°C. Plates coated with DNA modified with the alkylating agent melphalan, which induces unstable guanine *N*-7 adducts, are stable for at least one year when stored in this way.
13. Incubations are at 37°C with lids on plates. All washings are with PBSTw.

References

1. Strickland, P. T. and Boyle, J. M. (1984) Immunoassay of carcinogen modified DNA. *Prog. Nucleic Acids Res. Mol. Biol.* **31,** 1–58.
2. Poirier, M. C., Yuspa, S. H., Weinstein, I. B., and Blobstein, S. (1977) Detection of carcinogen-DNA adducts by radioimmunassay. *Nature* **270,** 186–188.
3. Van Schooten, F. J., Kriek, E., Steenwinkel, M-J. S. T., Noteburn, H. P. J. M., Hillebrand, M. J. X., and Van Leeuwen, F. E. (1987) The binding efficiency of polyclonal and monoclonal antibodies to DNA modified with benzo[a]pyrene diol epoxide is dependent on the level of modification. Implications for quantitation of benzo[a]pyrene-DNA adducts *in vivo. Carcinogenesis* **8,** 1263–1269.
4. West, G. J., West, I. W.-L., and Ward, J. F. (1982) Radioimmunoassay of a thymine glycol. *Radiat. Res.* **90,** 595–608.

5. Levine, L., Seaman, E., Hammerschlag, E., and Van Vunakis, H. (1966) Antibodies to photoproducts of deoxyribonucleic acids irradiated with ultraviolet light. *Science* **153,** 1666,1667.
6. Wakizaka, A. and Okuhara, E. (1979) Immunochemical studies on the correlation between conformational changes of DNA caused by ultraviolet irradiation and manifestation of antigenicity. *J. Biochem.* **86,** 1469–1478.
7. Plescia, O. J. (1968) Preparation and assay of nucleic acids as antigens, in *Methods in Enzymology,* Vol. 12 (Grossman, L. and Moldave, K., eds.), Academic, New York and London, p. 893–899.

Chapter 35

Sensitive Competitive Enzyme-Linked Immunoassay for Quantitation of Modified Bases in DNA

Michael J. Tilby

1. Introduction

A widely used configuration for competitive enzyme-linked immunosorbent assay (ELISA) of modified regions in polymeric DNA involves competition between a standardized quantity of antigen bound to the assay well and a variable quantity and/or quality of antigen in solution. These are competing for a limited standardized quantity of antibody in solution. The amount of antibody that binds to the immobilized antigen is measured and expressed as a percentage of the amount that binds in the absence of competing dissolved antigen. In this situation the amount of immobilized antigen is not generally known, but this is of no consequence, the essential feature being that the amount is highly uniform from well to well (*see* ref. *1* for a treatise on enzyme immunoassays).

This type of assay is valuable for characterizing the immunoreactivity of various antigens because accurately known amounts of competing antigen can be included in the assay well with no dependency on their ability to bind to plastic surfaces. It is also important because the quantity of analyte is not limited to what can be adsorbed to plastic, so that small amounts of modified base in many micrograms af DNA can be determined.

In order to maximize the sensitivity of the assay, the smallest possible amount of antibody should be included in each assay well. This amount is

From: *Methods in Molecular Biology, Vol. 10: Immunochemical Protocols*
Ed.: M. Manson ©1992 The Humana Press, Inc., Totowa, NJ

determined by the sensitivity of the method for detecting the antibody that binds to the immobilized antigen, the affinity constant of the antibody/antigen interaction, and the background signal generated as the result of nonspecific binding of immunological reagents to the assay wells. An extremely sensitive detection method, detailed below, involves the enzyme β-galactosidase and a substrate, 4-methylumbelliferyl β-D-galactoside, which becomes enzymatically hydrolyzed to yield 4-methylumbelliferone. This product can be detected with great sensitivity (about 50 pmol per assay well) through its strong fluorescence. Under ideal circumstances, the system is capable of detecting a few amol of antibody bound to each well *(2)*.

Unfortunately, the sensitivity of detection of the competing antigen generally appears to be less than that of the immobilized antigen. Using this assay system, we have achieved sensitivities (50% inhibition) of 30 and 2 fmol per assay well for melphalan *(3)* and platinum *(4)* adducts on DNA, respectively. The properties of the antibodies being used are clearly of major importance for determining the sensitivity that can be achieved in an assay.

2. Materials

All the following solutions 1–5 contain sodium azide (0.02% w/v) as an antimicrobial agent (CARE–TOXIC).

1. BSA coating solution: 1% Solution of bovine serum albumin (BSA) in 1 *M* sodium bicarbonate, pH 9.6.
2. PBS: 10 m*M* Na+K phosphates, 140 m*M* NaCl, pH 7.5.
3. PBSTw: PBS containing Tween-20 detergent at 0.1% v/v.
4. PB: 50 m*M* Sodium phosphate, pH 7.0
5. Stop buffer: 0.1 *M* Glycine adjusted to pH 10.3 with 10 *M* NaOH solution.
6. Multiway pipets (5–50 and 50–250 µL ranges).
7. 96-well plates: Ordinary flat and V-bottomed for diluting and mixing solutions and flat-bottomed immunosorbent assay plates for coating with DNA.
8. Solution A: PBS containing Tween-20 (0.2% v/v) and BSA (1% w/v). Phenol red (10 µg/mL) can be included to improve visibility of solutions in the wells of the microtiter plates.
9. Solution B: Solution A containing the primary antibody at double the desired final concentration (*see* Note 1).
10. Enzyme conjugated second antibody: We routinely use affinity purified $F(ab')_2$ species specific antibody conjugated to β-galactosidase (from Amersham, Slough, UK). This is supplied as a solution that should typically be diluted 1 in 500 into PBSTw containing 1% (w/v) BSA, 1 m*M* β-mercaptoethanol, and 10 m*M* $MgCl_2$.

11. Substrate solution: Dissolve 4-methyl-umbelliferyl β-D-galactoside (0.2 mg/mL) in PBS by gentle warming. Filter through filter paper and then add β-mercaptoethanol to 1 m*M* and $MgCl_2$ to 10 m*M* final concentrations.
12. Fluorescence plate reader capable of reading fluorescence with excitation and emission wavelengths of 380 and 450 nm, respectively.

3. Methods

3.1. Coating Assay Plates

The method described below is adopted in order to overcome problems of poor-uniformity of DNA binding to immunoassay wells (*see* Note 2). We do not use the outer wells of the plates because these give poorer uniformity than the 60 internal wells. The outer wells are left empty or, in order to improve uniformity of conditions on the internal wells, contain 50 µL of water.

1. Saturate the surface of each internal well with BSA by adding 50 µL of BSA coating solution using a multiway pipet. Ensure that the bottom of each well is completely covered with the solution by tapping the plate or briefly agitating it on a vortex-type plate mixer. Then place the plates, with lids on, in an incubator for at least 5 h (*see* Note 3).
2. Wash the plates three times (1 min per wash) with distilled water, to remove unbound protein.
3. Into each well place 50 µL of a solution of highly modified DNA at 0.2 µg/mL in PB, ensuring that the wells are properly covered with liquid as in Step 1.
4. Dry the solutes onto the bottoms of the wells by leaving the plates, without lids, overnight in an incubator (37°C), the door of which is left slightly open.
5. Next, crosslink the DNA to the protein by irradiating the plates with UV light *(5)* of about 254 nm wavelength such that each plate receives a total uniform exposure of about 8 min at about 800 µW/cm^2 (*see* Note 4). These plates can be stored until required. We store all DNA coated plates, sealed in polythene bags at –20°C, where they have been stable for at least one year.
6. Before use, soak the plates in PBSTw for about 15 min and then wash twice in PBSTw. This is to minimize the amount of DNA that is insecurely attached.

3.2. Competitive Assay

Each assay plate must include, in addition to wells containing various concentrations of competing antigen, wells with no competitor, and wells with no primary antibody. These indicate the maximum and minimum signals, respectively, to be expected. It is important to determine the maximum signal accurately. Therefore, we routinely use columns 5 and 11 for zero competitor with one well in each of these columns containing no primary antibody. This layout minimizes the effects of any systematic variation across the plate in level of antigen coating.

1. Serially dilute competing antigens. For accurate assays, the one in two dilution steps described here are appropriate. Larger steps could be used for preliminary tests. Place 120 µL of PBS in columns 3 to 11 of a flat-bottom 96-well microtiter plate, using a multiway pipet. Place the initial dilutions of the competing antigens in the wells of column 2. Serially dilute 120 µL vol across the plate using the multiway pipet, but do not dilute into columns 5 and 11, as these correspond to the no-competitor wells.
2. For each row of serial dilutions, transfer 55 µL vol to two rows of wells of a V-welled plate that has had the wells D5 and D11 ringed with marker pen to identify them as wells for no primary antibody.
3. The next task is to add to each well, 55 µL of solution A or B (i.e., without or with antibody). Since the accuracy of the assay depends on the accuracy of pipeting, some care is advisable, especially since solutions A and B contain protein and detergent that cause the solution to wet the pipet tips (*see* Note 5).
4. Incubate the antibody and competing antigens together for 30 min before transferring to the DNA-coated plates. This step appears to give a slight increase in assay sensitivity.
5. From each of the V-wells, transfer 50 µL, in a row by row fashion, to wells of duplicate DNA-coated assay plates. Thus, four assay wells, two on each assay plate, are used for each dilution of competing antigen. These plates are incubated for 1 h.
6. Then wash the assay plates five times with PBSTw.
7. Add the enzyme-conjugated second antibody reagent (30 µL per well) and incubate the plates for about 90 min.
8. Wash the plates extensively with PBSTw, at least five times. For the most sensitive assays we wash 10 times, although such a precaution may not be essential.
9. Add substrate solution (50 µL per well) and incubate the plates for the appropriate period, which will be dependent on the chosen dilution of the primary antibody (*see* Note 1).
10. Read the plates directly in the fluorescence plate reader. To terminate the reaction and increase the fluorescence intensity fourfold, add 150

μL of stop buffer to each well. The plates should then be read straightaway since the fluorescence intensity of 4-methyl umbelliferone is dependent on the pH of the solution and, being alkaline, the pH of the stopped wells will tend to fall as atmospheric CO_2 is absorbed. Addition of stop buffer is not always necessary.

3.3. Analysis of Data

1. Express the fluorescence intensity (V) for each dilution of antigen as a percentage of the maximum signal

$$V = \frac{(S\text{-}BG)}{(MAX\text{-}BG)} \times 100\%$$

where BG = mean value for wells without primary antibody;
MAX = mean value for wells with no competing antigen;
S = mean value for sample wells containing a given concentration of competing antigen.

2. Plot these values against concentration of antigen to generate a sigmoid curve from which the concentration necessary to give 50% inhibition can be read as a measure of immunoreactivity of a particular antigen, or as an indication of the assay sensitivity. Having plotted such a curve for a standard antigen, it is possible to read off single values to determine the antigen concentration in unknown samples. However, we find it preferable to generate, for each unknown sample, data from a dilution series that can be fitted to a curve to determine the 50% inhibition dilution. Not only is this more accurate, but also, the slope of the curve can yield information about the quality of the antigen and show up spurious results.
3. We use our own curve fitting software to fit the sets of data to an equation and to plot the results. The simplest fitting procedure uses the log–logit equation, but this is not very satisfactory and more accurate fits are obtained by using the logistic equation. This generates sigmoid curves using just two fitted parameters, namely, the competitor concentration that causes 50% inhibition, and the slope. The fitting procedure for the logistic equation requires starting estimates for the parameters which are calculated making a preliminary fit using the log–logit equation. Program listings and explanations of actual curve fitting procedures have been provided by Barlow *(6)*.

4. Notes

1. The final dilution at which the primary antibody is used must be determined for each antibody. It should be as high a dilution as possible, consistent with giving an assay signal that, in the absence of competing antigen, is at least 10 times the background signal observed in the wells from which primary antibody was omitted. For our monoclonal antibodies that recognize DNA adducts induced by melphalan and cisplatin, we use hybridoma culture supernatants that are diluted 2000 and 20,000 times, respectively, and we incubate the plates with substrate for 3 and 18 h, respectively.
2. Immunoassay plates appear to be manufactured and quality controlled to ensure uniform binding of protein antigens. We have experienced poor reproducibility of binding DNA directly to a variety of these plates. The method described here provides good uniformity, covalent coupling of the DNA, and low backgrounds. We have found that methods based on coating wells with polycationic substances, such as DEAE-dextran, are less satisfactory in all three aspects.
3. Incubation is at 37°C. Plates should not be stacked in the incubator, as this will result in nonuniform warming of the wells.
4. We rotate six plates mounted on a disc beneath two short wavelength UV lamps (type UVG, peak intensity at about 254 nm, from UV Products, San Gabriel, CA). UV irradiation boxes designed for crosslinking DNA to nylon blotting membranes should be an ideal alternative.
5. The best way to overcome this problem is to use an electronically controlled multiway pipet such that, for example, 230 µL of solution can be drawn up and then 4 × 55 µL dispensed, leaving 10 µL in the tips. The plates are then mixed on a vortex-type plate mixer. An alternative method using a conventional multiway pipet is to draw up 55 µL of solution A or B, dispense it in the appropriate wells, mix the contents of tips and wells by pipeting up and down a few times, and then use new tips for the next row of wells.

References

1. Maggio, E. T., ed. (1980) *Enzyme-Immunoassay,* CRC Press, Boca Raton, FL.
2. Ishikawa, E. and Kato, K. (1978) Ultrasensitive enzyme immunoassay. *Scand. J. Immunol.* **8 (Suppl. 7),** 43.
3. Tilby, M. J., Styles, J. M., and Dean, C. J. (1987) Immunological detection of DNA damage caused by Melphalan using monoclonal antibodies. *Cancer Res.* **47,** 1542–1546.

4. Tilby, M. J., Johnson, C., Knox, R., Cordell, J., and Dean, C. J. (1991) Sensitive detection of DNA damage induced by cisplatin and carboplatin *in vitro* and *in vivo* using monoclonal antibodies. *Cancer Research* **51**, 123–129.
5. Braun, A. and Merrick, B. (1975) Properties of the ultraviolet-light-mediated binding of bovine serum albumin to DNA. *Photochem. Photobiol.* **21**, 243–247.
6. Barlow, R. B. (1983) *Biodata Handling with Microcomputers*, Elsevier, Amsterdam.

CHAPTER 36

An Immunochemical Assay for Detecting Transition of B-DNA to Z-DNA

T. J. Thomas

1. Introduction

The enzyme immunoassay technique described below is a simple and sensitive method to detect left-handed Z-DNA formation in synthetic polynucleotides, recombinant plasmids, and native DNAs. This method utilizes the differences in the antigenic properties of right-handed B-DNA and left-handed Z-DNA.

In 1972, Pohl and Jovin *(1)* showed that the circular dichroism (CD) spectrum of poly(dG-dC).poly(dG-dC) inverted at >3*M* NaCl in comparison to its CD spectrum at <2*M* NaCl, thereby indicating the formation of a novel conformation of this polynucleotide at high salt conditions. Single crystal X-ray crystallographic and Raman spectroscopic studies by Rich and coworkers *(2,3)* established that the high salt form of poly(dG-dC).poly(dG-dC) existed in a left-handed Z-DNA conformation. Behe and Felsenfeld *(4)* showed that the B→Z transition could be provoked in poly(dG-dC).poly(dG-dC) at ionic conditions close to the physiologic cation concentrations when the polynucleotide was methylated at the C-5 position. Bromination of the polymer stabilized the Z-DNA form at a physiologically compatible ionic concentration of 0.15*M* NaCl *(5)*. These results and the demonstration of Z-DNA in chromosomal preparations *(6)* implied that Z-DNA might be involved in a number of biological functions, including gene regulation and DNA

From: *Methods in Molecular Biology, Vol. 10: Immunochemical Protocols*
Ed.: M. Manson ©1992 The Humana Press, Inc., Totowa, NJ

recombination. Convincing evidence for the existence of Z-DNA in plasmids in *E. coli* has also been provided recently *(7)*.

B-DNA and Z-DNA differ in their physical and chemical properties, interactions with ligands, and immunogenicity *(5,8)*. In contrast to the poor immunogenicity of B-DNA, Z-DNA is immunogenic and produces anti-Z-DNA antibodies in experimental animals. In addition, Z-DNA has high affinity toward anti-DNA autoantibodies present in systemic lupus erythematosus and rheumatoid arthritis *(9–11)*. Stollar and collaborators, as well as other investigators, used the immunogenic property of Z-DNA to produce polyclonal and monoclonal anti-Z-DNA antibodies *(6,12–14)*. These antibodies were used to detect the presence of Z-DNA segments in polytene chromosome, rat tissue, and other native DNAs *(15–18)*. In these studies, nitrocellulose filter binding assay, immunofluorescence microscopy, and electron microscopy were used to detect the presence of Z-DNA.

The enzyme immunoassay technique for the detection of Z-DNA formation was introduced to circumvent some of the difficulties associated with the use of spectroscopic techniques in studying the B→Z transition of polynucleotides and recombinant plasmids with potential Z-DNA forming inserts *(19,20)*. For example, the Z-DNA conversion of polynucleotides is associated with a condensation and/or aggregation of the polynucleotide *(21)*. The most widely used technique in B→Z transition, CD spectroscopy, is very sensitive to aggregation of DNA because aggregated particles cause differential scattering of circularly polarized light that complicates the interpretation of CD spectral changes *(22)*. In addition, the spectral properties of certain drugs overlap with those of DNA bases and pose a major problem in studying the interactions of drugs and environmental carcinogens with DNA *(23,24)*.

The enzyme immunoassay technique consists of five sequential steps:

1. Interaction of the Z-DNA inducing agent (ligand) with DNA. This step is achieved by incubating various concentrations of the ligands with the desired polynucleotide, plasmid, or native DNA.
2. Immobilization of the DNA conformation on a microtiter plate.
3. Titration of the Z-DNA conformation on the microtiter plate using a monoclonal or polyclonal anti-Z-DNA antibody.
4. Treatment of the monoclonal or polyclonal antibody with an enzyme-conjugated affinity purified antimouse or antirabbit immunoglobulin.
5. Quantitation of the amount of enzyme on the microtiter plate by monitoring the enzyme–substrate reaction with a microplate autoreader.

This method was originally developed to detect the induction of Z-DNA in polynucleotides by polyamines and inorganic cations *(19,24)*. The midpoints of transition determined from ultraviolet and CD methods were similar to the values obtained from ELISA *(19,20)*. Whereas the spectroscopic techniques record the average of global perturbations in all molecules of the sample, ELISA will detect changes in a small percentage of the sample and hence it is a much more sensitive probe to ascertain the formation of Z-DNA. Recently, we used this method for the detection of Z-DNA in small stretches of alternating purine-pyrimidine (APP) sequences inserted in recombinant plasmids *(25)*. The method is sensitive enough to detect Z-DNA formation in a mixture of polynucleotides and calf thymus DNA even when the nucleotide concentration is very small (<1%). Similarly, the salt-dependent variation of the inducibility of Z-DNA by polyamines has been documented by the enzyme immunoassay.

2. Materials

1. 50 m*M* NaCl buffer: 50 m*M* NaCl, 1 m*M* Na cacodylate, and 0.15 m*M* EDTA, pH 7.4. This buffer is used for dissolving DNA and/or diluting it to the required concentration.
2. Washing buffer: 0.05% Tween-20 and 0.02% sodium azide in phosphate buffered saline ([PBS]: 145 m*M* NaCl, 12 m*M* Na_2HPO_4, and 36 m*M* NaH_2PO_4, pH 7.4).
3. Conjugate buffer: Washing buffer with 1% bovine serum albumin (BSA).
4. Carbonate buffer: 50 m*M* $NaHCO_3$, 1 m*M* $MgCl_2$, and 0.02% NaN_3, pH 9.0.
5. 1*M* NaOH.
6. Polynucleotides. Poly(dG-m^{5}dC).poly(dG-m^{5}dC). This polynucleotide is available from Pharmacia, Inc., Piscataway, NJ or other biochemical companies. Dissolve the polynucleotide at a concentration of 0.5–1 mg/mL in 50 m*M* NaCl buffer. Dialyze the polynucleotide solution with this buffer three times to remove the excess salt and $MgCl_2$ that are usually present in commercial preparations.
7. Z-DNA-inducing ligand. Spermidine is used here as an example. Prepare a solution of spermidine in double-distilled deionized (dd) water at a concentration of 20 m*M* (*see* Note 1).
8. Anti-Z-DNA antibody. Monoclonal or polyclonal anti-Z-DNA antibody can be used in this assay. The method described here uses a monoclonal antibody Z22 that was prepared and characterized by the research group of David Stollar, Department of Biochemistry, Tufts University, Boston, MA.
9. Protamine sulfate from Salmon, Grade X (Sigma Chemical Co., St. Louis, MO). Prepare a solution in dd water at a concentration of 1 µg/mL (0.0001%) immediately before use.
10. Alkaline phosphatase-conjugated, affinity-purified polyvalent goat antimouse immunoglobulins for use with a mouse monoclonal in Step 8 above.
11. Phosphatase substrate: *p*-nitrophenyl phosphate at 1 mg/mL in dd water. Prepare freshly as required.

Table 1
Concentration of Spermidine Used in the B-DNA to Z-DNA Conformational Transition Study of poly(dG-m^{5}dC).poly(dG-m^{5}dC)

Test tube	Volume and concentration of spermidine	Final concentration
a	None	0
b	5 µL 1.25 m*M*	5 µ*M*
c	10 µL 1.25 m*M*	10 µ*M*
d	10 µL 2.5 m*M*	20 µ*M*
e	7.5 µL 5 m*M*	30 µ*M*
f	5 µL 10 m*M*	40 µ*M*
g	6.25 µL 10 m*M*	50 µ*M*
h	6.25 µL 20 m*M*	100 µ*M*
j	9.4 µL 20 m*M*	150 µ*M*

12. Microtiter Plate: 96-Well plates from Costar, Cambridge, MA are good for use with protamine sulfate as the linking agent for DNA to the plates (*see* Note 2).
13. Single and multichannel micropipets.
14. Microplate Autoreader.

3. Method

1. Add 300 µL/well of protamine sulfate solution to each of the inner 60 wells of the microtiter plate (*see* Notes 3 and 4). Incubate the plate at room temperature (≈22°C) for 90 min.
2. Dilute poly(dG-m^{5}dC).poly(dG-m^{5}dC) solution to a concentration of 10 µg/mL in 50 m*M* NaCl buffer. Pipet 1.25 mL of this solution into each of nine 5 mL glass test tubes. Label the test tubes a to h and j to identify them.
3. Dilute the spermidine solution with dd water to make four different concentrations: 10, 5, 2.5, and 1.25 m*M*. Add the spermidine solution to test tubes a to h and j as shown in Table 1. Incubate the polynucleotide–spermidine complex for 1 h at room temperature.
4. After 90 min incubation, remove the protamine sulfate solution from the microtiter wells. Wash the wells three times with 300 µL/well of dd water. The last wash should last at least 5 min (*see* Note 5).
5. Remove the washing buffer from the microtiter wells. Coat the microtiter wells with 200 µL/well of polynucleotide–spermidine solution as shown in Fig. 1. Incubate the plate for approx 16 h at 4°C.

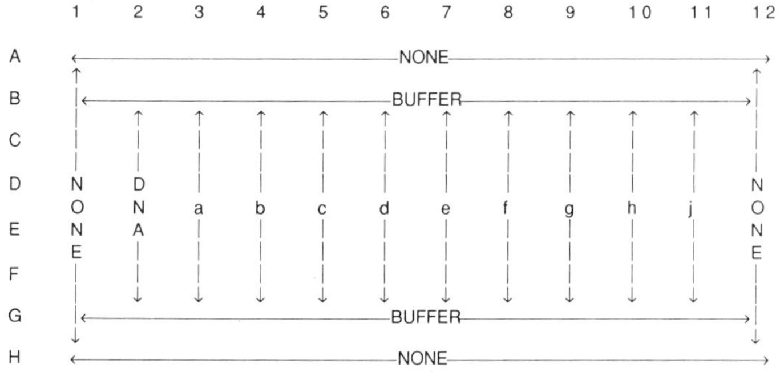

Fig. 1. Configuration of a microtiter plate used to study the induction of B-DNA to Z-DNA transition in poly(dG-m⁵dC).poly(dG-m⁵dC) by spermidine. The concentration of spermidine added to the polynucleotide (designated a to h and j inside the plate) is described in Table 1. Column 2 (DNA) contains poly(dG-m⁵dC).poly(dG-m⁵dC) similar to column 3. However, these wells will not be treated with the monoclonal antibody.

6. Remove the polynucleotide solution from the microtiter wells. Wash the wells three times with 300 µL/well of washing buffer as in Step 2.
7. Dilute the monoclonal antibody (Z22) solution to a concentration of 0.35 µg/mL in washing buffer (*see* Note 6). Add 200 µL/well of the monoclonal antibody solution to wells 3–11 in rows B→G. Add 200 µL/well of washing buffer to wells B2, C2, D2, E2, F2, and G2. Incubate the microtiter plate at 37°C for 1 h.
8. Remove the monoclonal antibody solution and wash the wells three times with washing buffer as in Step 3.
9. Dilute alkaline phosphatase-conjugated polyvalent immunoglobulin solution 1:350 in washing buffer containing 1% BSA. Add 200 µL/well of the substrate solution to all 60 inner wells that were treated with protamine sulfate. Incubate the microtiter plate at 37°C for 1 h.
10. Remove the enzyme-conjugated immunoglobulin solution and wash the wells three times with washing buffer as in Step 3.
11. Add 200 µL/well of the phosphatase substrate solution to all 60 inner wells as in Step 9. Incubate the plate at room temperature in the dark for 30 min.
12. Stop the enzyme–substrate reaction by adding 50 µL of 1*M* NaOH to each well.
13. Read the optical density at 405 nm using a Microplate Autoreader.
14. Calculate the mean optical density values of quadruplicate wells containing polynucleotide solution with the same spermidine concentration, for example, wells C4, D4, E4, and F4.

15. Calculate the mean background reading for each spermidine concentration. For example, background reading for wells identified in Step 14 is that from wells B4 and G4 (*see* Note 7).
16. Subtract the mean background values from the mean values for wells containing polynucleotide–spermidine complex. In the example given, subtract the mean obtained in Step 15 from that in Step 14. The result is the mean optical density owing to the binding of monoclonal anti-Z-DNA antibody to the polynucleotide.
17. In order to determine the efficacy of spermidine to provoke the B→Z transition, plot the net optical density values against the concentration of spermidine. The midpoint concentration is determined from this plot at 50% increase in net optical density.

4. Notes

1. The enzyme immunoassay method is described here to detect the induction of Z-DNA conformation in poly(dG-m^{5}dC).poly(dG-m^{5}dC) in the presence of spermidine. The polynucleotide and/or spermidine can be substituted for the system under investigation. We have successfully used this method to detect the induction of Z-DNA conformation in polynucleotides by NaCl *(19)*, $Co(NH_3)_6^{3+}$, and $Co(en)_3^{3+}$ *(24)*. In addition, we found that polyamines could provoke the Z-DNA conformation in recombinant plasmid DNAs with as little as 0.5% insert of a $(dG\text{-}dC)_n.(dG\text{-}dC)_n$ sequence by this technique *(25)*. In this method, we used the plasmid DNAs at a concentration of approx 10 µg/mL and incubated the ligand-DNA solutions for 1 h at room temperature to attain equilibrium. The ability of the monoclonal antibody to recognize the Z-DNA conformation rather than specific base sequences or poly-nucleotide–ligand combinations amply demonstrates that the immunological method described here is quite general in nature and is adaptable to recognize the presence of Z-DNA segments in native DNAs also as demonstrated in the case of recombinant plasmids *(25)*.
2. Costar #3590 microtiter plates were used in our experiments with protamine sulfate as the linking agent. From a comparative evaluation of various microtiter plates, Eaton et al. *(26)* found these plates to be the most satisfactory in enzyme immunoassays with anti-DNA antibodies. Our experience with various microtiter plates is consistent with their finding.
3. We have not used the outer wells in our assay because of the difficulty of achieving thermal equilibrium in these wells.
4. We routinely use protamine sulfate as a linking agent to facilitate the adsorption of DNA to the microtiter plate. In our experiments, the

background reading in wells that contained all reagents except the DNA was 0.1–0.2 compared to a reading of 2.2 in wells treated with poly(dG-m^5dC).poly(dG-m^5dC) + 50 μM spermidine or brominated poly(dG-dC).poly(dG- dC). Zouali and Stollar *(27)* found that UV irradiation is an effective method to adsorb DNA to the microtiter plate.

5. The plate washing can be conducted automatically if a microplate washing system is available. We used the Perkin Elmer-Cetus Pro/Pette instrument with a Pro/Wash head and the Beckman Biomek 1000 Automated Laboratory Workstation in our enzyme immunoassays. If automated instrumentation is not available, liquid transfers and plate washing can be carried out by a 12-channel micropipet.
6. The method described here uses a monoclonal anti-Z-DNA antibody Z22 *(12)*. This can be substituted by other monoclonal or polyclonal antibodies. Rabbits or goat could be immunized with brominated poly(dG-dC).poly(dG-dC) to produce polyclonal anti-Z-DNA antibody. This polynucleotide exists in the Z-DNA conformation at 150 m*M* NaCl.
7. If the background values are high, a solution of gelatin (1 mg/mL) in distilled water can be used to block nonspecific binding of antibodies to the microtiter plate *(28)*. In this case, add 300 µL of gelatin solution per well to the microtiter plate after the polynucleotide incubation step. Incubate the plates with gelatin for at least 4 h (preferably overnight at 4°C) before adding the monoclonal/polyclonal antibody.
8. The configuration of the microtiter plate given in Fig. 1 is only one way of adding the DNA solutions. Instead of quadruplicate wells with DNA, duplicates may suffice if the standard deviation is relatively small (<10%).
9. Keeping a column of wells with DNA poly ([dG-m^5dC].poly[dG-m^5dC]) (B2 to G2) will show the extent of enzyme–substrate reaction in the absence of the antibody.

Acknowledgment

This work was supported by grants from the National Institutes of Health (RO1 AR 39020), the New Jersey Chapter of the Arthritis Foundation, and the Foundation of the University of Medicine and Dentistry of New Jersey.

References

1. Pohl, F. M. and Jovin, T. M. (1972) Salt-induced co-operative conformational change of a synthetic DNA: Equilibrium and kinetic studies with poly(dG-dC). *J. Mol. Biol.* **67**, 375–396.
2. Wang, A. H-J., Quigley, G. J., Kolpak, F. J., Crawford, J. L., van Boom, J. H., van der Marel, G., and Rich, A. (1979) Molecular structure of a left-handed helical DNA fragment at atomic resolution. *Nature* **282**, 680–686.

3. Thamann, T. J., Lord, R. C., Wang, A. H-J. and Rich, A. (1981) The high salt form of poly(dG-dC). poly(dG-dC) is left-handed Z-DNA: Raman spectra of crystals and solutions. *Nucleic Acids Res.* **9**, 5443–5457.
4. Behe, M. and Felsenfeld, G. (1981) Effects of methylation on a synthetic polynucleotide: The B-Z transition in poly(dG-m^{5}dC). poly(dG-m^{5}dC). *Proc. Natl. Acad. Sci. USA* **78**, 1619–1623.
5. Lafer, E. M., Möller, A., Nordheim, A., Stollar, B. D., and Rich, A. (1981) Antibodies specific for left-handed Z-DNA. *Proc. Natl. Acad. Sci. USA* **78**, 3546–3550.
6. Arndt-Jovin, D. J. Robert-Nicoud, M., Zarling, D. A., Greider, C., Weimer, E., and Jovin, T. M. (1983) Left-handed Z-DNA in bands of acid-fixed polytene chromosomes. *Proc. Natl. Acad. Sci. USA* **80**, 4344–4347.
7. Jaworsky, A., Hsieh, W. T., Blaho, J. A., Larson, J. E., and Wells, R. D. (1988) Left-handed DNA in vivo. *Science* **238**, 773–777.
8. Rich, A., Nordheim, A., and Wang, A. H-J (1984) The chemistry and biology of left-handed Z-DNA. *Ann. Rev. Biochem.* **53**, 791–846.
9. Lafer, E. M., Valle, R. P. C., Möller, A., Nordheim, A., Schur, P. H., Rich, A., and Stollar, B. D. (1983) Z-DNA specific antibodies in human systemic lupus erythematosus. *J. Clin. Invest.* **71**, 314–321.
10. Thomas, T. J., Meryhew, N. L., and Messner, R. P. (1988) DNA sequence and conformational specificity of lupus autoantibodies. Preferential binding to the left-handed Z-DNA form of polynucleotides. *Arthritis Rheum.* **31**, 367–377.
11. Sibley, J. T., Lee, J. S., and Decoteau, W. E. (1984) Left-handed "Z" DNA antibodies in rheumatoid arthritis and systemic lupus erythematosus. *J. Rheumatol.* **11**, 633–637.
12. Möller, A., Gabriels, J. E., Lafer, E. M., Nordheim, A., Rich, A., and Stollar, B. D. (1982) Monoclonal antibodies recognize different parts of Z-DNA. *J. Biol. Chem.* **257**, 12081-12085.
13. Zarling, D. A., Arndt-Jovin, D. J., Robert-Nicoud, M., McIntosh, L. P., Thomae, R., and Jovin, T. M. (1984) Immunoglobulin recognition of synthetic and natural left-handed Z-DNA conformations and sequences. *J. Mol. Biol.* **176**, 369–415.
14. Pohl, F. M., Thomae, R., and DiCapua, E. (1982) Antibodies to Z-DNA interact with form V DNA. *Nature* **300**, 545,546.
15. Nordheim, A., Pardue, M. L., Lafer, E. M., Möller, A., Stollar, B. D., and Rich, A. (1981) Antibodies to left-handed Z-DNA bind to interband regions of Drosophila polytene chromosomes. *Nature* **294**, 417–422.
16. Lancillotti, F., Lopez, M. C., Alonso, C., and Stollar, B. D. (1985) Locations of Z-DNA in polytene chromosomes. *J. Cell Biol.* **100**, 1759–1766.
17. Arndt-Jovin, D. J., Robert-Nicoud, M, Baurschmidt, P., and Jovin, T. M. (1985) Immunofluorescence localization of Z-DNA in chromosomes: Quantitation by scanning microphotometry and computer-assisted image analysis. *J. Cell Biol.* **101**, 1422–1433.
18. Viegas-Pequignot, E., Derbin, C., Tailandler, E., Leng, M. and Dutrilaux, W. (1983) Z- DNA immunoreactivity in fixed metaphase chromosomes of primates. *Proc. Natl. Acad. Sci. USA* **80**, 5890–5894.
19. Thomas, T. J., Baarsch, M. J., and Messner, R. P. (1988) Immunological detection of B-DNA to Z-DNA transition of polynucleotides by immobilization of the DNA conformation on a solid support. *Anal. Biochem.* **168**, 358–366.

20. Thomas, T. J. and Messner, R. P. (1988) Structural specificity of polyamines in left-handed Z-DNA formation. Immunological and spectroscopic studies. *J. Mol. Biol.* **201**, 463–467.
21. Thomas, T. J. and Bloomfield, V. A. (1985) Toroidal condensation of Z DNA and identification of an intermediate form in the B to Z transition of poly(dG-m^{5}dC). poly(dG-m^{5}dC). *Biochemistry* **24**, 713–719.
22. Bustamante, C., Tinoco, I. Jr., and Maestre, M. F. (1983) Circular differential scattering can be an important part of the circular dichroism of macromolecules. *Proc. Natl. Acad. Sci. USA* **80**, 3568–3572.
23. Thomas, T. J. and Messner, R. P. (1986) Effects of lupus-inducing drugs on the B to Z transition of synthetic DNA. *Arthritis Rheum.* **29**, 638–645.
24. Thomas, T. J. and Thomas, T. (1989) Direct evidence for the presence of left-handed conformation in a supramolecular assembly of polynucleotides. *Nucleic Acids Res.* **17**, 3795–3810.
25. Thomas, T. J., Gunnia, U. B., and Thomas, T. (1989) A facile transition of recombinant plasmids to the left-handed Z-DNA form by polyamines. A possible mechanism for anti-DNA antibody production in lupus. *Arthritis Rheum.* **32**, S143. Abstract.
26. Eaton, R. B., Schnneider, G., and Schur, P. H. (1983) Enzyme immunoassay for antibodies to native DNA. Specificity and quality of antibodies. *Arthritis Rheum.* **26**, 52–62.
27. Zouali, M. and Stollar, B. D. (1986) A rapid ELISA for measurement of antibodies to nucleic acid antigens using UV-treated polystyrene microplates. *J. Immunol. Methods* **90**, 105–110.
28. Rubin, R. L., Joslin, F. G., and Tan, E. M. (1983) An improved ELISA for anti-native DNA antibodies by elimination of interference by anti-histone antibodies. *J. Immunol. Methods* **63**, 359–367.

CHAPTER 37

Cell Sorting Using Immunomagnetic Beads

Eddie C. Y. Wang, Leszek K. Borysiewicz, and Anthony P. Weetman

1. Introduction

Immunomagnetic beads are uniform, polymer particles coated with a polystyrene shell that provides both a smooth hydrophobic surface to facilitate physical absorption of molecules, such as antibodies, and surface hydroxyl groups that allow covalent chemical binding of other bioreactive molecules, such as streptavidin, lectins, and peptides. Iron (III) oxide (Fe_2O_3) deposited in the core gives the beads superparamagnetic properties that lead to consistent and reproducible reactions to a magnetic field without permanent magnetization of the particles. These are the two qualities on which immunomagnetic separation (IMS) depends.

IMS is a fast, simple method for separating a range of targets, and principally involves the removal of an indirectly magnetized target by a permanent magnet. "Magnetization" is achieved using beads coated with a target-specific, bioreactive molecule, and removal occurs by the application of a neodynium-iron boron or similar magnet. With respect to cell sorting, "target-specific, bioreactive molecules" consist of monoclonal antibodies (MAb) specific to certain cell subsets. Thus, OKT8 and OKT4 (used in this chapter) are mouse antihuman MAbs to the T-cell markers CD8 and CD4 that, respectively, target the cytotoxic/suppressor and helper T-cell subsets of the immune system.

From: *Methods in Molecular Biology, Vol. 10: Immunochemical Protocols*
Ed.: M. Manson ©1992 The Humana Press, Inc., Totowa, NJ

Although this chapter will only discuss IMS in relation to cell sorting (also termed MACS or magnetic-affinity cell sorting) and T-cell subset separation and selection, the technique has also been adapted for DNA sequencing *(1)*, purification of DNA binding proteins *(2)*, immobilization and isolation of nucleic acids *(3)*, tissue-typing *(4,5)*, quantification of lymphocyte subsets directly from blood *(6)*, bone marrow T-cell depletion *(7)*, depletion of malignant neuroblastoma cells from bone marrow *(8,9)*, and the selective enrichment of microorganisms *(10)*, and other cell types, such as Langerhans cells *(11)* or antigen-specific B-cells *(12)*. IMS is, therefore, a highly versatile procedure.

The advantages of this technique over other separation methods can be summarized:

1. The procedure is very rapid.
2. Beads are easy to work with under sterile conditions.
3. The procedure is cheaper than FACS (fluorescence-activated cell sorting).
4. The magnetic particles, once coated, may be stored for at least several months at 4°C.
5. Both positive and negative selection of cells can be achieved with attainment of positively selected cells by overnight incubation.
6. Unlike FACS, viability can be measured during the separation procedure using ethidium bromide and acridine orange.
7. The method can be used in large-scale isolations. However, the full effects of having polymer beads in close association with cells is still unknown, and isolation still involves the use of a MAb, which may react with target cells. IMS has enormous potential, but a fuller understanding of bead–cell interactions is currently needed.

IMS itself can be split into the following major procedures:

1. Preparation of beads;
2. Preparation of cells for use with immunomagnetic beads coated with sheep antimouse IgG;
3. Immunomagnetic separation using immunomagnetic beads coated with sheep antimouse IgG;
4. Immunomagnetic separation using uncoated immunomagnetic beads; and
5. Positive selection using immunomagnetic beads.

2. Materials

1. Immunomagnetic Beads: Dynabeads uncoated, or coated as M-280 or M-450 sheep antimouse IgG (available from Dynal, Wirral, Merseyside, UK). Stored at 2–8°C, they will remain stable for up to 1 yr. Thiomerosal

(0.01%) or sodium azide (0.02%) is use as preservative. Hazard Warning: Both sodium azide and thiomerosal are irritants to the skin and eyes, and are very toxic if inhaled or ingested. Handle with care.

N.B. Immunomagnetic particles developed by Biolab, Belgium are also available from Metachem Diagnostics.

2. Phosphate buffered saline (PBS), pH 7.4: 8.0 g of NaCl, 0.2 g of KH_2PO_4, 2.9 g of Na_2HPO_4. $12H_2O$, 0.2 g of KCl, and 0.2 g of NaN_3. Make up to 1 L with sterile, distilled water. Store at room temperature.
3. PBS/BSA: PBS plus 0.1% bovine serum albumin. Store at 4°C. Addition of 0.02% azide is optional.
4. PBS/FCS: PBS plus 1.0% fetal calf serum. Store at 4°C. Addition of 0.02% azide is optional.
5. Hank's balanced salt solution (HBSS), pH 7. 4: Make to 0.05*M*. Store at 15-30°C.
6. Permanent Magnet: Dynal Magnetic Particle Concentrator (MC 1 or MPC 6) (Dynal, UK) or equivalent magnet (Biolab, Belgium also supply magnets).
7. Tris-HCl, 0.05*M*, pH 9.5. Store at 4°C.
8. Specific monoclonal or polyclonal antibody to cell subsets. OKT8 (Leu 2) recognizes CD8 marker on cytotoxic/suppressor T-cells. OKT4 (Leu 3) recognizes CD4 marker on helper T-cells. Store at 4°C.
9. Heparin: Make to 50 U/mL with PBS. Use 5 U/mL of blood. Store at 4°C.
10. Lymphoprep: Store at 2–4°C protected from light.
11. Ethidium Bromide/Acridine Orange Mix: Stock solutions consist of 50 mg of ethidium bromide and 15 mg of acridine orange, dissolved in 1 mL of 95% ethanol and made up to 50 mL with distilled water. Store at –20°C.

    For use in viability counts, stocks are diluted 1:100 with PBS. Stored at room temperature in the dark, the solution has a shelf life of at least 2 mo. Hazard Warning: Both ethidium bromide and acridine orange have mutagenic properties, are combustible, toxic, and irritants to the skin, eyes, and respiratory system. Handle with care.

3. Methods

3.1. Preparation of Beads

Immunomagnetic beads can be obtained from Dynal, UK with sheep antimouse IgG attached (Section 3.1.1.) or uncoated for physical absorption of IgM antibodies or chemical coupling of IgG antibodies (as recommended by the producers) (Section 3.1.2.).

3.1.1. Removal of Preservatives

For immunomagnetic beads with sheep antimouse IgG attached, the following protocol should be used to remove preservatives, such as thiomerosal or sodium azide, that may affect cell viability.

1. Suspend the immunomagnetic beads well by pipeting up and down.
2. Take an appropriate aliquot of beads. For depletion, recommended bead:target cell ratios range from 10:1 to 40:1. For positive selection, a bead:target cell ratios of 3:1 is sufficient.
3. Make up to 10 mL with PBS/FCS.
4. Collect beads using a strong permanent magnet (the magnets suggested in the Materials section are supplied with a useful holding device for test tubes/universals). This process takes approx 2 min.
5. Discard the supernatant while the tube is still held against the magnet.
6. Repeat 10 times. The beads are now ready for use.

3.1.2. Coating the Beads

Uncoated immunomagnetic beads can be coated with IgM using the following procedure *(13)*. Note that Metachem Diagnostics supply immunomagnetic particles with antimouse IgM attached.

1. Make a homogenous suspension of beads by light shaking and pipet off the desired amount of beads.
2. Dissolve the IgM in 0.05*M* Tris-HCl, pH 9. 5. Usually, 1–5 µg of IgM per mg of immunomagnetic beads give optimal coating.
3. Remove the supernatant from uncoated immunomagnetic beads using the magnet for two minutes and either pouring or pipeting off the supernatant.
4. Add the IgM in Tris-HCl and mix well. The recommended concentration of beads to antibody is 50 mg beads/mL of IgM in Tris-HCl.
5. Incubate overnight at 4°C on a rotator at approx 20 rpm.
6. Remove the supernatant (as before) and wash four times for 5 min each time, in PBS/BSA or PBS/FCS with a final wash overnight at 4°C.
7. The beads are now ready for use.

3.2. Preparation of Cells

Preparation of cells for use with immunomagnetic beads coated with sheep antimouse IgG:

1. Prepare peripheral blood mononuclear cells (PBMC) from blood anticoagulated using heparin, by Ficoll-Hypaque density (Lymphoprep) centrifugation *(14)*. The process involves carefully layering anticoagulated

blood onto Lymphoprep (in a 4:3 ratio of blood to Lymphoprep), such that there is no mixing of blood and separation fluid. After spinning at 800*g* for 15 min without brake, mononuclear cells form a distinct, cloudy band at the interface between the sample layer and Lymphoprep solution. Erythrocytes aggregate and sediment to the bottom of the tube. PBMCs are harvested using a Pasteur pipet, avoiding removal of the upper layer.

2. Wash PBMCs three times with 10 mL of PBS by spinning down first at 800*g* for 10 min, then twice at 400*g* to remove excess Lymphoprep and platelets from the sample.
3. Make viability counts by taking 50 µL and adding an equal volume of ethidium bromide/acridine orange mix. Viable cells stain green, non-viable cells red *(15)*.
4. Resuspend PBMCs at 4×10^7 cells/mL in PBS.
5. Lay down 100 µL aliquots in 96-well U-shaped plates and add 50 µL of the chosen monoclonal antibody at an appropriate concentration. Usually, the antibody is used according to the manufacturer's instructions.
6. Incubate at room temperature for 25 min.
7. Wash with PBS by spinning down at 600*g* for 5 min to pellet cells, aspirating off the supernatant (care is needed to avoid removing pellet) and resuspending the cells in 150 µL of PBS. Repeat this twice to remove excess antibody.
8. Transfer cells to a test tube or universal bottle. The cells are now ready for separation.

3.3. Immunomagnetic Separation with Coated Beads

Immunomagnetic separation using immunomagnetic beads coated with sheep antimouse IgG.

1. Add an appropriate number of prepared beads (*see* Section 3.1.) to the cells. For depletion, a bead:target cell ratio of 10:1–40:1 is recommended (*see* Notes 1 and 2 for discussion on separation efficiency). For positive selection, bead:cell ratios of 3:1 are sufficient.
2. Incubate at 4–20°C for 30–60 min, depending on individual systems, with mild agitation every 5–10 min (*see* Notes 1 and 2).
3. Make up to 5 mL with HBSS.
4. Remove beads by applying the magnet for 2 min to the side of the tube and pouring off the supernatant. To obtain higher yields, wash the beads twice by adding another 5 mL of HBSS and repeat the separation.

5. Perform the separation procedure on the supernatant to remove as many carried over beads as possible.
6. A small sample should be passed through a flow cytometer to check the efficiency of depletion.
7. Sequential depletions may be more efficient. This involves two or more separation procedures using decreased bead:target cell ratios and shorter incubation periods, e.g., a 3:1 bead:cell ratio repeated three times with 20 min incubation periods at 4–20°C. This gives around 95% or greater depletion and use of lower bead:cell ratios may decrease carryover effects.

3.4. Immunomagnetic Separation with Uncoated Beads

1. Uncoated immunomagnetic beads can be coated with IgM as described in Section 3.1.
2. Coated beads can then be used for separation experiments by direct incubation with cells at similar bead:target cell ratios and with the same protocol as in Section 3.3.

3.5. Positive Selection Using Immunomagnetic Beads

Positively selected cells attached to the beads after incubation can be collected by overnight incubation at 37°C and then application of a magnet to remove the shed beads. For a discussion on problems with positive selection, *see* Notes 3–5.

4. Notes

The general consensus is that ideal conditions for separation using immunomagnetic beads vary from system to system and must, therefore, be investigated by the individual using the technique. However, there are a number of useful points that should be considered.

1. A major problem when using IMS to deplete cell subsets is the loss of cells owing to nonspecific binding and carriage with the beads during separation. This can be anything up to 70% if incubating at room temperature. Nonspecific binding is kept to a minimum by low temperatures, i.e., 4°C, though incubation periods have to be increased correspondingly to allow for the slower kinetics of interaction (*see* Figs. 1 and 2). However, higher temperatures favor depletion of cells expressing very low levels of the specific marker targeted (in this case, CD8), which remain in the supernatant when using incubations at 4°C, inde-

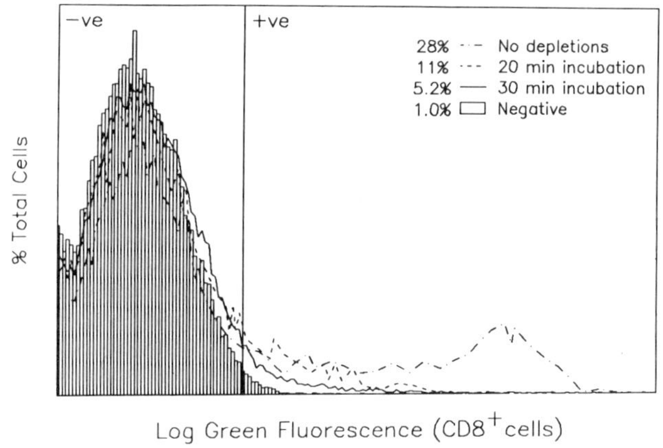

Fig. 1. Effect of incubation time on deletion of CD8+ cells. PBMCs were prepared from anticoagulated peripheral blood and stained with OKT8, as in Section 3.2. Beads were added at a 20:1 bead:target cell ratio. The mixture was left for varying incubation times (10, 20, or 30 min) at room temperature, after which samples were made up to 10 mL in HBSS and a magnet applied for 2 min to gather bead:CD8+ cell rosettes. The supernatant was collected, cells spun down, resuspended in 150 µL of PBS, and stained with antimouse Ig fluorescein isothiocyanate (FITC)-conjugated secondary antibody. After 25-min incubation at room temperature, samples were washed three times with PBS and passed through an EPICS-Profile flow cytometer (Coulter) to measure the CD8+ cells remaining. Negative controls included an unstained sample (negative) and a sample stained with the antimouse Ig FITC conjugate only (not shown). The positive control was a sample that had not been passed through a depletion cycle (no depletions). Results show the total depletion of cells expressing high levels of CD8 independent of the incubation period, but increased depletion of cells expressing low levels of CD8 with the 30-min incubation (solid line) compared to 20-min (dashed line). However, a small percentage of low-expressing CD8+ cells still remain.

pendent of the number of separations carried out. However, incubations at 37°C lead to high levels of nonspecific binding. The "best" temperature for separation is dependent on the individual system used and the goal of any particular experiment.

Carriage of a percentage of cells with beads is almost inevitable. This can be kept to a minimum by diluting mixtures before each separation procedure. Washing and application of the magnet on bead–cell mixtures several times after initial separation also help recapture as many carried cells as possible. Even with these precautions, between 30 and 50% of the original number of cells is usually lost.

2. With regard to depletion followed by cell culture after separation, there is the potential problem of leftover beads. A small number of beads

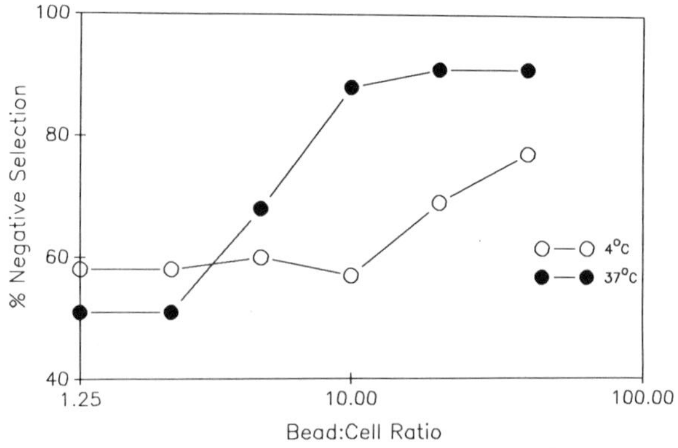

Fig. 2. Effect of temperature and bead:cell ratio on negative selection. PBMCs were prepared from anticoagulated peripheral blood as described in Section 3.2. Cells were either unstained, incubated with excess OKT8, or excess antimouse Ig FITC-conjugated secondary antibody (control for nonspecific binding of this Ab). Beads were added at varying bead:cell ratios (from 40 to 1.25:1). Mixtures were left for 20 min with agitation every 5 min, at 4°C (open circle), or 37°C (filled circle). Samples were then made up to 10 mL in HBSS and a magnet applied for 2 min to gather bead:CD8+ cell rosettes. The supernatant was collected, cells spun down, resuspended in 150 µL and stained with antimouse Ig FITC-conjugated secondary antibody, using the same procedure as with OKT8. Samples were washed three times with PBS and passed through an EPICS-Profile to measure the CD8+ cells remaining. Percentage negative selection was calculated using the formula: (%CD8+ cells in depleted samples – %+ve cells in antimouse Ig FITC control)/(%CD8+ cells in undepleted samples – %+ve cells in antimouse Ig FITC control) × 100. Results show a marked increase in depletion as the bead:cell ratio and the temperature increases.

always remain uncaptured by the magnet and stay with the supernatant. These seem to be harmless to cells in low numbers (and are usually phagocytosed by the macrophages in culture), but it is best to keep these numbers to a minimum by applying the separation procedure to the supernatant a number of times. Using HBSS for these washes (rather than PBS) has been reported to give significantly better yields, though the reasons for this are unknown. The use of lower bead:cell ratios (Fig. 2) with a greater number of separation procedures should be considered (Fig. 3). Although this would increase total loss of cells by nonspecific binding and cell carriage with the beads, there would be less beads carried over with the supernatant resulting in "cleaner" cultures.

3. With respect to positive selection of cells, it is claimed that cells shed beads with overnight incubation at 37°C, and therefore, removal of beads after such an incubation results in purified, functionally active, bead-free cell subsets. This is partially true, but about 50% of positively

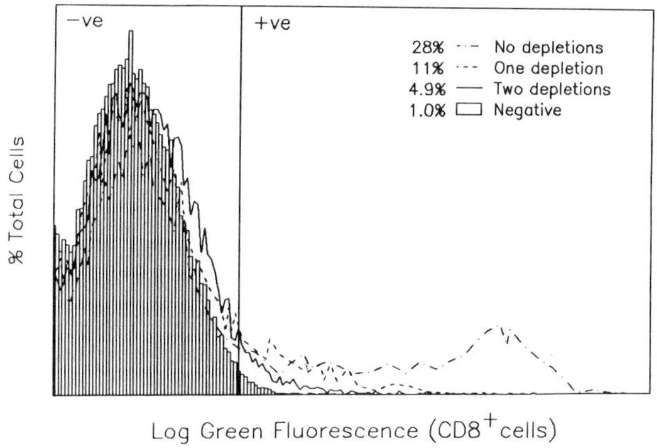

Fig. 3. Effect of number of cycles of depletion on CD8+ cells. PBMCs were prepared from anticoagulated peripheral blood as described in Section 3.2. Cells were either unstained, incubated with excess OKT8, or excess antimouse Ig FITC-conjugated secondary antibody (control for nonspecific binding of this Ab). Beads were added at a 20:1 bead:cell ratio. Samples were incubated for 20 min at 4°C with agitation every 5 min, and then made up to 10 mL in HBSS and a magnet applied for 2 min to gather bead:CD8+ cell rosettes. The supernatant was collected, cells spun down, resuspended in 150 µL, and beads added to the two depletion samples for a second 20-min incubation period and washes (as above). Antimouse Ig FITC-conjugated secondary antibody was then added and incubated at room temperature for 25 min. Samples were washed three times with PBS and passed through an EPICS-Profile to measure the CD8+ cells remaining. Negative controls included an unstained sample (negative) and an antimouse Ig FITC-conjugated secondary antibody stain only (not shown). Positive control was an OKT8 stained undepleted sample (no depletion). Results show that two cycles of depletion (solid line) reduce the numbers of low expressing CD8+ cells remaining compared with one cycle (dashed line).

selected cells still remain with the beads when separation is attempted, giving rise to low yields of bead-free cells.

4. Furthermore, there can be problems with cell viability following close contact with the beads. In this laboratory, it has been found that positively selected T-cells often do not survive longer than 48 hours in culture after shedding of beads, although use of apparently the same protocol has also yielded T-cells capable of good, though possibly reduced, proliferative responses to stimulants, such as OKT3 and phytohaemagglutinin (PHA). Thus, using CD4+ positively selected T-cells, stimulation indices (measured by ^{3}H-thymidine uptake) ranging from 12 to 2500 have been recorded after PHA stimulation at 1 µg/mL from separate cultures. The responses of positively selected cells have not been thoroughly investigated here, but a number of groups have had no problems with cell

function and viability after positive selection *(16)*.

5. The use of enzymes, such as papain, to cleave bead:cell rosettes has also been investigated *(17)*. Chymopapain, used at 200 U/10^7 cells per mL of TC199 tissue culture medium for 10 min at 37°C, was found to be most effective and least toxic to KG1a human leukemia cells and gave 100% recovery of viable cells.

References

1. Hultman, T., Stahl, S., Hornes, E., and Uhlen, M. (1989) Direct solid phase sequencing of genomic and plasmid DNA using magnetic beads as solid support. *Nucleic Acids Res.* **17,** 4937–4946.
2. Gabrielsen, O. S., Hornes, E., Korsnes, L., Ruet, A., and Oyen, T. B. (1989) Magnetic DNA affinity purification of yeast transcription factor T — a new purification principle for the ultrarapid isolation of near homogeneous factor. *Nucleic Acids Res.* **17,** 6253-6267.
3. Uhlen, M. (1989) Magnetic separation of DNA. *Nature* **340,** 733,734.
4. Hansen, T. and Hannestad, K. (1989) Direct HLA typing by rosetting with immunomagnetic beads coated with specific antibodies. *J. Immunogenet.* **16,** 137–139.
5. Laundy, G. J., Peberdy, M., Klouda, P. T., and Bradley, A. (1989) Applications of automated simultaneous double fluorescence (SDF). II. HLA class I phenotyping using immunomagnetically separated T-lymphocytes. *J. Immunogenet.* **16,** 141–148.
6. Brinchmann, J. E., Leivestad, T., and Vartdal, F. (1989) Quantification of lymphocyte subsets based on positive immunomagnetic selection of cells directly from blood. *J. Immunogenet.* **16,** 177–183.
7. Gee, A. P., Mansour, V., Weiler, M. (1989) T-cell depletion of human bone marrow. *J. Immunogenet.* **16,** 103–115.
8. Combaret, V., Favrot, M. C., Chauvin, F., Bouffet, E., Philip, I., and Philip, T. (1989) Immunomagnetic depletion of malignant cells from autologous bone marrow graft: From experimental models to clinical trials. *J. Immunogenet.* **16,** 125–136.
9. Kemshead, J. T., Heath, L., Gibson, F. M., Katz, F., Richmond, F., Treleaven, J., and Ugelstad, J. (1986) Magnetic microspheres and monoclonal antibodies for the depletion of neuroblastoma cells from bone marrow: Experiences, improvements and observations. *Br. J. Cancer* **54,** 771–778.
10. Lund, A., Helleman, A. L., and Vartdal, F. (1988) Rapid isolation of K88+ *Escherichia coli* by using immunomagnetic particles. *J. Clin. Microbiol.* **26,** 2572–2575.
11. Schmitt, D. A., Hanau, D., and Cazenaze, J.-P. (1989) Isolation of epidermal Langerhans cells. *J. Immunogenet.* **16,** 157–168.
12. Ossendorp, F. A., Bruning, P. F., Van den Brink, J. A. M., and de Boer, M. (1989) Efficient selection of high affinity B cell hybridomas using antigen-coated magnetic beads. *J. Immunol. Methods* **120,** 191-200.
13. Guadernack, G., Leivestad, T., Ugelstad, J., and Thorsby, E. (1986) Isolation of pure functionally active CD8+ T cells: Positive selection with monoclonal antibodies directly conjugated to monosized magnetic microspheres. *J. Immunol. Methods* **90,** 179–187.
14. Timonen, T., Ortaldo, J. R., and Herberman, R. H. (1981) Characteristics of human

large granular lymphocytes and relationships to natural killer and K cells. *J. Exp. Med.* **153,** 569–582.

15. Parks, D. R., Bryan, V. M., Oi, V. I., and Herzenberg, L. A. (1979) Antigen-specific identification and cloning of hybridomas with a fluorescent-activated cell sorter (FACS). *PNAS, USA* **76,** 1962–1966.
16. Leivestad, T., Guadernack, G., Ugelstad, J., Vartdal, F., and Thorsby, E. (1987) Isolation of pure and functionally active T4 and T8 cells by positive selection with antibody-coated monosize magnetic microspheres. *Transplantation Proceedings* **19,** 265–267.
17. Civin, C. I., Strauss, L. C., Fackler, M. J., Trischmann, T. M., Wiley, J. M., and Loken, M. R. (in press) Positive stem cell selection: Basic science. *Bone Marrow Transplantation.*

CHAPTER 38

Cell Preparation for Flow Cytometry

Michael G. Ormerod

1. Introduction

For flow cytometry, a suspension of single cells, free of large clumps and excess debris, is essential. The quality of the data obtained depends as much on the quality of the preparation as on that of the flow cytometer.

Cells taken from an animal are sometimes already in a single cell suspension, for example, in blood or bone marrow. However, analysis is usually easier if the erythrocytes are removed, and two methods for doing this are given (Sections 3.1. and 3.2.).

The preparation of a suspension of leucocytes from solid tissues is comparatively straightforward (Section 3.3.). The preparation of other cells from solid tissues is more difficult; the procedures depend on the tissue and the type of cell required *(1)*. Most methods involve the use of proteolytic enzymes that may destroy surface antigens. Frequently, better results are obtained if the cells are put in short-term culture. Section 3.4. gives a method for preparing cells grown on a plastic surface.

Cells prepared by these techniques are suitable for use in the procedures described in Chapters 41 and 42.

2. Materials

1. Ammonium chloride lysis buffer: 8.29 g of ammonium chloride, 1 g of potassium hydrogen carbonate, 37 mg of EDTA disodium salt in 1 L of distilled water. Make up fresh, adjust pH to 7.2 and pass through a 0.47-µm membrane filter before use.
2. Bovine serum albumin (BSA).

From: *Methods in Molecular Biology, Vol. 10: Immunochemical Protocols*
Ed.: M. Manson ©1992 The Humana Press, Inc., Totowa, NJ

3. 0.1% BSA–DAB solution (BSA–DAB): 500 mL of Dulbecco's formulas A and B phosphate buffered saline (DAB) pH 7.3 (use tablets, Oxoid Ltd., Basingstoke, UK, as instructed). Add 5 mL of 10% BSA in distilled water to 500 mL of DAB. Pass through a 0.47-μm membrane filter before use.
4. Phosphate buffered saline (PBS): 8.5 g of sodium chloride, 1.07 g of disodium hydrogen orthophosphate (or 2.7 g of $Na_2HPO_4 \cdot 12H_2O$), 0.39 g of sodium dihydrogen orthophosphate (or 0.51 g of $NaH_2PO_4 \cdot 2H_2O$). Make up to 1 L with distilled water.
5. Soybean trypsin inhibitor.
6. Leibowitz (L-15) culture medium.
7. Fetal calf serum (FCS).
8. Crystalline trypsin (chymotrypsin free).
9. Histopaque 1077 (Sigma Chemical Co. Ltd.)
10. Lymphoprep (Nyegaard AS, Oslo, Norway).
11. Deoxyribonuclease, bovine pancreatic Type 1 (DNAase).
12. Tubes for the collection of blood containing an anticoagulant (either EDTA or heparin).

3. Methods

3.1. Preparation of Leucocytes from Peripheral Blood by Red Blood Cell Lysis (2,3)

1. Collect 8 mL of blood into a tube containing an anticoagulant (EDTA or heparin). Place 2-mL aliquots into each of four 50-mL tubes.
2. To one tube at a time, rapidly add 45 mL of ammonium chloride lysis buffer and mix immediately.
3. Incubate at room temperature for 10 min to lyse the red blood cells.
4. Centrifuge at 400*g* for 10 min at 4°C. Decant the supernatant. Add 5 mL of BSA–DAB to each pellet, resuspend the cells and combine into one tube.
5. Centrifuge at 400*g* for 5 min at 4°C. Decant the supernatant and wash with a further 25 mL of BSA–DAB.
6. Adust the cell concentration to approx 5×10^6/mL with BSA–DAB and store on ice.

3.2. Preparation of Human Peripheral Blood Mononuclear Cells Using Density Gradient Centrifugation (2,4) (see *Note 1*)

1. Collect 10 mL of peripheral blood into a tube containing anticoagulant. Dilute with an equal volume of PBS.

2. Underlay with 5 mL of a high density medium (either Histopaque 1077 or Lymphoprep; density 1.077) (*see* Note 2).
3. Centrifuge at 700*g* for 30 min at room temperature. The erythrocytes and most of the granulocytes will have pelleted. The other mononuclear cells will have banded at the interface between the high density medium and the blood.
4. Carefully aspirate the layer of cells from the interface.
5. Make up to 20 mL with BSA–DAB and centrifuge at 400*g* for 10 min at 4°C.
6. Wash twice (10 mL BSA–DAB, centrifuge at 400*g*, 5 min, 4°C).
7. Resuspend the final pellet to a cell concentration of 5×10^6/mL in BSA–DAB.

3.3. Preparation of a Single Cell Suspension from Spleen, Thymus, and Lymph Node (3)

1. Place the tissue in a Petri dish with 15 mL of BSA–DAB. Cut into small pieces, 3–4 mm in size, and carefully tease apart with fine forceps.
2. Pass through a fine mesh stainless steel tea-strainer and then wash with BSA–DAB through a single sheet of Whatman lens tissue.
3. Underlay with Histopaque or Lymphoprep and proceed as for peripheral blood mononuclear cells, as in Section 3.2.

3.4. Preparation of Cultured Cells Grown on Plastic

1. Remove the culture medium from the dish and wash with PBS, 0.25% (w/v) EDTA.
2. Incubate with 0.25% (w/v) crystalline trypsin in PBS, 0.25% EDTA for 2–6 min at 37°C to give a suspension of predominantly single cells. Terminate the enzymatic action by adding L-15 plus FCS and 1% (w/v) soybean trypsin inhibitor.
3. Wash the cells by centrifugation in L-15 plus FCS and finally resuspend in the same medium at a dilution of about 10^6 cells/mL (*see* Note 3).

4. Notes

1. Methods similar to that in Section 3.2. can be used to prepare nucleated cells free of erythrocytes from bone marrow and from pleural and serous effusions.
2. Histopaque and Lymphoprep consist of a mixture of Ficoll and sodium metrizoate. In cell mixtures, the density can be adjusted to enrich one cell type at the expense of another.

3. Occasionally, the cells prepared according to the method in Section 3.4. become badly clumped by DNA that has been released from dying cells. If this is a problem, incubate the cells for 10 min at 37°C in culture medium containing 4 µg/mL of DNAse.

Acknowledgment

I thank N. P. Carter (University of Cambridge) and M. J. O'Hare for their assistance. This work was supported by a program grant from the Cancer Research Campaign.

References

1. Freshney, R. I. (1987) Culture of animal cells. A manual of basic technique. Liss, New York.
2. Boyle, W. (1968) An extension of the 51-Cr release assay for the estimation of mouse cytotoxins. *Transplantation* **6,** 761–764.
3. Carter, N. P. (1990) Measurement of cellular sub-sets using antibodies, in *Flow Cytometry* (Ormerod, M. G., ed.), Oxford University Press, Oxford, pp. 45–67.
4. Boyum, A. (1968) Separation of leukocytes from blood and bone marrow. *Scand J. Clin. Lab. Invest.* **21 (Suppl. 97),** 77–89.

CHAPTER 39

Preparation of Rat Lung Cells for Flow Cytometry

Janet Martin and Ian N. H. White

1. Introduction

Morphologically, the lung is a complex organ containing over 40 different cell types *(1)*. Although species and strain dependent, in the Fischer rat, the most common cell types include: endothelial, 33%; Type I, 6.5%; Type II, 12%; macrophages, 8%; ciliated and nonciliated bronchiolar epithelial (Clara) cells, 1.3 and 0.7%, respectively *(2,3)*. The lungs from an "adult" rat comprise about 4×10^8 cells *(3)*. Unlike the liver, there is no one enzyme, such as collagenase, universally used for cell dispersion. Instead, a wide variety of proteases, such as Protease I or XIV *(4,5)*, have been recommended. Enzyme cocktails have often been adjusted to suit the cell type being isolated: e.g., collagenase/trypsin/elastase for Type I cells *(6)*, elastase/trypsin for Type II cells *(7)*, and hyaluronidase/cytochalasin for tracheal epithelial cells *(8)*. However, such proteolytic enzymes may lead to the loss of specific cell surface markers. There is general agreement that DNAse is necessary to minimize cellular reaggregation once the lung cells have been isolated. This procedure describes the use of the proteolytic enzyme subtilisin, recently introduced for the isolation of lung cells *(9)*, which has been particularly effective in the preparation of functional Clara cells following their separation by flow cytometry *(10)*. Type II cells have also been isolated using similar procedures (unpublished), and macrophages are lavaged from the lung during the perfusion process and can be recovered if required.

From: *Methods in Molecular Biology, Vol. 10: Immunochemical Protocols*
Ed.: M. Manson ©1992 The Humana Press, Inc., Totowa, NJ

2. Materials

1. Pentobarbitone sodium (Sagatal): Obtained as a solution of 60 mg/mL (May and Baker, Dagenham, Essex, UK).
2. 70% Industrial methylated spirit (IMS).
3. Heparin: 1000 U/mL.
4. Cannula: bulbous ended 20 gage (*see* Note 1).
5. Recirculating perfusion apparatus.
6. Cylinder of 95% O_2/5% CO_2.
7. Hanks balanced salt solution (HBSS), pH 7.4. Prepare a 1X solution from the 10X concentrate (obtained from Gibco, Paisley, Scotland) and adjust the pH with 1.76 mL of 7.5% $NaHCO_3$/100 mL. Warm to 37°C (*see* Note 3).
8. Plastic cannula: Butterfly-19 winged needle infusion set (od 1.1 mm; id 0.8 mm: Abbott, Sligo, Rep. of Ireland). Cut off the winged needle and discard, leaving 5 cm of tubing with an angled cut open end.
9. Phosphate buffered saline (PBS), pH 7.3: 8 g of NaCl; 0.2 g of KCl; 1.15 g of $Na_2HPO_4 \cdot 2H_2O$; and 0.2 g of KH_2PO_4, made up to 1 L with distilled water.
10. Subtilisin: 15 mg in 10 mL of HBSS containing 1 mM EDTA (*see* Method). Filter through a 0.2 µm filter.
11. HBSS (1X) containing 10% fetal calf serum (FCS) and 0.05 mg/mL of DNAse. Prepare from the HBSS 10X concentrate and adjust the pH with 0.5 mL of 7.5% $NaHCO_3$/100 mL.
12. Male Fischer F344/N rat, 170–180 g.
13. 2 × 10 cm diameter plastic Petri dishes.
14. Nybolt gauze: 125 µm mesh (John Stanier and Co. Ltd., Manchester, UK). Place a square over a 250 mL plastic beaker and retain with an elastic band.
15. Dissection instruments, 20 cm of strong polyester thread, small clip (3-cm long), universal tubes, plastic pipets, pipetor, 0.2 µm filter, plastic syringes, 250-mL glass beaker.

3. Method

3.1. Vascular Perfusion

1. Anesthetize a male Fischer F344/N rat with pentobarbitone sodium (250 mg/kg, ip) (*see* Note 4).
2. To increase sterility, wet the fur of the ventral surface of the animal with 70% IMS.
3. Open the chest cavity, cutting through the diaphragm and up through the sides of the rib cage, deflect the sternum by using the weight of artery forceps clamped to the xiphisternum. Carefully cut the pericardium.
4. Apply 0.5 mL of heparin solution to the surface of the heart and lungs.

5. Insert a bulbous-ended 20-gage cannula (attached to a syringe containing 1.5 mL of heparin) through a small incision made in the right ventricle, and hence, into the pulmonary artery. Hold in place with a small clip having only the bulb projecting into the artery.
6. Cut the left auricle to allow the perfusate to escape.
7. Inject the heparin rapidly but steadily through the cannula. Remove the syringe and attach the cannula to the outlet of the perfusion apparatus containing 500 mL of oxygenated HBSS at 37°C. Perfuse at 50 mL/min. Run the effluent to waste (*see* Note 5).
8. During maintenance of the vascular perfusion for approx 10 min, cut away the liver. Cut up either side of the trachea. Remove tissue and membrane overlying the trachea. Insert and tie in the plastic tracheal cannula just below the larynx.
9. Lavage the lungs through this cannula using three successive 9-mL vol of PBS at room temperature (*see* Note 6). Inflate and deflate the lungs twice with each volume. The combined PBS lavages may be centrifuged at 300*g* for 4 min at room temperature to recover macrophages, although further purification may be required to remove contaminating erythrocytes.
10. Inflate the lungs with a further 9 mL of PBS and, holding both cannulae, cut the lungs and heart free. Remove the vascular cannula and lay the organs in a plastic Petri dish with the tracheal cannula still in place. Carefully cut away the heart and as many blood vessels as possible without damaging the trachea. Rinse the lungs with PBS (*see* Note 7).

3.2. Subtilisin Digestion

1. Switch the perfusion apparatus over to recirculate and oxygenate 150 mL of HBSS at 37°C (containing 55.8 mg of EDTA, disodium salt) in a 250-mL beaker while the lungs are being trimmed.
2. Fill the lungs with 9 mL of HBSS/EDTA through the tracheal cannula using a syringe. Immerse lungs in the beaker, attach the perfusion apparatus outlet to the cannula and add the subtilisin solution to the beaker.
3. Continue the enzyme recirculating perfusion for 15 min. The temperature can be measured by use of a thermistor probe in the beaker of perfusate and should be reading 37°C.
4. After 15 min, detach the perfusion outlet from the tracheal cannula and allow the perfusate to drain out. Flush the lungs with 12 mL of HBSS containing FCS and DNAse.
5. Remove the lungs to a Petri dish, cut away the cannula and major airways, and shred the lungs in the dish using two pairs of fine forceps (*see* Note 8).
6. Further disrupt the tissue by passage (3×) through a 10-mL plastic pipet, followed by filtration through 125 µm Nybolt gauze into the beaker.

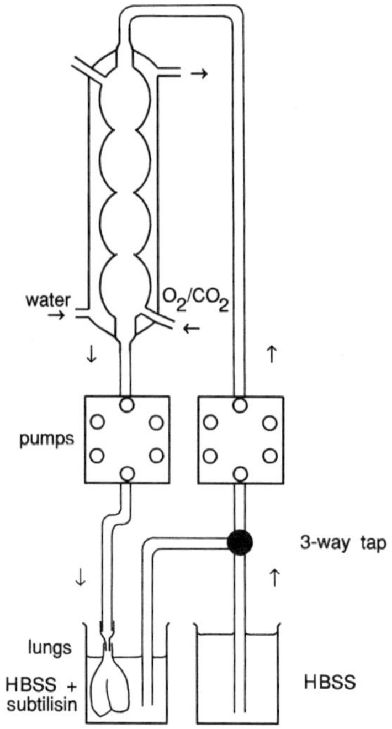

Fig. 1. Diagram of the perfusion apparatus used for the enzymic digestion of rat lungs.

Wash the Petri dish with further HBSS containing FCS and DNAse and pipet this onto the debris on the gauze, to a total of approx 30 mL of filtrate in the beaker.

7. Centrifuge this suspension at 300*g* for 4 min at room temperature. Aspirate to waste the supernatant containing broken cells and resuspend the pellet in 10 mL of HBSS containing FCS and DNAse and centrifuge as before.
8. Resuspend the pellet in 5 mL of the same solution and run through a small piece of Nybolt gauze into a Universal tube.

4. Notes

1. The vascular cannula is made by adding solder around the end of a 4-cm long 20-gage needle to give a bulb 2 mm in diameter, and extending 3 mm onto the needle.
2. The recirculating perfusion apparatus consists of a double pump (Masterflex with 7016 heads, Cole-Parmer, Chicago, IL) to and from a falling film oxygenator (Fig. 1). One inlet tube comes from a bottle of

HBSS (500 mL) used for the vascular perfusion, and the other from a beaker of HBSS/EDTA (150 mL) to which is added the subtilisin for the enzyme perfusion. These are switched to a common tube via the pump to the top of the oxygenator. The oxygenator is enclosed in a water jacket filled from the same water bath at 37°C in which the solutions are maintained. Oxygenated effluent from the bottom of the oxygenator goes via the second pump to be connected to either the arterial or the tracheal cannula.

3. Deionized, filtered water is used throughout for preparation of solutions.
4. Rapid induction of anesthesia within 5 min minimizes blood clotting in the periphery of the lungs during subsequent perfusion.
5. The solutions being recirculated are conveniently immersed in a heated water bath. The bath temperature should be high enough to maintain 37°C at the outlet, as the solutions will cool during their circulation. HBSS is gassed with O_2/CO_2 at a flow rate of 500 mL/min, starting 5 min prior to perfusion.
6. The open end of the plastic cannula should be in the trachea and must not enter the bronchi. The inflation and deflation of the lungs during lavage helps the clearance of the vascular system.
7. The lungs should be completely blanched at this stage. However, any small areas of clots around the periphery can be cut away after enzyme perfusion.
8. The lung lobes are shredded against each other using 200 strokes of two pairs of closed fine curved forceps. A dark background is advantageous, and tissue can also be pulled free of remaining airways.

References

1. Breeze, R. G. and Wheeldon, E. B. (1977) The cells of the pulmonary airways. *Am. Rev. Respir. Dis.* **116,** 705–777.
2. Pinkerton, K. E., Barry, B. E., O'Neil, J. J., Raub, J. A., Pratt, P. C., and Crapo, J. D. (1982) Morphologic changes in the lung during the lifespan of Fischer 344 rats. *Am. J. Anat.* **164,** 155–174.
3. Lehnert, B. E., Valdez, Y. E., and Holland, L. M. (1985) Pulmonary macrophages: alveolar and interstitial populations. *Exp. Lung Res.* **9,** 177–190.
4. Devereux, T. D. and Fouts, J. R. (1981) Isolation of pulmonary cells and use in studies of xenobiotic metabolism, in *Methods in Enzymology,* vol. 77 (Jakoby, W. B., ed.), Academic, NY, pp. 147-154.
5. Dawson, J. R., Norbeck, K., and Moldeus, P. (1982) The isolation of lung cells for the purpose of studying drug metabolism. *Biochem. Pharmacol.* **31,** 3549–3553.
6. Weller, N. K. and Karnovsky, M. J. (1986) Isolation of pulmonary alveolar Type I cells from adult rats. *Am. J. Pathol.* **124,** 448–456.
7. Leary, J. F., Finkelstein, J. N., Notter, R. H., and Shapiro, D. L. (1982) Isolation of Type II pneumocytes by laser flow cytometry. *Am. Rev. Resp. Dis.* **125,** 326–330.

8. Johnson, N. F., Wilson, J. S., Habbersett, R., Thomassen, D. G., Shopp, G. M., and Smith, D. M. (1990) Separation and characterisation of basal and secretory cells from the rat trachea by flow cytometry. *Cytometry* **11,** 395–405.
9. Adam, A. (1989) PhD Thesis, Faculty of Science, University of London.
10. Martin, J., Legg, R. F., Dinsdale, D., and White, I. N. H. (1990) Isolation of Clara cells from rat lung using flow cytometry. *Biochem. Soc. Trans.* **18,** 664.

Chapter 40

The Isolation of Rat Hepatocytes for Flow Cytometry

Reginald Davies

1. Introduction

The isolation and subsequent study of hepatocytes in in vitro conditions was first transformed by Berry and Friend *(1)*, who developed a collagenase liver perfusion method, allowing the isolation of large numbers of cells with high viability.

Numerous research groups have modified the procedure to their own needs, and a detailed investigation was carried out by Seglen *(2)*. Our routine method involves the perfusion of an excised liver with a recirculating collagenase solution. However, if the aim is the subsequent separation of hepatocytes according to phenotype or difference in ploidy *(3)* by centrifugal elutriation or flow cytometry, it is essential to get a hepatocyte preparation with a greater proportion of single cells. This is best achieved by the inclusion of ethyleneglycol *bis*-(aminoethylether) tetraacetic acid (EGTA) in the *in situ* perfusion step, together with the addition of calcium chloride to the collagenase perfusion step.

We routinely obtain about 500×10^6 cells from a rat weighing 200 g. The viability is about 90%. In the absence of DNAse in Step 8, the viability is nearer 80%. To obtain a hepatocyte population with a viability above 95%, the hepatocyte preparation can be centrifuged in a Percoll gradient *(4)*.

From: *Methods in Molecular Biology, Vol. 10: Immunochemical Protocols*
Ed.: M. Manson

2. Materials

1. Pentobarbitone sodium (Sagatal): 60 mg/mL.
2. Industrial methylated spirit (IMS) (70%).
3. Heparin: 1000 U/mL.
4. Intravenous cannula (*see* Note 1).
5. Recirculating perfusion apparatus (*see* Note 2).
6. Cylinder of 95% O_2/5% CO_2.
7. Calcium- and magnesium-free Hanks balanced salt solution (HBSS), obtained as a 10X solution from Gibco, Paisley, Scotland, and diluted to 1X, plus 1.76 mL of 7.5 Na_2HCO_3/100 mL, pH 7.4 (HBSSB). Warmed to 37°C and gassed with O_2 and CO_2.
8. Collagenase A: 25 mg (0.22 U/mg) in 10 mL of HBSSB. Filter through a 0.2-µm filter.
9. HBSS containing 0.01 m*M* HEPES buffer, pH 7.4 and 0.02% DNAse (HBSSHD). Kept at 4°C.
10. Male Fischer F344/N rat weighing about 200 g.
11. Sterile 50 mL measuring cylinder, sterile plastic filter funnel.
12. Nybolt gauze: 125-µm Mesh. A square is placed over a 250-mL plastic beaker, retained with an elastic band, and sterilized.
13. Dissection instruments, 20 cm of strong polyester thread, universal tubes, plastic pipets, pipetor, 0.2-µm filter.
14. Williams Medium E, containing 5% fetal calf serum (FCS).
15. 0.2% Trypan blue in 0.15*M* NaCl.

3. Method

1. Anesthetize a male Fischer F344/N rat with pentobarbitone sodium (200 mg/kg, ip).
2. Wash the abdomen with alcohol and carefully open up the abdominal cavity, cut the ligaments connecting parts of the liver to the stomach, and gently push the liver up against the diaphragm.
3. Expose the hepatic portal vein and place forceps under the bile duct and portal vein above the mesenteric vein. Gently pull through two threads, and tie loose knots around the vein.
4. Fill a cannula (Note 1) with heparin (1000 U/mL) and cannulate the hepatic portal vein about 1 inch below the point of branching into liver. Ease the cannula up to the point of branching and secure in place using the two threads, making sure that the knots are well away from the tapered tip of the cannula. If the cannulation is successful, blood will back-fill the cannula.

5. Cut the abdominal vena cava above the kidney and clear the blood from liver by *in situ* perfusion with HBSSB at 37°C, gassed with 95% O_2 and 5% CO_2 (Note 3). The flow rate of the perfusate is 50 mL/min.
6. Carefully cut the ligaments holding the liver. Start at the stomach and work anticlockwise around the liver. Gently lift the liver using the cannula to cut remaining ligaments and place the excised liver onto a plastic spade and place in the sterile filter funnel over the bottle containing 40 mL of HBSSB.
7. Add 10 mL of collagenase (25 mg, final concentration 0.025%) and circulate for 30 min (*see* Note 4). Monitor the temperature of the liver by placing the thermocouple of a digital thermometer under the perfusing liver. Temperature should be 37°C ± 1°C.
8. Stop the perfusion and hold liver by the cannula in a 50-mL glass sterile measuring cylinder. Make several cuts in the liver surface (about 5-mm long and 2-mm deep) using a sterile scalpel, and cut off the connective tissue surrounding the cannula releasing the liver into the cylinder. Gently agitate with a sterile glass rod in cold HBSSHD until the hepatocytes are released from the liver. Pour the released cells onto 125-µm Nybolt into the beaker.
9. Sediment the hepatocytes by centrifugation at 100*g* for 4 min. Remove the supernatant, and wash the resuspended hepatocytes in 20 mL of HBSSHD and repeat the centrifugation step. Repeat the washing and centrifugation step again.
10. Finally, resuspend the hepatocytes in 25 mL of Williams medium E containing 5% FCS (approx 15×10^6 cells/mL).
11. Determine the viability of the hepatocytes using 0.2% trypan blue in saline (Note 5).
12. Hepatocytes are now ready for cell culture or for incubation with antibodies for cell sorting.

4. Notes

1. For rats below 120 g wt, use a 1.3-mm outside diameter (od) cannula, for rats above this wt, use a 1.7-mm od cannula.
2. The recirculating perfusion apparatus consists of a double pump (Masterflex with 7016 heads, Cole-Parmer, Chicago, IL) to and from a falling film oxygenator (*see* Fig. 1 in Chapter 39). One inlet tube comes from a bottle of HBSSB (500 mL) used for the *in situ* perfusion, and the other from a bottle of 40-mL HBSSB, to which is added the collagenase for the enzyme perfusion. These are switched to a common tube via the pump to the top of the oxygenator. The oxygenator is enclosed in a

water jacket filled from the same water bath at 37°C in which the solutions are maintained. Oxygen- and CO_2-saturated effluent from the bottom of the oxygenator goes via the second pump to be connected to the cannula.

3. Seglen *(2)* showed that oxygenation was unnecessary for the preparation of viable hepatocytes for in vitro cultures. If EGTA is included in the perfusion, then the *in situ* perfusion is continued for 5 min with HBSSB containing 0.5 m*M* EGTA. Before the collagenase perfusion, a short perfusion with HBSSB alone is carried out to clear the liver of EGTA.
4. Some researchers double the collagenase concentration and halve the perfusion time, which they claim leads to preparation of hepatocytes with higher viabilities. If the *in situ* perfusion solution contained EGTA, then $6.8 \times 10^{-5} M$ $CaCl_2$ should be present during the collagenase perfusion step.
5. If most of the hepatocytes are in clumps, this suggests that the breakup of the liver by collagenase has proceeded too quickly. Try lowering the collagenase concentration during perfusion.

References

1. Berry, M. N. and Friend, D. S. (1969) High-yield preparation of isolated rat liver parenchymal cells. *J. Cell. Biol.* **43,** 506–520.
2. Seglen, P. O. (1976) Preparation of isolated rat liver cells, in *Methods in Cell Biology,* vol. 13 (Prescott, D. M., ed.), Academic, New York, pp. 29–83.
3. Davies, R., Cain, K., Edwards, R. E. Snowden, R. T., Legg, R. F., and Neal, G. E. (1990) The preparation of highly enriched fractions of binucleated rat hepatocytes by centrifugal elutriation and flow cytometry. *Anal. Biochem.* **190,** 266–270.
4. Meredith, M. J. (1988) Rat hepatocytes prepared without collagenase: Prolonged retention of differentiated characteristics in culture. *Cell Biol. Toxicol.* **4,** 405–425.

CHAPTER 41

Flow Cytometric Analysis of Cells Using an Antibody to a Surface Antigen

Michael G. Ormerod

1. Introduction

Identifying a subset of cells, particularly peripheral blood lymphocytes, by means of a fluorescently-tagged antibody is one of the commonest applications of flow cytometry. There are many methods, all of them basically similar, for labeling antigens on cell surfaces. Two slightly different procedures are given in this chapter; these can be adapted to meet most needs.

The first procedure uses a directly conjugated antibody. It has the advantage that it is simple and fast — only one incubation is involved. Its disadvantage is that a conjugated reagent has to be made or purchased for every antibody studied. It is also less sensitive. The second procedure describes a sandwich technique with the use of a primary reagent (a mouse monoclonal antibody) followed by a sheep antimouse Ig conjugated to fluorescein. Many different monoclonal antibodies can be used, unpurified, with the single-labeled secondary antibody. This procedure is also more sensitive, since it increases the number of fluorescein molecules at each site of reaction of the primary antibody.

The choice of method depends on the availability of reagents and whether a higher sensitivity is required.

The methods described assume that antibodies tagged with fluorescein will be used. This is by far the most widely used label, and all the commercially

From: *Methods in Molecular Biology, Vol. 10: Immunochemical Protocols*
Ed.: M. Manson ©1992 The Humana Press, Inc., Totowa, NJ

available flow cytometers are set up to measure fluorescein. Because of the differences between flow cytometers, only an outline of the method of analysis can be given. Details of the configuration of the instrument, which ports to use, and so on, must be selected by the reader depending on the instrument being used.

A concise, but comprehensive, outline of the principles of flow cytometry can be found in ref. *1*.

2. Materials

1. Bovine serum albumin (BSA).
2. 0.1% BSA–DAB solution: Prepare 500 mL of Dulbecco's formulas A and B phosphate buffered saline (DAB) pH 7.3 (use tablets, Oxoid Ltd., Basingstoke, UK, as instructed). Add 5 mL of 10% BSA in distilled water to 500 mL of DAB. Pass through a 0.47-µm membrane filter before use.
3. 2% Formaldehyde–BSA–DAB fixative: Add 5 mL of 37–40% formaldehyde solution to 100 mL of 0.1% BSA–DAB solution. Pass through a 0.47-µm membrane filter and store at 4°C. Filter again just prior to use.
4. Either, primary antibody labeled with fluorescein or primary antibody (mouse monoclonal) and sheep antimouse Ig, fluorescein-labeled.
5. Leibowitz culture medium plus 10% fetal calf serum (L-15 plus FCS).
6. Propidium iodide (PI) solution: Dissolve 1 mg of PI in 10 mL of distilled water. Store at 4°C in the dark (*see* Note 1).

3. Methods

3.1. Analysis of Human Peripheral Blood Mononuclear Cells Using a Directly Conjugated Antibody (2)

3.1.1. Staining the Cells

1. Carefully dispense 100 µL of primary, fluorescein-labeled, antibody, appropriately diluted in BSA–DAB, into the bottom of labeled test tubes. Use 100 µL BSA–DAB in place of diluted antibody for the negative control. Place the tubes on ice until the cells are ready. This can be done while the cells are being centrifuged (*see* Notes 2 and 3).
2. Add 100 µL of cell suspension (5×10^6/mL), prepared as described in Chapter 38, to the bottom of each tube, being careful not to contaminate the tube walls. Use a fresh pipet tip for each tube to avoid the possibility of carryover from one tube to the next.
3. Mix by vortexing the tubes briefly, then incubate for 30 min on ice.

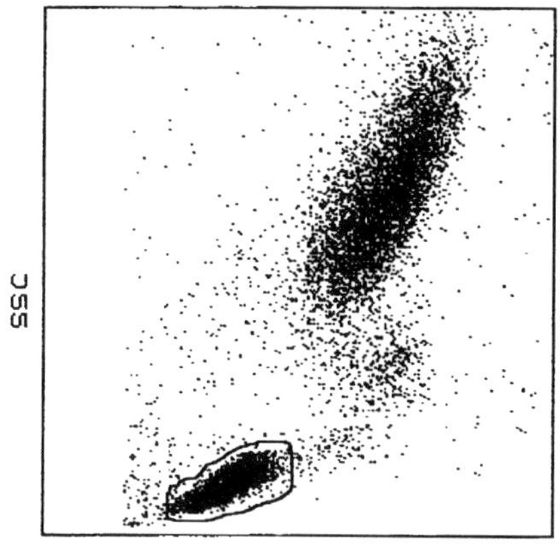

Fig. 1. Orthogonal light scatter (side scatter, SSC) (*y*-axis) displayed against forward scatter (FSC) (*x*-axis) for the author's peripheral blood leucocytes. This is in the form of a "dot plot." Each dot represents one cell. A gate has been set on the lymphocyte cluster. If there is any doubt, this cluster can be positively identified using a pan-T antibody (anti-CD3) and "back-gating," as described in Section 3.2.2. Data were recorded using a Becton-Dickinson FACSCAN.

4. Add 1.5 mL of BSA–DAB to each tube and centrifuge at 400*g* for 5 min at 4°C. Discard the supernatant and repeat the wash.
5. Either resuspend the pellet in 450 µL of BSA–DAB plus 0.02% azide and keep on ice or, if there may be some delay before analysis, resuspend in 450 µL of formalin fixative and store at 4°C in the dark.

3.1.2. Flow Cytometric Analysis

1. Use an argon-ion laser tuned to 488 nm. Record the light scattered in a forward direction and orthogonally, and green (>520 nm) fluorescence.
2. Display a cytogram (bivariate histogram or "dot plot") of orthogonal vs forward scattered light. Place a gate on the lymphocyte cluster to exclude debris, any remaining erythrocytes, any dead lymphocytes (which have less forward scatter than live cells), monocytes, and granulocytes. (A typical cytogram with the lymphocytes gated is shown in Fig. 1) (*see* Note 4).
3. Record green fluorescence of the particles lying within the gate. The method for adjusting the gain depends on the type of amplifier used. If a logarithmic amplification is employed, run the negative control and adjust the gain so that fluorescence lies in the lowest quartile of the fluorescence range. With linear amplification, select the sample that it is

anticipated will give the brightest fluorescence. Adjust the gain so that no more than 5% of the cells are off-scale on the bright end of the range.

3.2. Flow Cytometric Analysis of Cells with a Murine Monoclonal Antibody Using Indirect Immunofluorescence

The method describes the analysis of mononuclear cells from human bone marrow. It is, however, generally applicable and can easily be adapted for other types of cell.

3.2.2. Staining the Cells (Notes 2–8)

1. Remove erythrocytes from the sample by density gradient centrifugation, as described in Chapter 38. Aliquot approx 10^6 cells into each labeled sample tube.
2. Centrifuge at 400*g* for 5 min. Discard the supernatant and resuspend the cells in 200 µL of L-15 plus FCS containing the appropriate dilution of the primary antibody. To the negative control sample, add L-15 plus FCS alone.
3. Incubate on ice for 45 min.
4. Wash the cells by centrifuging them at 400*g* for 5 min, discarding supernatant and resuspending in L-15 plus FCS. Repeat. Pellet the cells again by centrifugation and discard the supernatant.
5. Resuspend in 200 µL of L-15 plus FCS containing fluorescein-labeled sheep antimouse Ig appropriately diluted (*see* Note 5).
6. Incubate on ice for 45 min.
7. Wash twice as in Step 4.
8. Finally, resuspend the cells either in 450 µL of BSA–DAB containing 0.02% azide and add 50 µL of propidium iodide (PI) solution, or in 500 µL of formalin fixative. Keep in the dark on ice or at 4°C until analysis.

3.2.3. Flow Cytometric Analysis

This description assumes that the cells have not been fixed in formalin and that PI has been added to the sample. Live cells exclude PI, whereas the DNA of the dead cells takes up the dye and fluoresces red (*see* Note 4).

1. Use an argon-ion laser tuned to 488 nm. Record the light scattered in a forward direction and orthogonally and green (antibody, use a 520 nm bandpass filter) and red (PI, >630 nm) fluorescence.
2. Display a histogram of red fluorescence and set a gate to exclude cells fluorescing red (these are dead).
3. Display a cytogram of orthogonal vs forward scattered light of the live cells and place a gate on the cell cluster of interest.

4. Display a histogram of the green fluorescence of these cells.

The protocol requires that a gate is set on the cytogram of scattered light to select the desired cell (Step 3). In peripheral blood, the lymphocyte cluster is easy to identify (Fig. 1). However, sometimes, in a complex mixture of cells, such as bone marrow, it is difficult to identify the light scatter of the cells of interest. In this case, include in the set of samples one that has been stained with an antibody that identifies the desired population of cells. Set a gate on the green fluorescence and display a cytogram of the light scatter of these cells. Set a gate on the cluster displayed and then use the identical gate on the cytogram of the light scatter of the whole cell population in the protocol described in the preceding paragraph. This procedure is sometimes called "back-gating."

This is illustrated in Fig. 2, which shows nucleated cells from the bone marrow of a patient with myeloma. The sample was labeled with an antibody to CD33, a marker of monomyeloid differentiation. The histogram of green fluorescence showed three cell populations, negative, weakly positive, and positive. By means of setting gates, the light scatter of the weakly positive and positive cells was displayed. The light scatter of the weakly positive cells fell in the region associated with granulocytes and granulocyte precursors (Fig. 2B). The positive cells fell in a cluster with less light scatter; this identified the myeloma cells (Fig. 2C). Having identified these regions, the correct gate could be set on the cytogram of light scatter from all the cells and the fluorescence of the granulocytes and the myeloma cells displayed separately (Fig. 3).

4. Notes

1. PI is a suspected mutagen and should be handled accordingly.
2. For both methods, if a series of different cells is being studied with one antibody, it is useful to include a sample of cells that is known to stain with the antibody in question. This could either be from the blood of a normal person or a cell line. This acts as a positive control to ensure that the method is working satisfactorily.
3. The appropriate dilution of primary antibody must be established by titration against a fixed number of cells (use doubling or tripling dilutions). Use the reagent at double the concentration at which increasing the concentration produces little increase in fluorescence intensity (as measured on the flow cytometer).
4. Dead cells (that is, cells whose plasma membranes have lost their integrity) may take up antibody nonspecifically. Using both the methods described, they should be excluded from the analysis. With peripheral

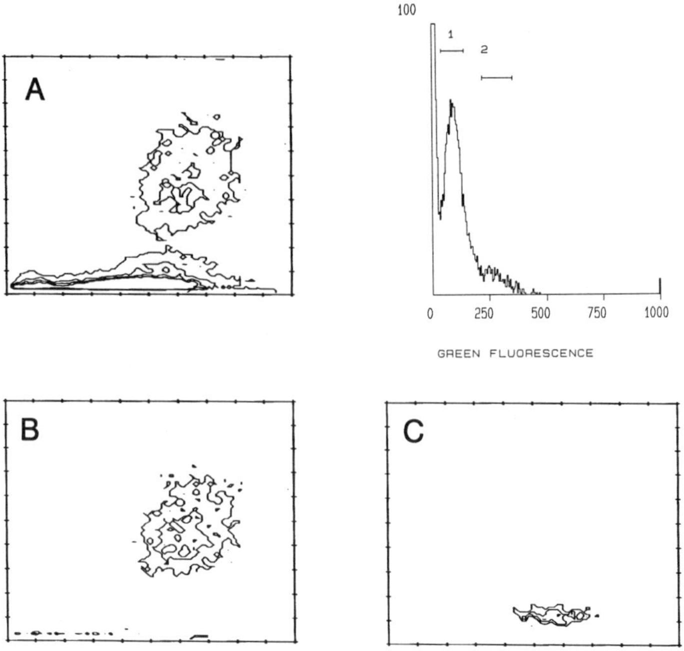

Fig. 2. Cytogram A shows orthogonal (*y*-axis) vs forward (*x*-axis) light scatter for bone marrow cells from a patient with myeloma. This is shown as a contour plot. Cytograms B and C show the light scatter from the cells selected by regions 1 and 2 set on the univariate histogram of green fluorescence of an antibody to CD33. Data were recorded on an Ortho Cytofluorograf 50 H.

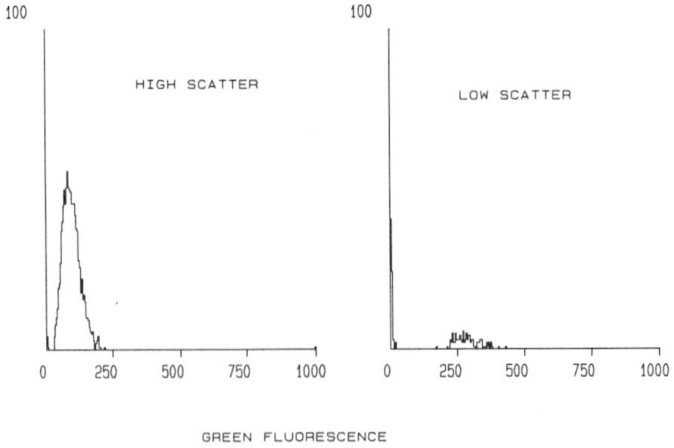

Fig. 3. Histograms of green fluorescence from CD33 expressed by the granulocytes and precursors (high light scatter) and by the myeloma cells (lower light scatter), using cytograms B and C in Fig. 2 to define the light scatter gates.

blood lymphocytes, this can be done using forward light scatter. The difference is preserved after fixation in formalin. For other cells, PI can be used, as described, as a vital stain; fixation is best avoided.

5. If a series of different antibodies is being used on a single type of cell, then an antibody that is known to give positive staining should be included.
6. For the indirect immunofluorescence, a sample that has been stained with the fluorescein-labeled second antibody alone should always be included in order to check for nonspecific or unwanted reaction of the second antibody.
7. The fluorescein-labeled antimouse Ig antibodies are obtainable from several manufacturers. Use either the dilution recommended by the supplier, or establish the correct dilution as described in Note 3.
8. The protocol in Section 3.2.1. can be used for other types of antibody, both polyclonal and monoclonal. The correct secondary antibody must be selected to match the species in which the primary one was made. The protocol can also be used for a variety of different types of cell.

Acknowledgments

I thank D. W. Hedley and P. R. Imrie for the data used to construct Figures 2 and 3. This work was supported by a programme grant from the Cancer Research Campaign.

References

1. Ormerod, M. G., ed. (1990) *Flow Cytometry: A Practical Approach.* Oxford University Press, Oxford.
2. Carter, N. P. (1990) Measurement of cellular subsets using antibodies, in *Flow Cytometry: A Practical Approach* (Ormerod, M. G., ed.) Oxford University Press, Oxford, pp. 45–67.

CHAPTER 42

Multiple Immunofluorescence Analysis of Cells Using Flow Cytometry

Michael G. Ormerod

1. Introduction

Prior to 1982, the measurement of more than one antigen simultaneously by flow cytometry required two lasers—an argon-ion laser to excite fluorescein (at 488 nm) and a krypton or a dye laser to excite rhodamine or one of its derivatives. The discovery of a naturally occurring fluorochrome, phycoerythrin (PE), changed this *(1)*. PE is a phycobiloprotein found in red algae. It can be excited efficiently at 488 nm (simultaneously with fluorescein) and has a peak fluorescence at 578 nm, sufficiently removed from the peak of 520 nm from fluorescein. There is some overlap in the emission spectra from the two dyes (in particular, there is still some emission from fluorescein above 580 nm) and this must be corrected, either electronically or by the computer software.

For labeling with three antibodies, Becton Dickinson have recently produced a compound, Duochrome, consisting of Texas Red (a rhodamine derivative) conjugated to PE. On excitation at 488 nm, the PE moiety is excited and transfers its energy to the Texas Red, which fluoresces with a peak emission at 620 nm. With the appropriate optical filters, this can readily be separated from the fluorescence from PE alone so that fluorescein, PE, and Duochrome can all be used together, excited by a single laser at 488 nm.

From: *Methods in Molecular Biology, Vol. 10: Immunochemical Protocols*
Ed.: M. Manson ©1992 The Humana Press, Inc., Totowa, NJ

Duochrome has yet to find widespread use, and much of the work with three antibodies is undertaken using two lasers *(2,3)*. An argon-ion laser is used to excite fluorescein and PE, whereas a second laser is used to excite the third dye. Either Texas Red is excited at 568 nm by a krypton or a dye laser, or allophycocyanin, another phycobiloprotein, is excited by a helium-neon laser at 630 nm.

Often, in a multiple labeling experiment, all the antibodies are mouse monoclonals. A two-step procedure using secondary antibody (as described in Chapter 41) cannot be used. The simplest way to overcome this problem is to use directly labeled primary antibodies, each conjugated to a separate fluorochrome. Unfortunately, these are not always available. An alternative with two antibodies is to use one directly labeled with fluorescein, and the other conjugated to biotin. The cells are incubated with the primary antibodies followed by an incubation with streptavidin conjugated to PE (streptavidin has a high affinity for biotin).

Some biotinylated antibodies will not react with some secondary reagents. If this is the case (which can only be established by experiment), it is possible to use one biotinylated primary antibody together with a second, unconjugated, primary antibody and a fluorescein-labeled anti-Ig. The cells are incubated with the biotinylated antibody followed by streptavidin–PE together with the second primary monoclonal antibody, followed by an antimouse Ig conjugated to fluorescein.

For some applications, such as that described here, a mixture of polyclonal and monoclonal antibodies raised in different species can be used. In this case, different species-specific secondary antibodies conjugated to different fluorochromes are employed. The method given is for three antibodies raised in different species. Human breast cells were stained with polyclonal rabbit antiepithelial membrane antigen (EMA, specific for luminal epithelial cells), monoclonal mouse anticommon acute lymphoblastic leukemia antigen (CALLA, specific in the breast for myoepithelial cells) and monoclonal rat anti-c-*erbB-2* (a putative growth factor receptor). The purpose of the experiment was to determine which cell type expressed the c-*erbB2* antigen.

Two of the primary antibodies were produced within the Institute of Cancer Research, but the method can easily be adapted for whichever set of antibodies is of interest to the reader. The procedure can also readily be modified for two antibodies. For directly labeled antibodies, use the procedure in Chapter 41, using two or three antibodies in place of one.

2. Materials

1. Primary antibodies: Polyclonal rabbit anti-EMA, monoclonal mouse anti-CALLA (CD10) (ascitic fluid, hybridoma PHM6) and monoclonal rat anti-c-*erbB2* (Institute of Cancer Research clone, BT1/83d).

2. Secondary antibodies: Donkey antirabbit-Texas Red, sheep antimouse-fluorescein, biotinylated sheep antirat (Amersham International plc, Aylesbury, UK).
3. Avidin–biotin–PE complexes (Vectastain ABC-phycoerythrin, Vector Laboratories, Burlingame, CA, USA). Prepare according to the manufacturer's instructions that describe mixing a solution of biotinylated PE with a solution of avidin in the correct ratio.
4. Leibowitz culture medium containing 10% fetal calf serum (L-15 plus FCS).

3. Methods

3.1. Staining the Cells

1. Aliquot approx 10^6 cells in L-15 plus FCS into each of 5 tubes. Centrifuge (2 min at 500*g*) and discard the supernatant.
2. To the "positive" sample, add 50 µL of rabbit anti-EMA antiserum diluted 1 in 40, 50 µL of mouse anti-CALLA monoclonal antibody diluted 1 in 60, and 50 µL of neat rat anti-c-*erbB2* monoclonal. Make all dilutions in L-15 plus FCS.
3. Set up four controls consisting of no first antibody and each antibody alone. Make up the volume with L-15 plus FCS.
4. Resuspend the cells and then incubate on ice for 45 min with rocking.
5. Pellet the cells by centrifugation and wash twice in L-15 plus FCS.
6. Add 50 µL of donkey antirabbit-Texas Red, sheep antimouse-fluorescein, and biotinylated sheep antirat, all to a final dilution of 1 in 40. Resuspend the cells and then incubate on ice for 45 min.
7. Wash twice as in Step 5.
8. Add avidin–biotin–PE complexes prepared according to the manufacturer's instructions. Resuspend the cells and then incubate on ice for 30 min.
9. Wash twice in L-15 plus FCS and finally resuspend in 500 µL of L-15 plus FCS.

3.2. Analysis on the Flow Cytometer

1. Analyze the cells using an argon-ion laser tuned to 488 nm (blue) and either a dye laser tuned to 590 nm or a krypton laser at 568 nm (yellow). Record scattered blue light (orthogonal, forward, or both, depending on instrument configuration), scattered yellow light and green (CALLA-fluorescein, 530 nm), orange (c-*erbB2*-PE, 578 nm), and red (EMA—Texas Red, >630 nm) fluorescences.
2. First, record data from the three control tubes containing cells labeled singly with each antibody. Use a cytogram of light scatter to exclude debris and clumps of cells, and to select single cells for further analysis.

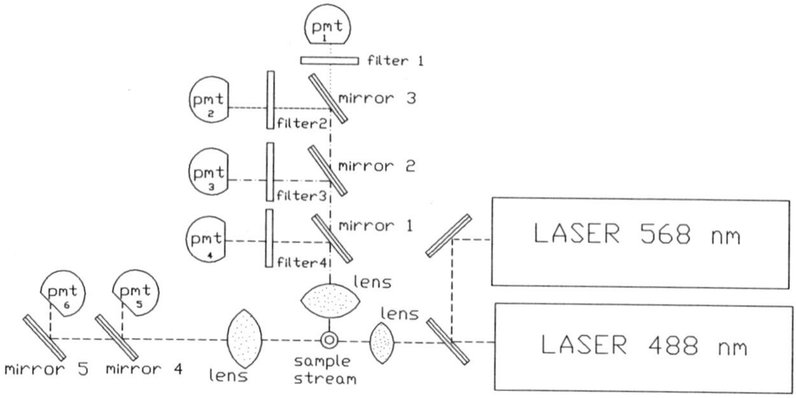

Fig. 1. A possible arrangement for a flow cytometer recording three immunofluorescences (*see* Table 1).

Set the amplifier gains so that the positive fluorescence from each color lies across the center of the range.

3. Either electronically or through the computer software, correct for spectral overlap between the ports (for example, for fluorescein fluorescence recorded in the PE channel).
4. Record data from the samples labeled with all three antibodies. Display a cytogram of green (CALLA) vs red (Texas Red) fluorescence. Set gates on the EMA and CALLA positive cells and display separate histograms of their orange (c-*erbB2*) fluorescence.

Figure 1 shows the layout of a hypothetical cytometer showing the arrangement of mirrors and filters. The optical properties of these components are shown in Table 1; by working through the system, the reader can see how the different colors are selected. The figure shows six ports in use. Few cytometers have as many ports as this and, in practice, one has to adapt the layout used to fit the limitations of the instrument.

Figure 2 shows a cytogram of EMA vs CALLA fluorescence recorded in the experiment described above together with the two univariate histograms of cell number (*y*-axis) vs c-*erbB2* fluorescence (*x*-axis). The EMA-positive and CALLA-positive cells are quite separate; none of the cells express both antigens, although a few cells express neither. From the histograms of c-*erbB2* fluorescence, it can be seen that the CALLA positive cells carried more c-*erbB2* on their surface than the EMA positive cells.

Table 1
Flow Cytometric Recording of Three Immunofluorescences

Dichroic mirrors	Filters	PMT records
1. 500 nm lp	520 nm bp	Green fluorescence
2. 600 nm sp	580 nm bp	Orange fluorescence
3. 560 nm sp	610 nm lp	Red fluorescence
4. 500 nm lp	488 nm	Blue scattered light
5. 560 nm sp		Blue scattered light
6.		Red scattered light

PXT = photomultiplier; lp = long pass; sp = short pass; bp = bandpass.

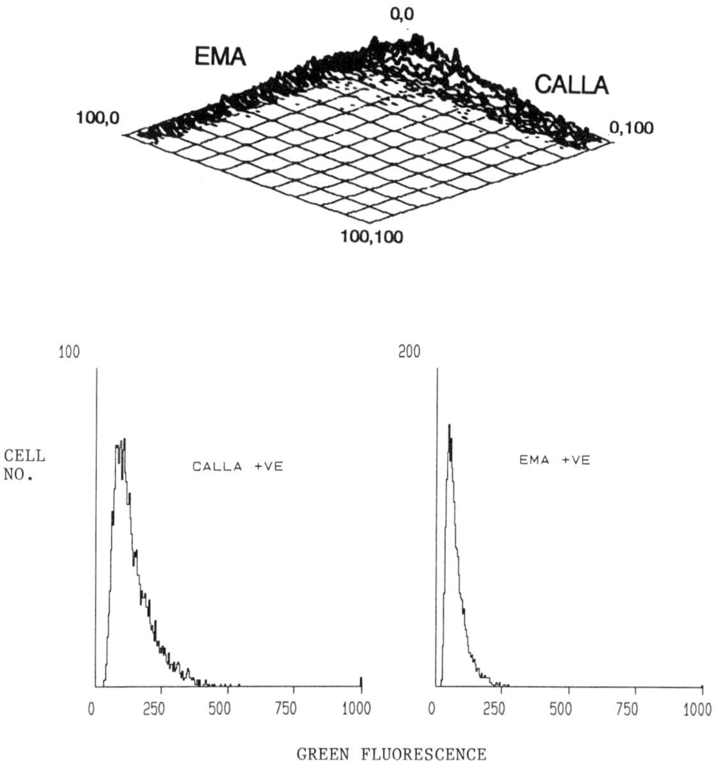

Fig. 2. A cytogram of red (EMA) vs green (CALLA) fluorescence from normal human breast cells displayed as an isometric plot. The vertical axis represents cell number. The histograms show the orange (c-*erbB2*) fluorescence from the CALLA and EMA positive cells.

4. Notes

The notes in the previous chapter on labeling with a single antibody are generally applicable.

Acknowledgment

This work was supported by a program grant from the Cancer Research Campaign.

References

1. Oi, V. T., Glazer, A. N., and Stryer, L. (1982) Fluorescent phycobiloprotein conjugates for analyses of cells and molecules. *J. Cell Biol.* **93**, 981–986.
2. Parks, D. R., Hardy, R. R., and Herzenburg, L. A. (1984) Three-color immunofluorescence analysis of mouse B-lymphocyte populations. *Cytometry* **5**, 159–168.
3. Loken, N. R., and Lanier, L. L. (1984) Three-color immunofluorescence analysis of Leu antigens on human peripheral blood using two lasers on a fluorescence activated cell sorter. *Cytometry* **5**, 151–158.

CHAPTER 43

Cell Kinetic Studies Using a Monoclonal Antibody to Bromodeoxyuridine

George D. Wilson

1. Introduction

The study of cell kinetics has traditionally involved the use of radioactive precursors of DNA, such as tritiated thymidine (^{3}HTdR), and autoradiography to detect their incorporation into DNA. These techniques have provided detailed knowledge of cell kinetics in both in vitro and in vivo experimental systems. The technique, however, is time consuming and arduous and is not readily applicable to human tumor research because of ethical problems involved in incorporation of a radioisotope into DNA.

The development of monoclonal antibodies to bromodeoxyuridine (BUdR) incorporated into DNA *(1)* and flow cytometric (FCM) techniques to simultaneously measure BUdR uptake and total DNA content *(2)* have led to renewed interest in cell kinetic studies. The speed and quantitative power of the flow cytometer, in conjunction with the specificity and sensitivity of monoclonal antibody techniques, provide the basis for the success of the BUdR technique.

BUdR/FCM offers several advantages over ^{3}HTdR/autoradiography in many cell kinetic studies.

1. By measuring DNA content and BUdR presence simultaneously, the technique is very sensitive in detecting cell cycle perturbations, owing to drugs

From: *Methods in Molecular Biology, Vol. 10: Immunochemical Protocols*
Ed.: M. Manson ©1992 The Humana Press, Inc., Totowa, NJ

or radiation, and tracing the lineage of a cell within the cell cycle. The technique can set "windows," using the computer facilities of the FCM, in any phase of the cell cycle, and is not restricted to the "mitotic window" as is the case with ^{3}HTdR/autoradiography.

2. The results from a cell cycle study can literally be obtained within a day using BUdR/FCM, whereas ^{3}HTdR/autoradiography may take several weeks to obtain an answer.
3. It is now possible to routinely study human tumor cell kinetic studies in vivo because BUdR does not have to be a radioisotope, as it is detected by a monoclonal antibody. BUdR shows no toxicity in the doses required for cell kinetic studies in man.

BUdR/DNA flow cytometry offers flexibility and diversity in the study of cell kinetics from cells in culture up to human tumors in vivo. Developments in monoclonals, staining techniques, and analysis of data are continually improving the sensitivity of the technique that is now replacing the outdated ^{3}HTdR/autoradiography procedure in many laboratories.

Several staining procedures have evolved to detect BUdR in different tissues or experimental systems. This chapter will describe the simplest procedure and point out variations that may be applicable in certain studies. The method requires ethanol-fixed cells or nuclei; these are partially denatured with HCl to allow access of the monoclonal antibody to its binding sites. A fluorochrome (FITC) is attached indirectly through a second antibody step to the BUdR. Total DNA content is stained using propidium iodide.

2. Materials

1. 5-Bromo-2-deoxyuridine: This can be obtained from Sigma Chemical Co. (Poole, UK) for experimental studies. A preparation suitable for in vivo use in humans can be obtained from the Investigational Drugs Branch of the National Cancer Institute, Bethesda, MD.
2. 70% Ethanol for fixation.
3. Phosphate buffered saline (PBS) pH 7.4.
4. 2*M* hydrochloric acid.
5. 0.4 mg/mL pepsin in 0.1*M* HCl (pH 1.5). The pepsin is dissolved in a small vol of PBS or water (1–2 mL) before addition of HCl.
6. Antibody incubation buffer: 0.5% normal goat serum (NGS) and 0.5% Tween-20 in PBS.
7. Monoclonal antibody against BUdR. Many preparations are currently available; the antibody used by our group is a rat IgG_{2a} (BUI/75 ICRl) obtained from Sera-Lab, Crawley Down, Sussex, UK.
8. Goat antirat IgG (whole molecule) fluorescein isothiocyanate conjugate.

9. Nylon filter (35-μm pore).
10. Conical bottom tube (12-mL).
11. Swinnex holder.
12. 10-mL syringe.
13. Propidium iodide.
14. A flow cytometer.

3. Methods

A prerequisite for flow cytometry studies is a single cell suspension. In vivo studies in solid tumors and tissues require disaggregation of the tissue into single cells or nuclei. The staining method described here includes a method for obtaining nuclei from solid tumors. However, it is not appropriate to describe detailed methods for obtaining cell suspensions from solid tissues in this chapter. Readers are directed to Chapters 38–40 and also *(3)* for a recent review of methods for obtaining single cell suspensions for flow cytometry studies.

3.1. BUdR Incorporation

1. To incorporate BUdR in vitro into monolayers or cell suspensions, the cells are incubated with 10 to 20 μ*M* BUdR for 10–20 min. Concentrations as small as 1 μ*M* can be detected, whereas concentrations above 50 μ*M* may cause cell cycle perturbations. The BUdR is thoroughly washed out by two washes in PBS or medium.
2. To incorporate BUdR in vitro into 1–2 mm^3 solid pieces of tissue in explant culture, use concentrations of 50 μ*M* or greater in the culture medium, for 1 h under high pressure oxygen, to ensure minimum diffusion of the DNA precursor into the core of the tissue *(4)*.
3. For in vivo incorporation of BUdR in experimental animals, inject 10–100 mg/kg of BUdR in 0.9% saline by the ip route. A concentration of 10 mg/mL is routinely used. The animal is sacrificed after 1 h for labeling index studies or at 1–2 hourly time intervals for cell cycle progression studies. In vivo incorporation of BUdR into humans, at Mount Vernon Hospital, involves iv injection of 200 mg of BUdR in 20 mL of normal saline as a single bolus over 3–5 min.

3.2. Fixation

Prior to the staining procedures, fix cells in 70% ethanol. Best results are obtained by resuspending the cell pellet in 1 mL of PBS and adding 9 mL of ice-cold 70% ethanol, while vortexing, to prevent agglutination. Solid pieces of tissue can be fixed as 5 mm^3 pieces directly in ice-cold 70% ethanol. Cells and tissues are usually left at least 1 h prior to staining. Fixed cells or tissues should be stored at 4°C and, in our experience, will remain suitable for BUdR staining studies for at least three years.

3.3. Staining Procedure

1. Cut solid tumors or tissues, removed from the fixative, into 1–2 mm^3 pieces and incubate in 5–10 mL of 0.1*M* HCl containing 0.4 mg/mL pepsin (pH 1.5) for 30 to 60 min at 37°C with constant agitation. The period of incubation varies from one tumor or normal tissue to another (*see* Notes 1 and 2).
2. When the tissue starts to break up, ensure complete dissociation into individual nuclei by pipeting, using a 5-mL automatic pipet.
3. Filter the suspension of nuclei into a 12-mL conical bottomed tube through 35-µm pore nylon filter, in a Swinnex holder and 10-mL syringe, to remove large debris and clumped nuclei.
4. Centrifuge the suspension at 700*g* for 5 min.
5. If cell suspensions are used as starting material, then omit Steps 1–4. Centrifuge the suspension from cultures or dissociated from solid tissues at 300*g* for 5 min and decant the fixative.
6. Resuspend the pellets in 2*M* HCl and incubate for 10–15 min in the case of nuclei, or 20–30 min in the case of whole cells, at room temperature with occasional mixing (*see* Note 3). This step achieves the partial unwinding of DNA into single strands to allow monoclonal antibody access to the binding sites.
7. Add 5 mL of PBS to the HCl and centrifuge the tubes at 700*g* for nuclei, or 300*g* for cells, for 5 min.
8. Add 5 mL of PBS to the pellets and repeat the centrifugation. The washes with PBS are sufficient to remove all the acid (*see* Note 4).
9. Resuspend the pellet in 0.5 mL of antibody incubation buffer and 25 µL of monoclonal antibody. The dilution of monoclonal will vary according to the preparation (*see* Notes 5 and 6). The monoclonal used in our studies is derived directly from the supernatant from the antibody producing cell line. Incubate the tubes for 1 h at room temperature with occasional mixing (*see* Note 7).
10. Add 5 mL of PBS and centrifuge the tubes at 700*g* for nuclei, or 300*g* for cells, for 5 min.
11. Resuspend the pellets in 0.5 mL of antibody incubation buffer containing 25 µL of goat antirat IgG (whole molecule) FITC conjugate. Allow binding of the second antibody to the monoclonal antibody to proceed for 1 h at room temperature with occasional mixing.

12. Repeat Step 10 (*see* Note 8).
13. Resuspend the resulting pellet in 1–3 mL of PBS containing 10 μg/mL of propidium iodide.
14. Analyze on a flow cytometer (*see* Notes 9 and 10). The flow cytometer used in these studies is an Ortho Systems 50-H Cytofluorograph with 2150 computer. Green (BUdR) and red (DNA) fluorescence should be collected at 90° to the incident light and separated from the exciting wavelength (488 nm) by a 510 or 520 nm long pass filter. Green fluorescence is detected with a 540 nm short pass band filter and the red fluorescence collected after passing through a 620 nm long pass filter. Data is routinely collected as linear signals, although it may be preferable to use log green fluorescence if the green signal is intense. As with all procedures measuring DNA, it is advantageous to discriminate cell doublets by processing the red fluorescence signal into height, area, and width. If this is not possible, gating on red fluorescence and forward or 90° scatter is advisable to exclude debris and cell clumps. Routinely, data is collected in list mode and 10,000 events recorded. However, in the case of slowly dividing tissues such as lung and kidney, up to 100,000 events may be necessary to register a significant number (100) of BUdR-labeled cells.

3.4. Examples of Data and Data Analysis

Cell kinetic data can be generated from a wide range of experimental conditions using BUdR/FCM. Figure 1 shows bivariate distributions obtained from a "pulse chase" experiment of V79 cells in vitro. The basic cell populations can be separated from each other by considering both their BUdR uptake and DNA content. Two populations show little BUdR uptake, those at channel 20 are G1 cells and those at channel 42 are G2+M cells. The S-phase cells lie between these two populations in terms of DNA content, but are separated from them by virtue of their BUdR uptake. Early S-phase cells can be seen arising from the G1 population, as can mid- and late-S-phase cells, owing to their increasing DNA content. The progression of cells through the cell cycle can be easily detected in the subsequent profiles at 2-h intervals. BUdR-labeled cells increase their DNA as they move through S-phase. They spend a short time in G2 before dividing and entering G1. Thus, at 3 h, the original BUdR-labeled population is clearly separated into two subpopulations. One cohort of BUdR-labeled cells are progressing through mid and late S-phase and G2, whereas the other comprises cells that have divided and now reside in G1. With time, this latter compartment increases as more of the original S-phase cells divide, whereas the former diminishes as the early S-phase cells reach late S and G2. At 7 h, virtually all the original S-phase cells have either divided or are in G2. In addition, some of these cells (those that

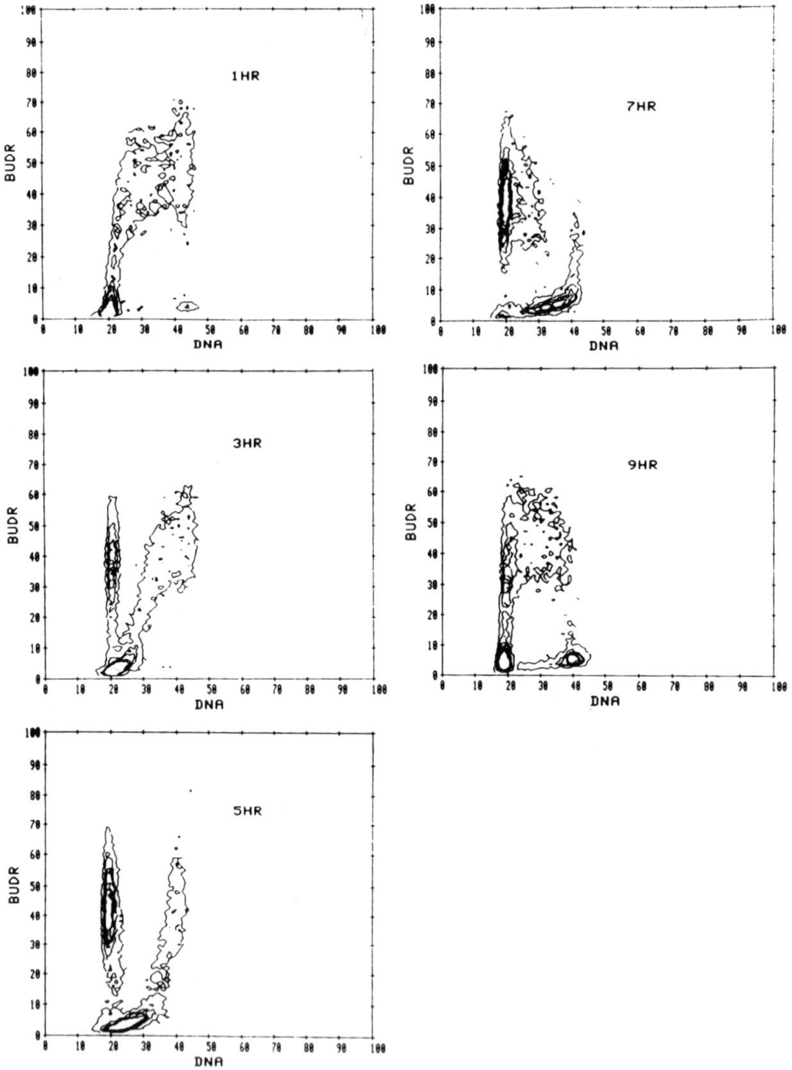

Fig. 1. Cell cycle progression of V79 cells. Bivariate distributions of DNA content (*x*-axis) vs BUdR uptake (*y*-axis) of V79 cells removed at time intervals after pulse labeling with BUdR. Cells, in exponential growth, were pulse-labeled with 10 μ*M* BUdR for 20 min. After washing out the BUdR, the cells were resuspended in fresh medium and harvested at 1 h intervals.

were in late S-phase at time zero) have progressed through G1 and entered early and mid S-phase. By 9 h the population has almost regained the 1-h profile as the cell cycle is almost complete.

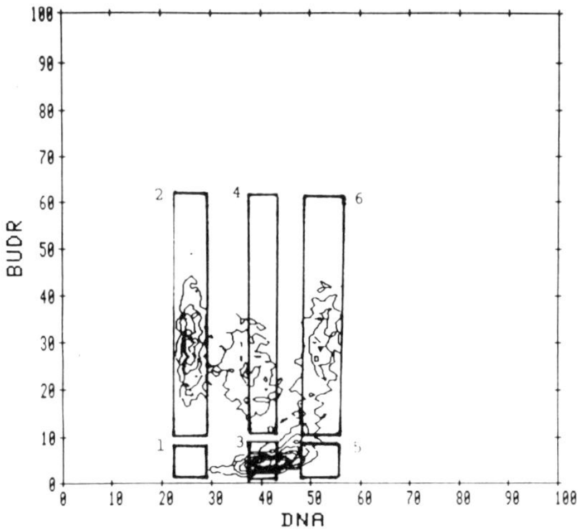

Fig. 2. Bivariate distribution of V79 cells perturbed by hydroxyurea. Cells were pulse-labeled with 10 μM BUdR and allowed to progress through their cycle for 3 h. 0.3 m*M* hydroxyurea was added for 5 h. The profile above was obtained 2 h after removal of hydroxyurea.

Analysis of this type of data is handled by the computer facilities of the flow cytometer. Basically, regions of interest or "windows" can be set in any phase of the cell cycle looking at both BUdR and non-BUdR-labeled cells. Figure 2 shows an example of data from a cell population perturbed by hydroxyurea in vitro. In this experiment, cells were pulse-labeled with BUdR and allowed to progress through their cycle for 3 h. Hydroxyurea was then added for 5 h. During this time, cells that are in early and mid S-phase stop DNA synthesis, while cells in late S and G2 can progress through to cell division. Cells in G1 are able to traverse this phase of the cycle but are blocked at the G1/S boundary. Once the drug is removed, there is a burst of cells moving into and through S-phase. Figure 2 shows a complex BUdR/DNA profile obtained 2 h after the removal of hydroxyurea, in which six regions were set to study this drug effect. Regions 1 and 2 are set in G1 to study non-BUdR and BUdR-labeled cells, respectively. Similarly, regions 3 and 4, and 5 and 6 are set in mid S and G2, respectively. These basic regions can give information on all cell cycle parameters and in most experimental conditions.

Regions 3 and 4 generate a curve analogous to a "pulse-labeled mitosis" curve from ^{3}HTdR/autoradiography. By plotting the mean green fluorescence in region 4 relative to the mean green fluorescence in both regions 3 and 4 or, alternatively, by plotting the number of cells in these regions, Fig. 3 can be generated. From this analysis, the mean green fluorescence should stay at its

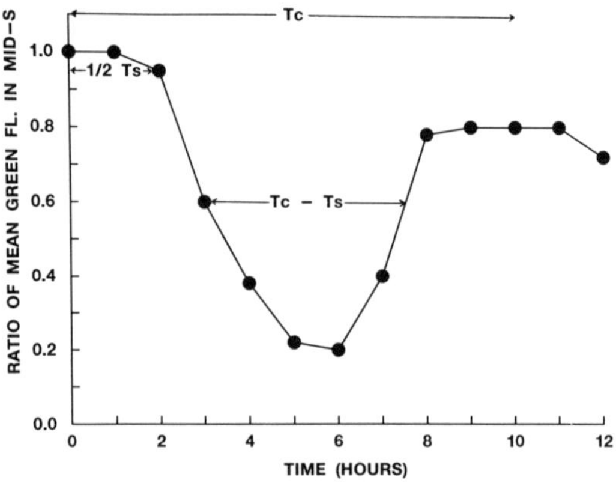

Fig. 3. Cell cycle analysis of BUdR/DNA profiles. The curve is generated from narrow regions set in mid-S phase. One region is set around the BUdR-labeled cells and the other encompasses all cells in mid-S. The *y*-axis is calculated from the ratio of mean green fluorescence in the BUdR-labeled compartment divided by that in the total mid-S region.

maximum for a time equal to half T_s (duration of S-phase). The mean fluorescence then falls and remains low for a period equal to the cell cycle time (T_c) minus T_s; this can be read off from the midpoint of the maximum and minimum values of fluorescence. The mean fluorescence then rises again and the cell cycle time can be read off from the midpoint of the second peak.

In human tumors, both the labeling index (LI) and T_s can be calculated from a single observation *(5)*. These two parameters can be used to calculate the potential doubling time (Tpot), which describes the shortest possible time a cell population can double its number in the absence of cell loss. This procedure is based on the measurement of the mean DNA content of the BUdR-labeled cohort of cells.

Immediately after labeling, as in the 1 h profile from Fig. 1, the mean DNA of the BUdR-labeled cells will be halfway between that of G1 and G2 cells, as there is a uniform distribution of cells throughout S-phase. If this were expressed as a function of the difference in DNA content between G1 and G2 cells, i.e., by subtracting the G1 mean DNA content from the mean S DNA content, and dividing by the G1 subtracted from the G2, then the starting value would be 0.5. As the BUdR-labeled cells progress through S-phase, then this value will increase (ignoring the cells that divide) until the only cells which are BUdR-labeled, and have not divided, are in G2. The value or relative movement will be 1.0. Assuming linearity of T_s then, the progression of 0.5 to

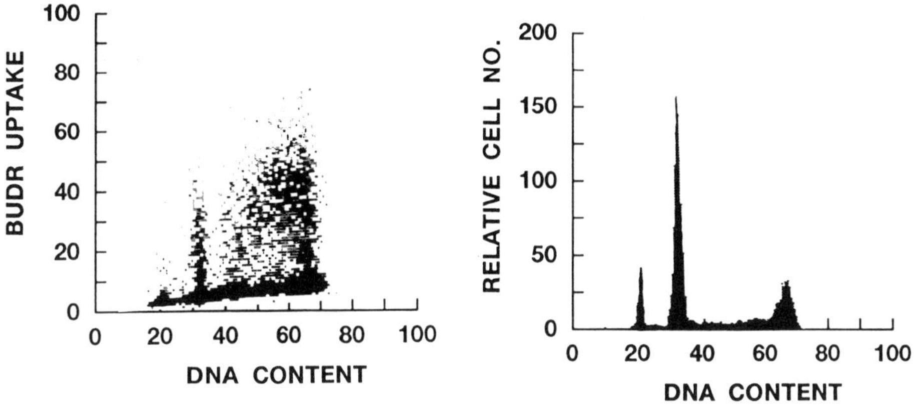

Fig. 4. BUdR/DNA profile of a human colonic adenocarcinoma. Bivariate BUdR/DNA distribution (right) and DNA profile (left) of a human colonic adenocarcinoma These profiles were obtained 6 h after injection of 200 mg of BUdR.

1.0 will describe the movement of early S-phase cells to G2, i.e., T_s Thus, by taking a biopsy several hours after an injection of BUdR, T_s can be calculated.

An example of a human tumor analyzed by this method is shown in Fig. 4. This was an adenocarcinoma of the colon removed 6 h after injection. The profile shows the complex nature of many human tumors that possess abnormal DNA content. The first major peak, in the DNA profile, is diploid G1 cells, followed by a peak of aneuploid G1 cells that have 1.6 times more DNA than the diploid cells. The bivariate distribution clearly shows the redistribution of BUdR-labeled cells through the cell cycle after 6 h. One population has divided and resides in G1, whereas the other is still progressing through mid and late S and G2+M. It is also clear from this profile that virtually all proliferation is associated with the cells with aneuploid DNA content. From the profile, the T_s was calculated to 16.6 h and the LI of the aneuploid cells was 12.5%. The LI has to be corrected for those cells that have divided and shared their BUdR between the two daughter cells. This is simply done by measuring the number of G1 cells with BUdR uptake, halving that number and recalculating a corrected LI by subtracting it from the total BUdR-labeled and total cell number. These values of LI and T_s compute a Tpot of 4.4 d. This technique has permitted routine analysis of proliferative characteristics of human solid tumors on a timescale useful for prognostic or diagnostic purposes *(6)*.

4. Notes

1. The concentration of cells or nuclei is important. The procedure outlined above is designed for a starting density of 2–3 million cells. The volume of HCl should be increased and the antibody dilution decreased, if more cells are used. A cell or nuclei count is usually performed prior to Step 6.
2. The procedure outlined above has been kept simple, with as few steps as possible. However, many variations exist that may have benefit in certain systems. For cell suspensions, digestion into nuclei, prior to staining, with 0.4 mg/mL pepsin in 0.1*M* HCl for up to 30 min at room temperature may improve results *(7)*. This step may be particularly useful if cells have a tendency to clump or have a large cytoplasm:nucleus ratio that may cause problems with nonspecific staining. The incubation time of cells or tissues in pepsin varies from one cell type to another. Solid tissues should be monitored throughout the incubation, and the digestion stopped when the large pieces disappear. Thirty minutes digestion is usually the optimum incubation period.
3. The denaturation step is perhaps the most crucial of the assay. Enough unwinding of DNA is required to ensure sufficient antibody binding, but not enough to disrupt the stoichiometry of propidium iodide binding, which requires double-stranded DNA. The denaturing agent, HCl, has proved the most reliable method with little loss of material. The incubation time in HCl varies according to the material being stained. Normal tissues require much less denaturation than tumors (10 min vs 15–20 min). Cells require much longer denaturation than nuclei (20–30 min vs 10–15 min). The optimum denaturation period should be characterized for each particular experimental system. The use of HCl, however, is relatively mild and does not result in extensive denaturation of DNA. If increased sensitivity of BUdR detection is required, an alternative method involving incubation of cells in 0.1*M* HCl containing 0.7% Triton X-100 at 0°C for 10 min, followed by boiling in 2 mL of distilled water for 10 min *(8)*, will increase sensitivity by a factor of 20–30. This procedure, and several like it, suffer the drawback that cell loss is a major problem; in the case of hematopoietic cells, up to 90% of the staining material may be lost during the process. Procedures have been developed to combine denaturation with digestion *(9)* using 0.2 mg/mL pepsin in 2*M* HCl for 30 min at room temperature. This technique has proved particularly useful for epidermal cells. One other potentially useful method for denaturing DNA is the use of restriction endonucleases

and exonuclease III *(10)*. In these procedures, nuclear protein is extracted with 0.1*M* HCl or citric acid prior to digestion with endonuclease *Bam* HI, *Dde* I, *Eco* RI, or *Hind* III followed by digestion with exonuclease III. This procedure may achieve the sensitivity of heat denaturation without severe cell loss, morphology change, or protein loss.

4. Many procedures employ 0.1*M* sodium tetraborate (pH 8.5) to neutralize after the acid denaturation. This does not appear to be necessary when large volumes of PBS are used as a washing medium.
5. Many monoclonal antibodies are commercially available to detect halogenated pyrimidines. Most antibodies will detect both iodo- and bromodeoxyuridine. However, Br3 shows extremely high specificity for BUdR *(11)*, whereas IU4 (both available from Caltag Inc.) shows much higher specificity for IUdR *(12)*. These antibodies have opened the pathway for double labeling experiments on the FCM. The dilution of the monoclonal will depend on its specificity and whether it is a purified, ascitic, or supernatant preparation; for instance, we routinely use IU4 at a dilution of 1:2000 compared to 1:20 for the Sera-Lab antibody.
6. Some workers prefer to carry out the antibody incubations at 4°C, as some monoclonal preparations were found to have DNAse activity from mycoplasma contaminants. This should not be a problem with commercial preparations.
7. There is no need to perform the assay in dark conditions. BUdR is photosensitive, but there is no advantage when incubations are carried out in the dark.
8. It has been reported that incubation of cells in 0.1% sodium borohydride in 1 mL PBS for 30 min at room temperature, after the antibody incubations, reduces autofluorescence and nonspecific staining. This has not proved to be advantageous in our hands.
9. We found that stained preparations can be kept for up to at least one month at 4°C with little loss of fluorescence.
10. If the staining procedure fails to detect BUdR incorporation, it is always worth repeating the antibody incubations. One of the most common reasons for failure is an excess of cellular or nuclear material.

References

1. Gratzner, H. (1982) Monoclonal antibody against 5-bromo and 5-iododeoxyuridine: A new reagent for detection of DNA replication. *Science* **218,** 474,475.
2. Dolbeare, F., Gratzner, H., Pallavicini, M., and Gray, J. W. (1983) Flow cytometric measurement of total DNA content and incorporated bromodeoxyuridine. *Proc. Natl. Acad. Sci. USA.* **80,** 5573–5577.

3. Pallavicini, M. G. (1987) Solid tissue dispersal for cytokinetic analyses, in *Techniques in Cell Cycle Analysis* (Gray, J. W. and Darzynewicz, Z., eds.), Humana, Clifton, NJ, pp. 139–162.
4. Wilson, G. D., McNally, N. J., Dunphy, E., Karcher, H., and Pfragner, R. (1985) The labeling index of human and mouse tumours assessed by bromodeoxyuridine staining in vitro and in vivo and flow cytometry. *Cytometry* **6,** 641–647.
5. Begg, A. C., McNally, N. J., Shrieve, D. C., and Karcher, H. (1985) A method to measure the duration of DNA synthesis and the potential doubling from a single sample. *Cytometry* **6,** 620–626.
6. Wilson, G. D., McNally, N. J., Dische, S., Saunders, M. L, des Rochers, C., Lewis, A. A., and Bennett, M. H. (1988) Measurement of cell kinetics in human tumours in vivo using bromodeoxyuridine incorporation and flow cytometry. *Br. J. Cancer* **58,** 423–431.
7. Schutte, B., Reynders, M. M. J., van Assche, C. L. M. V. J., Hupperets, P. S. G. J., Bosman, F. T., and Blijham, G. H. (1987) An improved method for the immunocytochemical detection of bromodeoxyuridine labeled nuclei using flow cytometry. *Cytometry* **8,** 372–276.
8. Beisker, W., Dolbeare, F., and Gray, J. W. (1987) An improved immunocytochemical procedure for high sensitivity detection of incorporated bromodeoxyuridine. *Cytometry* **8,** 235–239.
9. van Erp, P. E. J., Brons, P. P. T., Boezeman, J. B. M., de Jongh, G. J., and Bauer, F. W. (1988) A rapid flow cytometric method for bivariate bromodeoxyuridine/DNA analysis using simultaneous proteolytic enzyme digestion and acid denaturation. *Cytometry* **9,** 627–630.
10. Dolbeare, F. and Gray, J. W. (1988) Use of restriction endonuclease and exonuclease III to expose halogenated pyrimidines for immunochemical staining. *Cytometry* **9,** 631–635.
11. Vanderlaan, M., Watkins, B., Thomas, C., Dolbeare, F., and Stanker, L. (1986) Improved high-affinity monoclonal antibody to iododeoxyuridine. *Cytometry* **7,** 499–507.
12. Dolbeare, F., Kuo, W. L., Vanderlaan, M., and Gray, J. I. V. (1988) Cell cycle analysis by flow cytometric analysis of the incorporation of iododeoxyuridine (IdUrd) and bromodeoxyuridine (BrdUrd). *Proc. Am. Assoc. Cancer. Res.* **29,** 1896–1901.

CHAPTER 44

Production and Use of Nonradioactive Hybridization Probes

Victor T.-W. Chan and James O'D. McGee

1. Introduction

1.1. Preparation of Nonradioactive Hybridization Probes

Molecular hybridization is a useful technique for identifying specific target sequences even when they are present as a single copy in a complex population. It can be performed either on a solid matrix on which pure DNA (or RNA) is bound (blot hybridization) or on tissue sections (*in situ* hybridization). Until recently, the probes used in hybridization were usually labeled with radioisotopes. However, the short half-life, disposal, and safety problems of radioactive probes stimulated the development of nonradioactive hybridization techniques. In these, the probes are labeled with nonradioactive reporter molecules, which can be haptens, proteins, digoxigenin, biotin, and so forth. These reporter molecules can then be detected by enzyme-labeled antibodies or streptavidin (in the case of biotinylated probes). Of these reporter molecules, biotin and digoxigenin have several advantages over the others because of their small size. Therefore, they minimally interfere with hybridization efficiency. In addition, the high affinity of the binding of biotin and streptavidin ($K_d = 10^{-15} M$) is almost equivalent to covalent bonds. In fact,

From: *Methods in Molecular Biology, Vol. 10: Immunochemical Protocols*
Ed.: M. Manson ©1992 The Humana Press, Inc., Totowa, NJ

the biotin system was the first nonradioactive hybridization technique sensitive enough for routine use on blot hybridization *(1)* and *in situ* hybridization *(2)*.

In the original system, biotin is attached to the deoxy analog of γ-UTP via a spacer arm. Biotinylated dUTP is incorporated into DNA strands by a conventional labeling reaction, nick translation, which is also widely used to prepare radioactive probes *(3)*. The presence of a spacer arm between biotin and dUTP separates these two molecules far apart, and thus, reduces the steric hindrance caused between them. Therefore, the efficiency of labeling, hybridization, and detection is greatly increased.

The principle of nick translation *(3)* is based on the ability of deoxyribonuclease I (DNase I) to attack each strand of the DNA molecule independently in a random fashion, resulting in single-stranded nicks at low enzyme concentration in the presence of Mg^{2+}. *E. coli* DNA polymerase I (Pol I) synthesizes DNA complementary to the intact strand in a 5'-3' direction using the 3' OH termini of the nicks as primers. At the same time, the 5'-3' exonuclease activity of Pol I removes nucleotides in the same direction. The result of these two enzyme reactions is the replacement of the original nucleotides with new nucleotides, and the movement of the nicks in a 5'-3' direction along the DNA strands whose nucleotide sequences are unchanged. If one of the deoxyribonucleotide triphosphates included in the reaction is labeled, the original nucleotide will be replaced by labeled counterparts and the DNA is thus labeled. Since DNase I introduces nicks to both strands in a random fashion, theoretically, the probe is uniformly and almost completely labeled. However, (practically?) nick translation is relatively less reproducible, simply because it depends on the combination of the reactions of two enzymes that may be selectively inhibited by different contaminants in the DNA samples. Under optimal conditions, this method can be used to label all the different forms of double stranded DNA molecules with very high sensitivity.

Biotinylated dUTP can also be used to label DNA probes by a different method, namely random-primed labeling *(4)*. The principle of this method is based on the reannealing of hexadeoxyribonucleotide primers, which have random specificity, to the denatured DNA strands. The DNA to be labeled has to be linearized and denatured before the strands are used as templates in the labeling reaction. The complementary strands are synthesized from the 3' OH termini of the reannealed hexanucleotides by the Klenow fragment of *E. coli* DNA polymerase I. The primers reanneal at random sites of the template strands, so that the synthesis of the complementary strands is primed at random sites. If one of the deoxyribonucleoside triphosphates present in the reaction mixture is labeled, the newly synthesized strands will become labeled by the incorporation of the labeled nucleotides. The end product of this reaction is a mixture of unlabeled (template) and labeled (synthesized) DNA strands. However, this method is more reliable and reproducible than nick translation, because only the Klenow fragment of Pol I is used, which is rather resistant to the inhibition caused by various contaminants in the DNA samples. This method can only be used to label linear DNA molecules. Supercoiled DNA has to be linearized before being labeled. In general, the sensitivity of probes labeled by this method is reasonably high.

Another development of the biotin system is the synthesis of biotinylated UTP that can be used to prepare single-stranded biotinylated RNA probes. The success of this technique is based on two important components: (1) the synthesis of a biotinylated derivative of γ-UTP, and (2) the fact that this derivative of γ-UTP is a good substrate for T3 and T7 RNA polymerase (but not SP6 RNA polymerase). The DNA to be transcribed is cloned into a vector containing phage transcription promoters (e.g., T3 and T7) that can initiate the in vitro transcription of the DNA insert in the presence of the respective phage RNA polymerase. In some cases, two different phage transcription promoters are placed on either side of the polylinker cloning region of the vector in opposite orientation, so they can initiate the transcription of the coding and noncoding strands of the DNA insert, respectively. In general, hybrids of target sequences and RNA probes have a higher melting temperature (T_m) than that of DNA probes. Furthermore, after hyridization, the nonspecifically bound RNA probes can be selectively removed by RNase treatment. All these turn out to be advantages of this method, which enhance its sensitivity and specificity.

Soon after the demonstration of the general utility of the biotin system in molecular hybridization, a photoactivatable biotin analog (photobiotin) was synthesized *(5)*. A cationic spacer arm is placed between a photoactivatable group and biotin to enhance the efficiency of labeling. The photoactivatable group of this molecule is extremely reactive upon exposure to light and with a wide range of chemical groups. Therefore, it can be used to label various biomolecules (e.g., DNA, RNA, and protein).

Besides biotin, a digoxigenylated derivative of dUTP was also synthesized. This derivative of dUTP can be incorporated into DNA by Pol I (or the Klenow fragment of Pol I). Therefore, digoxigenin-labeled DNA probes can be prepared by standard nick translation and random primed-labeling reactions developed for the biotin system. It is almost certain that more nonradioactive alternatives to biotin and digoxigenin will be developed in the near future. Chemiluminescent methods for nonradioactive probe detections are now being introduced.

Chemically modified DNAs can also be used as hybridization probes, provided that the modification does not interfere with the formation of hybrid DNA molecules. Of the various chemical methods, acetylaminofluorene (AAF)-modified DNA *(6)* became more widely used in molecular hybridization. The labeling reaction is simple and straightforward. However, the potent genotoxic effect of AAF (that is, the consequence of its interaction with DNA) is a major factor that limits its general utility as a nonradioactive label.

2. Materials

2.1. Nick Translation

1.a. Nucleotides/buffer mix I (for dUTP analog): 0.2 m*M* dATP, 0.2 m*M* dCTP, 0.2 m*M* dGTP in 500 m*M* Tris-HCl, pH 7.5, containing 50 m*M* $MgCl_2$, 10 m*M* β-mercaptoethanol; or

b. Nucleotides/buffer mix II (for dATP analog): 0.2 m*M* dTTP, 0.2 m*M* dCTP, 0.2 m*M* dGTP in 500 m*M* Tris-HCl, pH 7.5, containing 50 m*M* $MgCl_2$, 10 m*M* β-mercaptoethanol.

2. Enzyme mix: 0.5 U of DNA polymerase 1/µL, 50 pg of DNase 1/µL in 100 m*M* $KHPO_4$, pH 6.5, containing 1 m*M* dithiothreitol (DTT), 500 µg of BSA/mL and 50% glycerol.
3. a. dUTP Analog: 0.2 m*M* Biotin dUTP, or 0.2 m*M* digoxigenin dUTP in 50 m*M* Tris-HCl, pH 7.5; or
   b. ATP Analog: 0.2 m*M* Biotin-7-dATP in 50 m*M* Tris-HCl, pH 7.5.
4. Template DNA.
5. TE Buffer: 10 m*M* Tris-HCl, 1 m*M* EDTA, pH 8.
6. Stop solution: 150 m*M* EDTA, pH 8.
7. 3*M* sodium acetate (NaOAc), pH 5.2.
8. Carrier DNA: Sonicated salmon sperm DNA, 10 mg/mL of distilled water.
9. Ethanol.

2.2. Random-Primed Labeling

10.a. Nucleotides/buffer mix I (for dUTP analog): 0.5 m*M* dATP, 0.5 m*M* dCTP, 0.5 m*M* dGTP, 50 U of $pd(N)_6$ (hexanucleotides)/mL in 2*M* Hepes, pH 6.6 containing 50 m*M* $MgCl_2$, 10 m*M* β-mercaptoethanol; or

b. Nucleotides/buffer mix II (for dATP analog): 0.5 m*M* dTTP, 0.5 m*M* dCTP, 0.5 m*M* dGTP, 50 U of $pd(N)_6$/mL in 2*M* Hepes, pH 6.6, containing 50 m*M* $MgCl_2$ and 10 m*M* β-mercaptoethanol.

11. Klenow fragment of Pol I: 0.5 U/µL in 100 m*M* $KHPO_4$, pH 6.5, containing 1 m*M* DTT, 500 µg of BSA/mL and 50% glycerol.

12.a. dUTP Analog: 0.5 m*M* Biotin dUTP, 0.5 m*M* digoxigenin-11-dUTP in 50 m*M* Tris-HCl, pH 7.5;
b. dATP Analog: 0.5 m*M* Biotin-7-dATP in 10 m*M* Tris-HCl, pH 7.5. Materials 4–9 are also required for this protocol.

2.3. *In Vitro Transcription*

13. Nucleotides/buffer mix: 4 m*M* γ-ATP, 4 m*M* γ-GTP, 4 m*M* γ-CTP in 400 m*M* Tris-HCl, pH 8.0, containing 80 m*M* $MgCl_2$, 20 m*M* spermidine and 500 m*M* NaCl.
14. 0.75*M* Dithiothreitol (DTT).
15. Human placental ribonuclease inhibitors.
16. T3 and T7 RNA polymerase: 10 U/μL in 20 m*M* $KHPO_4$, pH 7.7, containing 100 m*M* NaCl, 0.1 m*M* EDTA, 1 m*M* DTT, 0.01% Triton X-100, and 50% glycerol.
17. γ-UTP Analog: 4 m*M* Biotin-11-γ-UTP in 10 m*M* Tris-HCl, pH 7.5.
18. DNase I (RNase free): 10 U/μL in 10 m*M* Tris-HCl, pH 7.6, containing 10 m*M* $CaCl_2$, 10 m*M* $MgCl_2$ and 50% glycerol.
19. Phenol/chloroform 1:1.
20. Chloroform.
21. DNA in a suitable vector.
Materials 4–9 are also required for this protocol.

2.4. *Photobiotinylation*

22. Photobiotin: 1 mg/mL in water (it should be kept in the dark all the time).
23. 100 m*M* Tris-HCl, pH 9.0.
24. 2-Butanol.
25. Sunlamp: For example, General Electric Model No. RSM, 275W.
Also required are materials 4, 5, 7–9.

2.5. *Acetylaminofluorene Labeling (AAF Labeling)*

26. *N*-acetoxy-*N*-2-aceylaminofluorene (AAF): 120 μg/mL in ethanol. *Caution:* This chemical is an extremely potent carcinogen. It should be handled with precautions accordingly.

27. 2 mM sodium citrate, pH 7.
28. 50 mM sodium borate, pH 9.
29. 100 mM Tris-HCl, pH 7.
30. Diethyl ether.

Also required are materials listed in 7 and 9.

3. Methods

3.1. Nick Translation (see Notes 1 and 2)

1. Transfer 1 µg of DNA (500 µg/mL in TE) to an Eppendorf tube.
2. Add 5 µL of nucleotides/buffer mix I or II and 5 µL of the appropriate labeled nucleotide. Add sterile distilled water to give a final vol of 45 µL.
4. Mix the sample thoroughly and keep it on ice.
5. Add 5 µL of enzyme mix and mix the sample gently.
6. Incubate at 14°C for 1–2 h (*see* Note 3).
7. Add 5 µL of stop solution and 5 µL of 3M NaOAc, pH 5.2.
8. Add 2 µL of carrier DNA.
9. Add 140 µL of ethanol to precipitate the DNA.
10. Dissolve the DNA pellet in 20 µL of TE.

3.2. Random-Primed Labeling (see Notes 4 and 5)

1. Transfer 0.05–1.0 µg of restricted DNA (2 µL) to an Eppendorf tube.
2. Add water to 35 µL.
3. Heat-denature the DNA sample at 100°C for 10 min.
4. Incubate the sample on ice for 5 min and then spin it briefly.
5. Add 5 µL of nucleotides/buffer mix.
6. Add 5 µL of labeled nucleotide and mix the sample thoroughly.
7. Add 5 µL of Klenow fragment of Pol I and mix the sample gently.
8. Incubate at room temperature for 6–16 h, or at 37°C for 1–4 h (*see* Note 6).
9. Add 5 µL of stop solution and 5 µL of 3M NaOAc, pH 5.2.
10. Add 2 µL of carrier DNA.
11. Add 140 µL of ethanol to precipitate the DNA.
12. Dissolve the DNA pellet in 20 µL of TE.

3.3. In Vitro Transcription

1. Transfer 1 µg of RNase free restricted DNA (100 µg/mL in TE) to an Eppendorf tube (*see* Note 7).
2. Add 3 µL of nucleotide/buffer mix and 3 µL of labeled nucleotide.
3. Add 1 µL of 0.75*M* DTT and 25 U of RNase inhibitor.
4. Add water to 29 µL and mix the sample thoroughly.
5. Add 1 µL of RNA polymerase and mix sample gently (*see* Note 8). Incubate at 37°C for 30 min.
6. Add 1 µL of DNase I and incubate the sample for another 15 min at 37°C.
7. Extract the sample with phenol/chloroform (1:1) once, then with chloroform once.
8. Add 1.5 µL of carrier DNA and 3 µL of 3*M* NaOAc, pH 5.2.
9. Add 85 µL of ethanol to precipitate the labeled RNA.
10. Dissolve the RNA pellet in 20 µL of TE containing 25 U of RNase inhibitor (*see* Note 9).

3.4. Photobiotinylation (see Notes 11 and 12)

1. Transfer the DNA or RNA sample (1 mg/mL in water) to be labeled to an Eppendorf tube.
2. Add an equal vol of photobiotin stock solution.
3. Mix thoroughly and keep the sample on ice.
4. Initiate the labeling reaction by irradiating the reaction mix 10 cm below the sunlamp for 15 min.
5. Add 150 µL of 100 m*M* Tris-HCl, pH 9 to the sample.
6. Extract the sample with an equal vol of 2-butanol.
7. Discard the upper phase and repeat the extraction twice.
8. Add 3 µL of carrier DNA and 1/10 vol of 3*M* NaOAc, pH 5.2.
9. Add 2.5× the vol of ethanol to precipitate labeled nucleic acids.
10. Dissolve the pellet in an appropriate amount of TE.

3.5. AAF Labeling (see Note 13)

1. Sonicate or digest DNA (or RNA) to approx 1 kb.
2. Transfer the DNA (or RNA) sample (100 µg/mL in 2 m*M* sodium citrate, pH 7.0) to an Eppendorf tube.
3. Heat-denature the sample at 100°C for 5 min.

4. Add 2.5 µL of AAF. Incubate at 37°C for 2 h in the dark.
5. Add 12.5 µL of 50 m*M* sodium borate, pH 9. Heat-treat at 100°C for 3 min.
6. Add 12.5 µL of 100 m*M* Tris-HCl, pH 7.0.
7. Extract the sample with an equal vol of cold diethyl ether five times.
8. Add 2 µL of carrier DNA and 4 µL of 3*M* NaOAc, pH 5.2.
9. Add 110 µL of ethanol to precipitate the modified nucleic acids.
10. Dissolve the pellet in 20 µL of TE.

4. Notes

4.1. Nick Translation

1. Nick translation is probably the most sensitive method to prepare non-radioactive DNA probes. In general, it is simple, easy, and reasonably reliable. However, in order to achieve reproducible results, use of reagents and DNA samples of the highest quality is strongly recommended.
2. The reaction can be either scaled up or scaled down at will.
3. The incubation time in Step 7 should be optimized experimentally for different DNA samples to give highest sensitivity.

4.2. Random-Primed Labeling

4. The end product of the random-primed labeling reaction is a mixed population of unlabeled and labeled strands. Even if the reaction goes to completion, the labeled (synthesized) strands only account for 50% of the DNA strands. This is probably the reason why probes labeled by this method are not as sensitive as optimally nick-translated probes. However, this method is relatively more reliable and reproducible than nick translation.
5. This method can only be used to label linear DNA molecules. It is particularly useful for the DNA samples extracted from agarose gels, especially short DNA fragments, for which nick translation usually gives poor results.
6. The incubation time in Step 8 should be optimized experimentally for different amounts of DNA, and different DNA samples to give highest sensitivity.

4.3. In Vitro Transcription

7. The vector should be linearized with a restriction enzyme before the transcription reaction in order to obtain transcripts of a defined length.

Using intact plasmid DNA as template for transcription will result in heterogeneous transcripts of multiple plasmid lengths.

8. Excessive amounts of RNA polymerase should not be used with the vector containing both T3 and T7 promoters, otherwise transcription may not be promoter-specific (strand-specific). Nonspecific initiation of RNA transcripts may also occur at the ends of the DNA template. This is most prevalent with a 3' protruding terminus. Nonspecific initiation may be reduced by increasing the final NaCl concentration in the transcription buffer to 100 m*M*. When possible, restriction enzymes that leave blunt or 5' protruding ends, should be used.
9. The addition of RNase inhibitor is to preserve the full-length single-stranded RNA transcripts.
10. Since SP6 polymerase cannot use bio-γ-UTP efficiently, it is not suitable for preparing biotinylated RNA probes.

4.4. Photobiotinylation

11. Photobiotinylation is a simple and inexpensive method to prepare large quantities of nonradioactive hybridization probes. The advantage of this method is that the same reagents can be used to label DNA as well as RNA molecules.
12. The sensitivity of probes labeled by this method is not as high as for those labeled by enzymatic methods. Furthermore, photobiotinylated probes usually give higher nonspecific background owing to the cationic spacer arm of photobiotin, which nonspecifically binds to anionic molecules (e.g., DNA and RNA).

4.5. AAF Labeling

13. AAF labeling can also be used to prepare large quantities of nonradioactive hybridization probes. It is simple and inexpensive. However, it is still not widely used. This is partly caused by the potent genotoxic effect of the carcinogen. In addition, the sensitivity of probes labeled by this method has to be confirmed. These two factors significantly limit the general usefulness of this method for the preparation of nonradioactive molecular probes.

Acknowledgments

Work done in this laboratory was supported by grants to J. O'D. McGee from the Cancer Research Campaign, UK.

References

1. Chan, V. T.-W., Fleming, K. A., and McGee, J. O'D. (1985) Detection of subpicogram quantities of specific DNA sequences on blot hybridzation with biotinylated probes. *Nucl. Acids Res.* **13,** 8083–8091.
2. Burns, J., Chan, V. T.-W., Jonasson, J. A., Fleming, K. A., Taylor, S., and McGee, J. O'D. (1985) A sensitive method for visualizing biotinylated probes hybridized *in situ*: Rapid sex determination on intact cells. *J. Clin. Pathol.* **38,** 1085–1092.
3. Rigby, P. W. J., Dieckmann, M., Rhodes, C., and Berg, P. (1977) Labelling deoxyribonucleic acid to high specific activity in vitro by nick translation with DNA polymerase I. *J. Mol. Biol.* **113,** 237–251.
4. Feinberg, A. P., and Vogelstein, B. (1983) A technique for radio-labeling DNA restriction endonuclease fragments to high specific activity. *Anal. Biochem.* **132,** 6–13.
5. Forster, A. C., McInnes, J., Skingle, D. C., and Symons, R. H. (1985) Nonradioactive hybridization probes prepared by chemical labeling of DNA and RNA with a modified novel reagent, photobiotin. *Nucl. Acids Res.* **13,** 745–761.
6. Tchen, P., Fuchs, R. P. P., Sage, E., and Leng, M. (1984) Chemically modified nucleic acids as immunodetectable probes in hybridisation experiments. *Proc. Natl. Acad. Sci. USA* **81,** 3466–3470.

CHAPTER 45

Cellular Human and Viral DNA Detection by Nonisotopic *In Situ* Hybridization

C. Simon Herrington and James O'D. McGee

1. Introduction

In situ hybridization may be defined as the detection of nucleic acids *in situ* in cells, tissues, chromosomes, and isolated cell organelles. The technique was described in 1969 by two separate groups, who demonstrated repetitive ribosomal sequences in nuclei of Xenopus oocytes using radiolabeled probes *(1,2)*. Refinements in recombinant DNA technology and the development of nonisotopic probe labeling and detection obviate the need for radiation protection and disposal facilities, and have converted nonisotopic *in situ* hybridization (NISH) from a purely research technique to one that can be used in routine laboratory testing.

Nonisotopic reporter molecules, such as biotin, digoxigenin, mercury, and acetylaminofluorene (AAF) can be detected using either affinity or immunocytochemical techniques. The reporters can be detected by fluorochromes and enzyme/chromogen combinations; the former requires expensive equipment and is less suitable for routine analysis. The immunoenzymatic procedures, however, can be combined with routine tissue and cytological staining, enabling simultaneous analysis of nucleic acid content within intact cells and tissues. By combining NISH and immunocytochemistry (ICC), individual cells can be genotyped and phenotyped *(3)*. Similarly, more than one nucleic acid can be detected in individual cells within tissue sections *(4)*. These

From: *Methods in Molecular Biology, Vol. 10: Immunochemical Protocols*
Ed.: M. Manson ©1992 The Humana Press, Inc., Totowa, NJ

techniques and their combination can be applied to a variety of experimental and clinical situations (for reviews, *see 5–7*).

NISH technology has four elements: (1) choice of probe/reporter and probe labeling, (2) pretreatment of cells/tissues, (3) denaturation/hybridization, and (4) detection of signal.

1. The ideal reporter molecule is determined by considerations of sensitivity, specificity, safety, and practicability. Sensitivity is dependent not only on the reporter, but also on all other components of the NISH procedure, particulary nucleic acid unmasking and probe detection. Specificity can be enhanced by using reporters that are not normally present in cells. For example, biotin-labeled probes produce unacceptable background noise in tissues containing high levels of endogenous biotin (e.g., liver). Biotin is the most widely used reporter in NISH and was originally chosen because of its high affinity for, and tetravalent binding to, avidin. Many alternatives have been formulated, including mercury, acetylaminofluorene, bromodeoxyuridine, sulphones (for review *see 8*) and digoxigenin *(9)*. Many have intrinsic disadvantages. For example, mercuric cyanide is toxic and acetylaminofluorene, carcinogenic. However, digoxigenin, a derivative of the therapeutic cardiac glycoside digoxin, is safe, produces low background, and can be incorporated into nucleic acids in the same way as biotin. These alternatives can be used in place of, or in addition to, biotin. The choice depends on the experimental situation.

   Multiple *in situ* nucleic acid detection can be achieved in two ways: by sequential hybridization with probes labeled with the same reporter, and by simultaneous hybridization using probes labeled with different reporters. Any of the reporters listed above can be combined to allow dual nucleic acid detection, but we have found biotin and digoxigenin the most generally useful, as they are safe and sensitive. Probes labeled with these reporters can be combined in NISH and detected differentially according to the scheme presented in Fig. 1 *(4)*.
2. Having chosen the reporter and incorporated it into a probe, the cell/tissue to be analyzed is prepared for hybridization. The following requirements must be met to allow a successful hybridization reaction to take place:
   a. The cell/tissue and its nucleic acid content must be fixed in such a way that cell/tissue morphology is preserved and nucleic acids retained during the reaction.
   b. The cell/tissue sample must be sufficiently permeable to allow the labeled probe to reach its target nucleic acid.

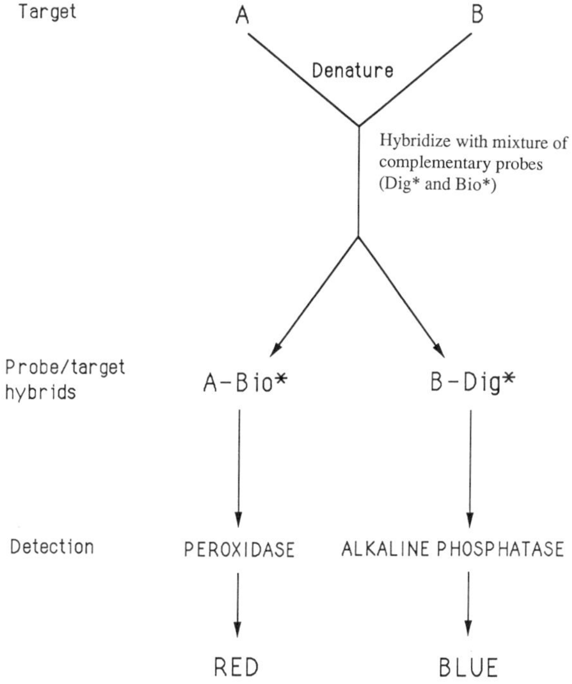

Fig. 1. Dual nucleic acid detection. The target DNAs (A and B) are denatured and hybridized with complementary DNA probes labeled with biotin (Bio*) or digoxigenin (Dig*). Biotin and digoxigenin residues are detected, respectively, with avidin peroxidase (red) and antibody to digoxigenin labeled with alkaline phosphatase (blue).

The latter is dependent on interplay between the degree of permeability of the cell/tissue and the size of the probe. In practice, probes with a median size of 200–400 bps are suitable for cells/tissues fixed in aldehyde and subjected to varying degrees of proteolysis (*see* Sections 3.1.1. and 3.1.2.). We have found that fixation of cells in aldehyde followed by mild proteolysis produces the best results.

3. Hybridization of probe to target requires that both probe and cellular target are denatured to form single-stranded molecules. Denaturation can be achieved by treatment with alkali or heat. The former is inappropriate for probes labeled with many nonisotopic labels, since the ester linkage between reporter and nucleotide undergoes alkaline hydrolysis. Hybridization occurs when the reaction temperature falls below the melting temperature (T_m) of the duplex formed between probe and target. Stringency conditions, which determine the degree to which the probe cross hybridizes with closely related sequences, can be varied according to individual requirements (for full discussion, *see* ref. 10).

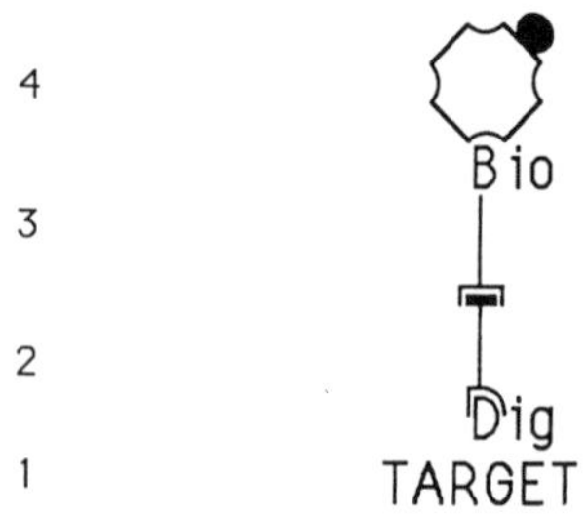

Fig. 2. Diagrammatic representation of three-step digoxigenin detection. The target nucleic acid (1) is labeled with digoxigenin (Dig). Monoclonal antidigoxigenin (2) is linked via biotinylated (Bio) rabbit antimouse (3) to avidin alkaline phosphatase (4).

4. The principles employed in the detection of nonisotopic probe labels are similar to those used for ICC; the main differences between NISH and ICC are the degree of unmasking required to render nucleic acids available for hybridization and the hybridization reaction itself. Affinity and immunohistochemical techniques are used for probe detection employing avidin conjugates for the detection of biotin-labeled probes and antibody based systems for the detection of digoxigenin. Combining affinity and immunohistochemical systems produces hybrid detection systems of high sensitivity. Thus, biotinylated linker antibodies are used as a bridge between monoclonal antibodies to the reporter molecule and an avidin conjugate (Fig. 2). Similarly, the alkaline phosphatase antialkaline phosphatase (APAAP) system can be applied to the detection of both biotin- and digoxigenin-labeled probes. The latter has the advantage that it is biotin-independent.

In this chapter, methods developed for the detection of human papillomaviruses and repetitive genomic DNA sequences (individually and in combination) in cultured cells and routinely processed surgical biopsies, are described.

2. Materials

1. Aminopropyltriethoxysilane (2%, v/v): Mix 12 mL of aminopropyltriethoxysilane with 588 mL of acetone immediately prior to use.
2. CaSki cells can be obtained from the American Type Culture Collection (ATCC, Rockville, MD).
3. Complete medium: Fetal calf serum (50 mL) is added to 440 mL of RPMI 1640 medium, together with 5 mL of penicillin/streptomycin solution (100X) and 5 mL L-glutamine (100X). This medium can be stored at 4°C.

4. Phosphate buffered saline (PBS): 10 m*M* phosphate, 150 m*M* NaCl, pH 7.4.
5. PBS glycine: Dissolve 0.2 g of glycine in 100 mL of PBS to give a 0.2% solution.
6. Tris buffered saline (TBS): 50 m*M* Tris-HCl, 100 m*M* NaCl, pH 7.2.
7. Tris, Bovine serum albumin (BSA), Triton (TBT): TBS containing 3% (w/v) BSA (Fraction V) and 0.05% (v/v) Triton X-100.
8. Alkaline phosphatase substrate buffer: 50 m*M* Tris-HCl, 100 m*M* NaCl, 1 m*M* $MgCl_2$, pH 9.5.
9. Tris-EDTA (TE) buffer; 10 m*M* Tris-HCl, 1 m*M* EDTA, pH 8.
10. Paraformaldehyde (4%, w/v): Boil 100 mL of PBS containing 4 g of paraformaldehyde in a fume cupboard. Cool on ice prior to use. The final pH of this solution should be 7.2–7.4 without adjustment.
11. Proteinase K/pepsin: The source used is important. The activity of these enzymes varies according to manufacturer and lot. We have found the proteinase K supplied by Boehringer Mannheim (FRG) the most consistent, and the details quoted in Methods refer to that source. We find that pepsin obtained from Sigma (Poole, Dorset, UK) is superior to that from Boehringer (FRG) and can be used at 0.1% (w/v) compared with 0.4% (w/v) in PBS.
12. Pepsin solution: Dissolve 0.1 g pepsin in 96 mL of distilled water prewarmed to 37°C and add 4 mL of 5*M* HCl slowly.
13. Standard saline citrate (SSC, 1X): 150 m*M* NaCl, 15 m*M* sodium citrate.
14. Hybridization mix (HM) is prepared by adding 1 mL of 50% (w/v) dextran sulfate in distilled H_2O and 1 mL of 20X SSC to 5 mL of deionized formamide. This mix is adjusted to pH 7.0 using 5*M* HCI and stored at 4°C. Under these conditions, it lasts up to one year.
15. Four-spot multiwell slides: In our laboratory, we use these for most *in situ* hybridizations. These allow the analysis of four samples (under identical conditions) on the same slide and require a smaller volume of reagent per section than conventional microscope slides. As the spot diameter is 12 mm, 14 mm cover slips are used to cover the section without adjacent aliquots of reagent running together. These can be obtained from Chance (Raymond Lamb, UK).
16. Microtiter or Terasaki plates (Gibco/Nunc, Paisley, UK). Proteolysis, denaturation, and hybridization are carried out in these plates. They can be floated in a 37°C water bath or incubated in a moist environment at 42 or 95°C. Two slides can be accommodated in each plate.
17. All antibody and conjugate dilutions are determined by experiment using reagents from Dako (High Wycombe, UK) unless otherwise stated: Monoclonal antibiotin; biotinylated rabbit antimouse [F(ab')$_2$ fragment] (this is used to further reduce background staining); monoclonal

antidigoxin, Sigma (UK); rabbit antimouse immunoglobulin; APAAP complex, Dako (UK); streptavidin–peroxidase; avidin (or streptavidin) alkaline phosphatase; antidigoxigenin alkaline phosphatase (Boehringer Mannheim, FRG).

18. Enzyme substrates. A variety of different substrates for both alkaline phosphatase and peroxidase are available. In this chapter, only the alkaline phosphatase substrate nitroblue tetrazolium/bromochloroindolylphosphatase (NBT/BCIP) and the peroxidase substrate 3-amino-9-ethyl carbazole (AEC) are described. Other substrates can be found in textbooks of histochemistry (*see also* Chapter 10).
    a. NBT/BCIP is made in advance and stored at –20°C. After equilibration of 30 mL of alkaline phosphatase substrate buffer at 37°C, 10 mg of NBT is dissolved in 200 µL of dimethylformamide (DMF) and added to 1 mL of the prewarmed substrate buffer. The mixture is added dropwise to the remaining substrate buffer. BCIP (5 mg), dissolved in 200 µL of DMF, is then added slowly and the whole preparation stored in 4 mL aliquots at–20°C.
    b. AEC is prepared fresh daily by dissolving 2 mg of AEC in 1.2 mL of dimethylsulfoxide in a glass tube. The mixture is added to 10 mL of 20 m*M* acetate buffer, pH 5.0–5.2. Immediately prior to use, 1 µL of 30% (v/v) hydrogen peroxide is added. The final mix may require filtration prior to use.
19. Glycerol jelly is used as an aqueous mountant (*see* Note 1). Ten grams of gelatin are dissolved in 60 mL of distilled water on a hot stirrer. Glycerol (70 mL) and phenol (0.25 g) are added, and the mountant is thoroughly mixed. Glycerol/gelatin can be stored at room temperature (solid) or at 42°C (liquid).

3. Methods

3.1. Preparation of Slides for Archival Material

1. Place the multiwell slides in a slide rack and immerse in 2% (v/v) Decon 90 at 60°C for 30 min.
2. Rinse thoroughly in distilled water, then acetone, and air-dry (*see* Note 2).
3. Immerse in 2% (v/v) aminopropyltriethoxysilane in acetone for 30 min.
4. Rinse in acetone, wash in distilled water, and air-dry at 37°C.
5. Slides prepared in this way can be stored indefinitely at 22°C.
6. Cut 5-µm sections from routine paraffin embedded blocks onto slides prepared as above.

7. Bake the sections either overnight at 60°C or for 45 min at 75°C.
8. Store the slides at 22°C.
9. Dewax the sections by heating to 75°C for 15 min (*see* Note 3).
10. Then plunge them into xylene and wash for 2× 5 min.
11. Remove xylene by washing in 99% ethanol (industrial grade) for 5 min at 22°C.
12. Wash in distilled water.

Unmasking of nucleic acids is achieved by using either proteinase K or pepsin HCl (*see* Note 4).

3.1.1. Proteinase K

13. Dilute lyophilized proteinase K to 500 μg/mL in PBS and spot onto the slides (100 μL/spot), put in Terasaki plates, and float in water bath at 37°C for 15 min.
14. Wash in distilled water and air-dry at 75°C.

3.1.2. Pepsin HCl

15. Incubate sections in pepsin solution in a coplin jar for 15 min at 37°C.
16. Wash in distilled water and air-dry at 75°C.

3.2. Preparation of Slides Using CaSki (or Other Adherent Cultured) Cells

1. Grow CaSki cells to mid-log phase in complete medium.
2. Subculture onto four-spot multiwell slides and grow almost to confluence.
3. Fix in 4% paraformaldehyde (*see* Note 5) for 15 min at 22°C.
4. Rinse in 0.2% (w/v) glycine in PBS (5 min), PBS (5 min), and air-dry.
5. Unmask nucleic acids using proteinase K. Dissolve 1 mg of proteinase K in 1 mL of PBS.
6. Add 100 μL to 100 mL PBS prewarmed to 37°C to give a final concentration of 1 μg/mL.
7. Incubate slides in this solution in a coplin jar for 1–5 min at 37°C.

3.3. Probe Preparation/Hybridization

This and the following Methods sections apply to slides prepared by either of the two methods described above.

For each spot:

1. Add 1 μL (10–20 ng) of each of biotin and/or digoxigenin labeled probe to 7 μL of HM.
2. Excess carrier DNA may be added at this stage (*see* Note 6).
3. Add TE to a total vol of 10 mL .

4. Spot 8 µL of the mixture onto each well (*see* Note 7).
5. Cover each spot with a 14-mm coverslip (*see* Note 8).
6. Place slides in Terasaki plates (2 slides/plate).
7. Denature in a convection hot air oven at 95°C for 15 min. A hot plate is probably also satisfactory, but we do not use them.
8. Hybridize at 42°C in a hot air oven for 2 h.

3.4. Stringency Washing and Blocking Procedure

1. Wash the slides in 4X SSC at room temperature for 2× 5 min.
2. Wash in the appropriate solution for stringency; e.g., 50% formamide/ 0.1X SSC, if required for discriminating closely homologous sequences (*see* ref. *10*). Adjust all washing solutions to pH 7.0 with 5*M* HCl. The temperature of the solution should be monitored directly using a mercury thermometer. Washing should be carried out for 30 min.
3. Wash in 4X SSC at 22°C for 5 min.
4. Incubate for 10–15 min in blocking solution (TBT).

All incubations in antibody/avidin/enzyme conjugates described below are carried out at 22°C for 30 min, unless otherwise stated. The substrate reactions are carried out at room temperature and signal development monitored by light microscopy. The substrate incubation times are, therefore, determined empirically for each experiment. All slides are finally washed in distilled water, air-dried at 42°C, and mounted in glycerol jelly.

3.5. Detection of Biotinylated Probes

3.5.1. Conventional Signal Detection

1. Incubate the slides in avidin alkaline phosphatase (or streptavidin alkaline phosphatase) diluted 1:100 in TBT.
2. Remove unbound conjugate by washing twice in TBS for 5 min.
3. Incubate in NBT/BCIP for 15–20 min.
4. Terminate the reaction by washing in distilled water for 5 min (*see* Note 9).

3.5.2. Amplified Signal Detection

1. Incubate the slides in monoclonal antibiotin diluted 1:50 in TBT.
2. Wash in TBS for 5 min twice.
3. Incubate in biotinylated rabbit antimouse [F(ab')$_2$ fragment] diluted 1:200 in TBT.
4. Wash in TBS for 5 min twice.
5. Incubate in avidin alkaline phosphatase diluted 1:50, or avidin peroxidase diluted 1:100 in TBT, containing 5% (w/v) nonfat milk.

6. Wash in TBS and develop the signal, using either NBT/BCIP or AEC/H_2O_2.

3.5.3. APAAP Detection System

1. Incubate the slides in monoclonal antibiotin diluted 1:50 in TBT.
2. Wash in TBS for 5 min twice.
3. Incubate in rabbit antimouse immunoglobulin diluted 1:50 in TBT and wash in TBS.
4. Incubate in APAAP complex diluted 1:50 in TBT.
5. Develop the signal using NBT/BCIP.

3.6. Detection of Digoxigenin Labeled Probes

3.6.1. Conventional Detection (see *Note 10*)

1. Incubate the slides in alkaline phosphatase conjugated antidigoxigenin diluted 1:600 in TBT.
2. Wash in TBS and develop the signal using NBT/BCIP as described for detection of biotinylated probes.

3.6.2. Amplified Signal Detection

1. Incubate the slides in monoclonal antidigoxin diluted 1:10,000 in TBT.
2. Wash in TBS.
3. Incubate in biotinylated rabbit antimouse [F(ab')$_2$ fragment] diluted 1:200 in TBT.
4. Wash in TBS (5 min) and incubate in avidin alkaline phosphatase diluted 1:50 or avidin peroxidase diluted 1:100 in TBT containing 5% (w/v) nonfat milk.
5. Wash in TBS (5 min) and incubate in either alkaline phosphatase or peroxidase substrate.

3.6.3. APAAP Detection System

1. Incubate in monoclonal antidigoxin diluted 1:10,000 in TBT.
2. Proceed as for APAAP detection of biotinylated probes from Step 2 (*see above*).

3.7. Double Probe Detection

1. Incubate the slides in a mixture of avidin–peroxidase conjugate diluted 1:100 and alkaline phosphatase conjugated antidigoxigenin diluted 1:600 in TBT; in practice, 1 µL of antibody and 6 µL of avidin are added to 600 µL of TBT.
2. Remove any unbound conjugate by washing twice in TBS for 5 min.
3. Incubate in AEC for 30 min at 22°C.
4. Terminate the reaction by thorough washing in TBS.

5. Wash in alkaline phosphatase substrate buffer for 5 min at 22°C.
6. Incubate in NBT/BCIP for 20-40 min.
7. Terminate the reaction by washing in distilled water for 5 min.

4. Notes

1. The NBT/BCIP and AEC reaction products are soluble, and/or crystallize out, in organic solvent based mountants.
2. It is important that the slides dry completely, since aminopropyltriethoxysilane is insoluble in water.
3. It is important to preheat the sections prior to dewaxing, as this leads to more effective dewaxing of nucleic acids.
4. We have noted that the proteolysis step appears to be the most critical in the whole technique. Variability in signal intensity is virtually always caused by incomplete or excessive proteolysis; the only remedy is repetition of the experiment! Silane is an effective adhesive for most cells and sections, although occasionally, high concentrations of proteinase K can cause repeated section dehiscence. Under these circumstances, reduction of the proteinase K concentration or, alternatively, the use of pepsin HCl, may prove more effective; these, however, reduce the sensitivity of the method.
5. The fixation of CaSki cells in aldehyde has been shown to enhance the sensitivity of the detection of human papillomavirus type 16 by *in situ* hybridization compared with fixation by acid/alcohol *(9)*.
6. Carrier DNA (e.g., salmon sperm) can be added to the reaction mix in excess to reduce nonspecific probe binding. We have not found this necessary for the detection of human papillomaviruses in archival surgical biopsies, but it is often necessary for cultured cells.
7. The volume of probe mixture required to cover the section without adjacent aliquots running together varies inversely with the size of the section. Thus, large sections require as little as 6 µL of mixture and small ones, up to 8.5 µL.
8. We have found that sealing the cover slips with rubber cement is not necessary.
9. The NBT/BCIP reaction can be terminated chemically by lowering the pH (e.g., by rinsing in TBS, pH 7.2), adding phosphate ions (e.g., PBS), or by chelating Mg ions using EDTA. However, we find that thorough rinsing in distilled water is as effective.
10. At the time of writing, the antidigoxigenin antibody is available conjugated only to alkaline phosphatase. Peroxidase detection cannot therefore be used with this method. The amplified methods, however, allow greater flexibility.

Acknowledgments

C.S.H. is a Cancer Research Campaign (UK) Clinical Research Fellow and holds a Junior Research Fellowship at Green College, Oxford. This work was supported by grants from the CRC (UK) to J. O'D. M. G.

Further Reading

Polak, J. and McGee, J. O'D., eds., (1990), In Situ *Hybridization: Principles and Practice* Oxford University Press, Oxford.

Hames, B. D. and Higgins, S. J., eds., (1985), *Nucleic Acid Hybridization: A Practical Approach* IRL Press, Oxford.

References

1. Gall, J. G. and Pardue, M. L. (1969) Formation and detection of RNA-DNA hybrid molecules in cytological preparations. *Proc. Natl. Acad. Sci. USA* **63,** 378–383.
2. John, H. A., Birnstiel, M. L., and Jones, K. W. (1969) RNA–DNA hybrids at the cytological level. *Nature* **223,** 582–587.
3. Graham, A. K., Herrington, C. S., and McGee, J. O'D. (1991) Sensitivity and specificity of monoclonal antibodies to human papillomavirus Type 16 Capsid protein: Comparison with simultaneous viral detection by nonisotopic *in situ* hybridisation. *J. Clin. Pathol.* **44,** *96–101.*
4. Herrington, C. S., Burns, J., Graham, A. K., Bhatt, B., and McGee, J.O'D. (1989) Interphase cytogenetics using biotin and digoxigenin labeled probes. II: Simultaneous detection of two nucleic acid species in individual nuclei. *J. Clin. Pathol.* **42,** 601–606.
5. Warford, A. (1988) *In situ* hybridisation: A new tool in pathology. *Med. Lab. Sci.* **45,** 381–394.
6. Herrington, C. S, Flannery, D. M. J., and McGee, J. O'D. (1990) Single and simultaneous nucleic acid detection in archival human biopsies, in In Situ *Hybridization: Principles and Practice* (Polak, J. and McGee, J. O'D., eds.), Oxford University Press, Oxford, pp. 187–215.
7. Herrington, C. S., Burns, J., and McGee, J. O'D. (1990) Nonisotopic *in situ* hybridisation in human pathology, in *Nonisotopic* In Situ *Hybridization and the Study of Development and Differentiation* (Harris, N. and Wilkinson. D., eds.), Cambridge University Press, Cambridge, pp. 241–269.
8. Matthews, J. A. and Kricka, L. J. (1988) Analytical strategies for the use of DNA probes. *Anal. Biochem.* **169,** 1–25.
9. Herrington, C. S., Burns, J., Graham, A. K., Evans, M. F., and McGee, J. O'D. (1989) Interphase cytogenetics using biotin and digoxigenin labeled probes 1: Relative sensitivity of both reporters for detection of HPV16 in CaSki cells. *J. Clin. Pathol.* **42,** 592–600.
10. Herrington, C. S., Burns, J., Graham, A. K., and McGee, J. O'D. (1990) Discrimination of closely homologous HPV types by nonisotopic *in situ* hybridisation: Definition and derivation of tissue Tms. *Histochem. J.* **22,** 545–554.

CHAPTER 46

Chromosomal Mapping of Genes by Nonisotopic *In Situ* Hybridization

Bhupendra Bhatt and James O'D. McGee

1. Introduction

With the advent of nonradioactive labels, *in situ* hybridization (ISH) has become a useful technique for the detection of viral DNA in infected tissue (*1*), mRNA expression (*2*), sex determination (*3*), human gene mapping (*4*), and interphase cytogenetics (*5–8*). For chromosomal mapping of genes by ISH, labeled DNA probe is hybridized to metaphase spreads. If the label is a radioisotope, the signal is detected by autoradiography (*9*). Nonradioactive labels, such as biotin (*4*), digoxigenin (*10*), or 2-acetylaminofluorene (AAF), are detected by immunocytochemistry. AAF is incorporated by chemical modification of the DNA probe. Biotin and digoxigenin are incorporated enzymatically by nick translation or by random primer extension (*11*), although both can be incorporated by chemical modification (*12*). In the authors' laboratory, biotin has been successfully used in mapping genes with probes of 0.8 Kb (*10*). The method described here, therefore, applies to biotin-labeled probes although, with minor modifications, can be adapted to probes labeled by digoxigenin or AAF. An example of chromosomal mapping of a unique sequence is shown in Fig. 1.

There are several steps involved in mapping genes by nonisotopic *in situ* hybridization (NISH).

1. Preparation of metaphase spreads from whole blood cultures.
2. Removal of endogenous RNA.

From: *Methods in Molecular Biology, Vol. 10: Immunochemical Protocols*
Ed.: M. Manson

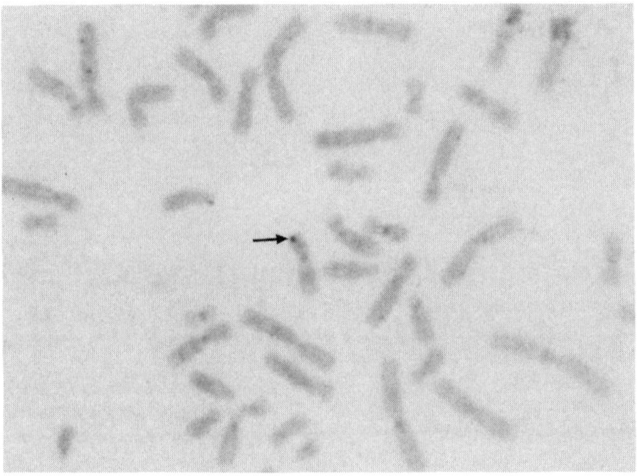

Fig. 1. Human chromosome spread probed with a biotinylated β-globin probe. There is a signal at band p1.5 (arrow) on chromosome 11.

3. Labeling the probe.
4. Denaturation and hybridization followed by stringency washes for the removal of nonspecifically bound probe.
5. Immunocytochemical detection of signal.
6 Chromosome banding.
7. Assignment of signal to specific chromosome band(s).

This technique is faster than radioisotopic *in situ* hybridization (RISH), which takes at least 1 wk to generate an interpretable signal, compared to 24 h for NISH. In addition, the technique puts gene mapping within reach of laboratories not equipped to handle radioisotopes. NISH complements other gene mapping techniques such as somatic cell hybrids (*12*) and linkage analysis. Apart from gene mapping, the method described here is potentially useful in the field of diagnostic molecular cytogenetics. For instance, biotin-labeled probes for the distal half of the q arm of the Y chromosome (α-satellite repeats) are used for sexing chorionic villus samples, amniotic cells, and embryos *(3,8)*. It can also be employed in the analysis of donor-recipient cell interactions in sex mismatched bone marrow transplants *(13–15)*. Additionally, probes for the α-satellite centromeric repeats specific for many human chromosomes are now available (*16*). These may be employed in prenatal diagnosis of numerical chromosomal aberrations, such as Down's syndrome (trisomy 21), as well as chromosomal aberrations in solid tumors *(8,17)*.

2. Materials

2.1. Tissue Culture

1. Complete Medium (CM): 76 mL of RPMI 1640, 20 mL of fetal bovine serum, 2 mL of phytohemagglutinin (M FORM), 1 mL of antibiotic/antimycotic solution (100X = 10,000 units penicillin, 10,000 µg Streptomycin and 25 µg fungizone) and 1 mL of 100X L-glutamine (100X = 30 mg/mL).
2. 5'-Bromo-2-deoxyuridine (BUdR): Dissolve 10 mg of BUdR per mL in distilled water. Sterilize by filtration with a 0.2-µm filter. Store aliquots at –20°C. Handle BUdR in subdued light.
3. Thymidine (dTTP): Dissolve 250 µg/mL in distilled water. Filter and store aliquots at –20°C.
4. Colcemid: Dissolve 5 µg in 1 mL of distilled water. Filter and store aliquots at –20°C.
5. Potassium chloride (KCl): Dissolve 560 mg of KCl (0.075*M*) in 100 mL distilled water. Prepare fresh and keep at 37°C.
6. Fixative: Mix three parts methanol with one part glacial acetic acid. Prepare fresh and keep on ice.
7. Probe DNA: Label 1 µg of probe DNA.
8. Nick translation kit.
9. Biotin-7-dATP or Biotin-11-dUTP.
10. Glycogen: Mussel glycogen (BCL), 20 mg/mL in sterile distilled water.
11. 20X SSC: 175 g of Sodium chloride, 88 g of trisodium citrate. Dissolve in 1 L of distilled water, adjust pH to 7.4 and autoclave.
12. Ribonuclease A (RNase): Dissolve 10 mg of pancreatic RNase A per mL of 20 m*M* sodium acetate (pH 5.0). Place in boiling water for 10 min. Store aliquots at –20°C. For a working solution dilute 1:100 in 2× SSC.
13. Hybridization Mixture (HM):
    a. Deionized formamide: Mix 5 g of mixed bed ion-exchange resin (Bio-Rad AG501-x8, 20–50 mesh) with 50 mL of formamide. Stir for 45 min. Filter and store at –20°C.
    b. 100X Denhardt's solution: 0.2 g of Ficoll, 0.2 g of polyvinyl pyrrolidone, 0.2 g of bovine serum albumin (BSA). Add sterile distilled water to 10 mL. Store aliquots at –20°C.
    c. Dextran sulfate: 50% (w/v) in sterile distilled water.
    d. Human or salmon sperm DNA: Dissolve 10 mg/mL in distilled water. Stir for 2–4 h. Shear DNA by passing it several times through an18 gage needle or by sonicating. Place in boiling

water for 10 min. Store aliquots at –20°C. Human DNA may also be extracted from placenta.

e. HM Mix: 6 mL of Deionized formamide, 1 mL of 100X Denhardt's solution, 2 mL of 50% dextran sulfate, 1 mL of 20X SSC. Store at 4°C.

14. PBT solution: Dissolve 5% BSA in phosphate buffered saline (PBS). Add Triton X-100 to 0.1% (v/v).
15. Primary antibody (Enzobiochem Inc., NY): This is a polyclonal antibody to biotin raised in rabbit (rabbit α-biotin). Use at 1:100 dilution in PBT. A mouse monoclonal to antibiotin is also available (Dako, High Wycombe, UK). Use at 1:50 dilution in PBT.
16. Second antibody (Dako): Goat antirabbit immunoglobulins conjugated to horseradish peroxidase. If the primary antibody is the mouse monoclonal antibiotin, the second antibody should be rabbit antimouse immunoglobulins (Dako).
17. 3,3'-diaminobenzidine tetrahydrochloride (DAB solution): Dissolve 2.5 mg of DAB powder in 5 mL of PBS. Add 2 mL of 30% hydrogen peroxide (H_2O_2). DAB is a suspected carcinogen. Handle with care in a fume cupboard.
18. Sodium chloroaurate (gold-sodium-chloride) (BDH Chemicals Ltd): Make up a 10% stock solution in distilled water. Store at 4°C. For the working solution, dilute the stock solution 1:100 (100 µg in 10 mL of distilled water), pH 2.2.
19. Sodium Sulfide (Na_2S): Dissolve 100 mg in 10 mL distilled water. Add 560 µL of 1 *M* HCl. Handle in a fume cupboard.
20. Silver solution, Intense M or Intense BL (Amersham, Aylesbury, UK): Follow the manufacturer's instructions.
21. Hoechst 33258: Prepare a 100X stock solution by dissolving 5 mg of dye in10 mL of methanol. Store at 4°C in the dark and make a working solution by diluting 1:100 in 2X SSC.
22. Giemsa: Mix a 5–10% Giemsa solution in PBS (v/v) at pH 6.8 (or pH 7.2).

3. Methods

3.1. Preparation of Metaphase Chromosomes

1. Add 0.8 mL of heparinized human venous blood to each culture bottle containing 10 mL of complete medium. Incubate at 37°C for 72 h (*see* Note 1).
2. Add 100 µL of BUdR to each culture. Return the cultures to 37°C for a further 16–17 h. (*see* Notes 2 or 3).

3. Centrifuge the cultures at 700*g* for 8–10 min. Discard the supernatant and resuspend the pellet in 10 mL of RPMI 1640 or in PBS.
4. Repeat Step 3.
5. Resuspend the pellet in 10 mL of fresh CM containing 2.5 µg/mL of dTTP. Incubate at 37°C for 6 h.
6. Add 100 µL of colcemid solution to each culture and incubate cultures at 37°C for a further 15–30 min.
7. Centrifuge the cultures and resuspend the pellet in fresh chilled fixative. Keep on ice for at least 20 min.
9. Wash the cells several times in fresh fixative by repeating Step 8.
10. Cells may be left indefinitely in 5–10 mL of fixative at –20°C.
11. Centrifuge and resuspend the pellet in 0.5–1 mL of fresh fixative and keep on ice.
12. Soak precleaned microscope slides in distilled water.
13. Place the wet slides on a hot plate maintained at 60°C.
14. Immediately drop 10 µL of cell suspension on each wet slide. Allow to dry.
15. Examine the slides under phase contrast. Select those that show a large number of well-spread metaphase chromosomes.

3.2. Removal of Endogenous RNA

1. Place 100 µL of RNase solution on each slide under a cover slip. Place the slides in Terasaki plates containing a few drops of 2X SSC. Incubate at 37°C for 60 min.
2. Allow the cover slips to float off in 2X SSC, and wash in two changes of 2X SSC (2 × 5 min).
3. Dehydrate through a series of 50, 70, 90%, and absolute ethanol.
4. Air-dry at 22°C; slides may be stored in a dessicator for a few weeks.

3.3. Labeling the DNA Probe

Label 1 µg of DNA probe with biotin-11-dUTP or biotin-7-dATP by nick translation, according to the following protocol (*see* Note 4).

Probe DNA (plasmid)	1 µg
dNTPs	5 µL
Biotin-11-dUTP or biotin-7-dATP	2.5 µL
DNA polymerase/DNase mixture	5 µL
10^{-2} *M* $MgCl_2$	50 µL

1. Incubate the complete nick translation mix at 15°C for 90 min.
2. Add 5 µL of 0.2*M* EDTA to stop the reaction.

3. Heat-inactivate at 70°C for 5 min.
4. Add 1 μL of mussel glycogen.
5. Add 0.1 mL vol of 3*M* sodium acetate, pH 5.6.
6. Add 2 vol of absolute ethanol.
7. Incubate at –70°C overnight or on dry ice for 2 h.
8. Centrifuge in a microfuge for 8–10 min.
9. Discard the supernatant and wash the pellet four times with 80% ethanol.
10. Dry the pellet in a freeze drier.
11. Dissolve in 50 μL of TE or distilled water.

3.4. Denaturation/Hybridization/Removal of Nonspecifically Bound Probe

1. Add 20–100 ng (1–5 μL) of labeled probe to 5 μL of sonicated human placental DNA or salmon sperm DNA. Bring the vol to 40 μL with HM (*see* Note 5).
2. Apply 30 μL of the mixture to chromosomal preparations under a cover slip. Seal the edges of the cover slip with rubber cement.
3. Place the preparations in Terasaki plates and incubate at 75°C for 7 min (*see* Notes 6 and 7).
4. Incubate at 37°C for 16 h or overnight.
5. Remove the cover slips and incubate the slides in 50% formamide in 2X SSC (pH 7.2) at 42°C for 20 min
6. Rinse in 2X SSC for 20 min at 42°C (*see* Note 8).
7. Rinse in 0.1X SSC at room temperature for 20 min.

3.5. Immunocytochemical Detection of Bound Probe

1. Incubate the slides in PBT for 15 min (*see* Notes 9 and 10).
2. Dry the back and edge of each slide, and place 100 μL of primary antibody on each slide.
3. Incubate at 37°C for 30 min (or at room temperature) for 45–60 min.
4. Rinse briefly in PBT.
5. Place 100 μL of second antibody on each slide.
6. Incubate as in Step 3.
7. Rinse in PBS for 3 × 5 min.
8. Prepare fresh DAB solution and immediately place a few drops of complete mixture on each slide. Incubate at 22°C for 5 min.
9. Rinse in distilled water for 3 × 5 min.
10. Add 100 μL of gold solution to each preparation. Incubate at 22°C for 5 min (*see* Note 11).

11. Rinse in distilled water for 3 × 5 min.
12. Place a few drops of sodium sulfide solution on each slide.
13. Incubate for 5 min at 22°C.
14. Rinse in distilled water for 3 × 5 min.
15. Place a few drops of silver solution on each preparation, and monitor under the microscope. Development of silver dots normally takes 5–18 min.
16. Wash in distilled water.

3.6. Replication Banding

After hybridization and signal development, proceed to replication banding as follows.

1. Stain the slides with 5 µg/mL Hoechst 33258 in 2X SSC for 20 min. Keep the preparations in the dark.
2. Rinse in distilled water, mount in 2X SSC, and seal the cover slips with rubber cement.
3. Expose the slides to UV light for 1–16 s.
4. Rinse in distilled water and stain with 10% Giemsa made in PBS (pH 6.8–7.2) for 10 min.
5. Rinse in distilled water, air-dry, and mount in DPX.

3.7. Chromosomal Gene Assignment

The signal appears as a black dot on banded chromosomes. With single copy DNA sequences, some background is inevitable. Therefore, up to 50 metaphase spreads should be analyzed and silver dots faithfully recorded on the respective chromosomes. The band on which there is a significant clustering of dots is judged as the gene under investigation (*see* ref. 4).

4. Notes

4.1. Preparations of Metaphase Spreads

1. There is no reason why chromosome spreads should not be prepared from cells other than PHA-activated lymphocytes (whole blood cultures). In practice, however, unless there is a specific reason for using other sources of cells, such as mapping translocations, deletions, and so on, blood cultures are easy to set up, and provide an ample source of metaphase spreads.
2. Synchronization with BUdR for replication banding permits direct detection of signal on banded chromosomes. If unsynchronized methotrexate cultures are used, or cultures are synchronized with 5-fluorodeoxyuridine, conventional G-banding will destroy the signal. It

may, therefore, be necessary to photograph the signal on unbanded Giemsa-stained chromosomes. The chromosome may be banded prior to or subsequent to *in situ* hybridization and photographed. The two sets of photographs are then compared to localize the signal to the chromosome band.

3. BUdR is sensitive to light. After addition of BUdR, the cultures should be handled in subdued light until the cells are fixed, to avoid extensive chromosomal damage.

4.2. Probe Storage

4. Biotin-labeled probes are stable for more than a year when kept at 4°C. Therefore, large amounts (up to 5 µg) can be nick-translated and stored at 4°C. Unlike tritium-labeled probes (5 ng), biotin-labeled probes need to be used in larger amounts (20–100 ng per slide). Single-stranded DNA and RNA probes in appropriate vectors are believed to be more promising, but have yet to be evaluated for gene mapping.

4.3. Denaturation/Hybridization and Stringency Washes

5. Inclusion of formamide in the HM mix and posthybridization washes lowers the melting temperature of DNA. Denaturation, hybridization, and stringency washes in formamide, therefore, helps preserve the architecture of the chromosomes, by lowering the denaturation and stringency washing temperature.
6. If separate denaturation of the probe and chromosomal DNA is desired, the probe may be denatured by boiling (in HM mix) at 100°C for 5 min and kept on ice. The DNA on microscope slides is denatured in 70% formamide in 2X SSC (pH 7.2) at 75°C in Terasaki plates.
7. The denaturation temperature (75°C) applies to slides kept in Terasaki plates. If other containers are used, both the temperature and time should be re-evaluated. Overheating damages chromosomes. If slides are denatured in coplin jars containing 70% formamide in 2X SSC (prewarmed to 70°C), denaturation at 70°C for 2 min is sufficient.
8. Reducing the salt concentration in SSC solution increases the stringency.

4.4. Detection

9. Tris buffered saline (TBS) may be used instead of PBS.
10. Nonspecific antibody binding sites may be blocked by incubation of slides

in PBS or TBS containing 1–5% of serum of the animal in which the antibody is raised, e.g., for goat antirabbit antibody, goat serum is used.

11. The use of gold-conjugated second antibody (Janssen EMGAR5, EMGAR10, or EMGAR15, equivalent antisera now available from Amersham) eliminates Steps 8–14 in Section 3.5. After incubation in the second antibody (gold-conjugate), slides must be rinsed thoroughly in distilled water before applying silver (Step 15). The following sequence of reagents are used. Rabbit antibiotin, biotin conjugated goat antirabbit Ig, streptavidin gold or avidin gold, and finally, silver.

Acknowledgments

This work was supported by grants to J. O'D. McGee from the Cancer Research Campaign.

Further Reading

Polak, J. M. and McGee, J. O'D, eds. (1990) In Situ *Hybridisation: Principles and Practice.* Oxford University Press, UK.

References

1. Burns, J., Graham, A. K., Frank, C., Fleming, K. A., Evans, M. P., and McGee, J. O'D. (1987) Detection of low copy human papilloma virus DNA and mRNA in routine paraffin sections of cervix by nonisotopic *in situ* hybridisation. *J. Clin. Pathol.* **40,** 858–864.
2. Bresser, J. and Eringer-Hodges, M. (1987) Comparison and optimization of *in situ* hybridization procedures yielding rapid, sensitive mRNA detections. *Gene. Anal. Tech.* **4,** 89–104.
3. Burns, J., Chan, V. T.-W., Jonasson, J. A., Fleming, K. A., Taylor, S., and McGee, J. O'D. (1985) Sensitive system for visualising biotinylated DNA probes hybridised *in situ*: Rapid sex determination of intact cells. *J. Clin. Pathol.* **38,** 1085–1092.
4. Bhatt, B., Burns, J., Flannery, D., and McGee, J. O'D. (1988) Direct visualisation of single copy genes on banded metaphase chromosomes by nonisotopic *in situ* hybridisation. *Nucleic Acids Res.* **16,** 3951–3961.
5. Cremer, T., Landegent, J., Bruckner, A., Scholl, H. P., Schardin, M., Hager, H. D., Devilee, P., Pearson, P., and van der Ploeg, M. (1986) Detection of chromosome aberrations in the human interphase nucleus by visualisation of specific target DNAs with radioactive and nonradioactive *in situ* hybridization techniques. Diagnosis of trisomy 18 with probe L1.84. *Human Genet.* **74,** 346–352.
6. Herrington, C. S., Burns, J., Graham, A. K., Evans MF and McGee, J. O'D. (1989) Interphase cytogenetics using biotin and digoxigenin labeled probes I. Relative sensitivity of both reporters for detection of HPV16 in CaSki cells. *J. Clin. Pathol.* **42,** 592–600.

7. Herrington, C. S., Burns, J., Graham, A. K., Bhatt, B., and McGee, J. O'D. (1989) Interphase cytogenetics using biotin and digoxigenin labeled probes II. Simultaneous detection of two nucleic acid species in individual nuclei. *J. Clin. Pathol.* **42,** 601–606.
8. Herrington, C. S. and McGee, J. O'D. (1990) Interphase cytogenetics: A review. *Neurochem. Res.* **15,** 467–474.
9. Buckle, V. and Craig, I. W. (1987) *In situ* hybridization, in *Human Genetic Diseases. A Practical Approach* (Davies, K. L., ed.), IRL Press, Oxford, UK.
10. Bhatt, B. and McGee, J. O'D. (1990) Chromosomal assignment of genes, in In situ *Hybridization: Principles and Practice* (Polak, J. M. and McGee, J. O'D., eds.), Oxford University Press, Oxford, UK, pp. 149–164.
11. Herrington, C. S., Burns, J. and McGee, J. O'D. (1990) Nonisotopic *in situ* hybridization in human pathology, in In Situ *Hybridization: Application to Developmental Biology and Medicine* (Harris, N. and Wilkinson, D. G., eds), Cambridge University Press, Cambridge, UK, pp. 241–269.
12. Forster, A. C., McInnes, J. L., Skingle, D. C., and Symons, R. H. (1985) Nonradioactive hybridization probes prepared by the chemical labeling of DNA and RNA with a novel reagent, photobiotin. *Nucleic Acids Res.* **13,** 745–761.
13. Reittie, J. E., Poulter, J. W., Hoffbrand, A. V., McGee, J. O'D., et al. (1988) Differential recovery of phenotypically and functionally distinct circulating antigen presenting cells after allogeneic bone marrow transplantation. *Transplantation* **45,** 1084–1091.
14. Lytletton, M. P. A., Browett, P. J., Bremner, M. K., Cordingley, F. T., Kohlman, J., McGee, J. O'D., Hamilton-Dutoit, S., Prentice, M. G., Hoffbrand, A. V. (1988) Prolonged remission of Epstein-Barr virus associated lymphoma secondary to T cell depleted bone marrow transplantation. *Bone Marrow Transplantation* **3,** 641–646.
15. Athanasou, N. A., Quinn, J., Brenner, M. K., Grant Prentice, H., Graham, A., Taylor, S., Flannery, D., and McGee, J. O'D. (1990) Origin of marrow stromal cells and hemopoietic chimerism following bone marrow transplantation determined by *in situ* hybridisation. *Br. J. Cancer* **61,** 385–389.
16. Hopman, A. H. N., Ramaekers, F. C. S., and Vooijs, G. P. (1990) Interphase cytogenetics of solid tumors, in In Situ *Hybridization: Principles and Practice* (Polak, J. M. and McGee, J. O'D., eds.), Oxford University Press, Oxford, UK, pp. 165–186.
17. Cremer, T., Lichter, P., Borden, J., Ward, D. C., and and Manueldis, L. (1988) Detection of chromosome aberrations in metaphase and interphase tumor cells by *in situ* hybridization using chromosome-specific library probes. *Hum. Genet.* **80,** 235–246.

CHAPTER 47

Nonisotopic *In Situ* Hybridization

Immunocytochemical Detection of Specific Repetitive Sequences on Chromosomes and Interphase Nuclei

John A. Crolla

1. Introduction

The term *in situ* hybridization describes a wide range of techniques concerned with the detection of DNA or RNA sequences within individual cells, tissues, or on identifiable regions of chromosomes. The technique utilizes an ability to label DNA or RNA probes so that, following hybridization with complementary sequences in the target tissues, the labeled DNA or RNA can be detected by various techniques.

Labeling of probes for *in situ* hybridization relies on the incorporation of either a radioisotopic dNTP (e.g., ^{3}H dCTP), or of a nonisotopic molecule, such as biotin-7-dATP or biotin-11-dUTP, by either nick translation or random priming. The site(s) of hybridization can then be seen using autoradiography with isotopic probes, or immunocytochemically if biotin is incorporated into the probe DNA. It is with the latter form of *in situ* hybridization methodology that this chapter is concerned.

Advances in molecular biology and immunocytochemistry *(1)* have resulted in the rapid growth of nonisotopic *in situ* hybridization, and although less sensitive than isotopic methods, the nonisotopic equivalents have the

From: *Methods in Molecular Biology, Vol. 10: Immunochemical Protocols*
Ed.: M. Manson ©1992 The Humana Press, Inc., Totowa, NJ

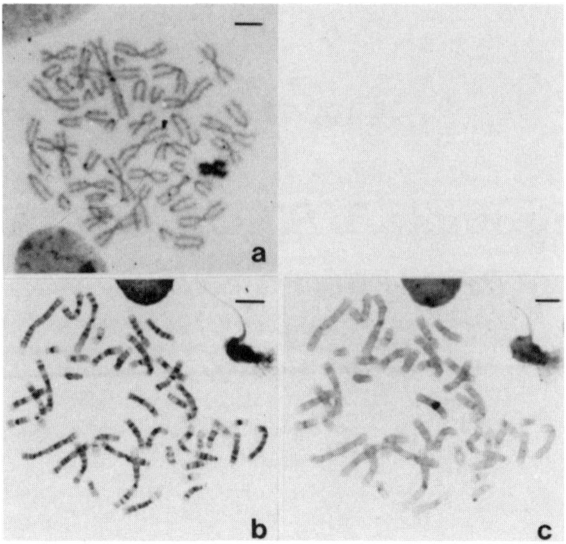

Fig. 1(**a**). A metaphase from the somatic cell hybrid HORL-9X that contains a single human chromosome in a rodent cell. The darkly staining chromosome is the human X following *in situ* hybridization with biotinylated human genomic DNA and immunocytochemical detection. (**b**). Normal human male metaphase (46,XY) following trypsin banding and prior to *in situ* hybridization. (**c**). The same cell as in 1(**b**) following *in situ* hybridization with a human X-centromere specific probe pSV2X5 (DXZ2) and immunocytochemical detection.

advantage of speed, safety, and relative simplicity in their use and applications. With methods similar to the one described here, applications have so far included screening of somatic cell hybrids with biotinylated human genomic DNA for the presence of whole or parts of human chromosomes (Fig. 1a and refs. *2,3*); the detection of centromere specific alphoid probes *(4)* on human chromosomes, and of RNA transcripts in the cytoplasm of differentiating embryonic tissues *(5)*.

This chapter describes the immunocytochemical detection of biotinylated human alphoid centromere specific probes on human chromosomes (Figs. 1b and c) and interphase nuclei. This is achieved by nick translating probe DNA with biotin and detecting the site(s) of hybridization using double antibody amplification steps followed by a biotin/streptavidin-horseradish peroxidase complex onto which diaminobenzidine is precipitated in the presence of hydrogen peroxide. This methodology has already been applied to human cytogenetics, in which conventional staining techniques had not been able to resolve the origin of small ring chromosomes *(6, 7)*. Similar methodologies have been applied to automated chromosome analysis for the diagnosis of Fragile X *(8)*, and for the detection of unique sequence genes on human chromosomes (*9–11* and *see also* this vol., Chapter 46). More recently, the "painting" of specific chromosomes with libraries of biotinylated chromosome specific sequences has been demonstrated *(12)*, which, in combination with the applications noted above, has opened the way for rapid developments in the application of nonisotopic *in situ* hybridization to not only large numbers of clinical cytogenetic problems, but also a wide range of problems in developmental biology and genetics.

2. Materials

The reagents and buffers required for *in situ* hybridization techniques are diverse, particularly for a method using immunocytochemical detection. The following sections have therefore been subdivided into prehybridization, hybridization, and posthybridization. Genomic human DNA was extracted from fetal placenta according to the method described in Maniatis *(13)* and the alphoid X-centromere specific probe was as described by Wolfe et al. *(14)*.

2.1. Prehybridization

1. Nick Translation Kit (Gibco BRL Ltd., UK).
2. Biotin-7-dATP (Gibco) or biotin-11-dUTP (Enzo, NY) (*see* Note 1).
3. 5% Sodium dodecyl sulfate (SDS) in distilled water.
4. G5O Sephadex mini-column.
5. Sodium hydroxide denaturing solution: 7 mL of 1*M* NaOH, (freshly prepared), 23 mL of distilled water, and 70 mL of industrial methylated spirit (IMS) (*see* Note 2).
6. 70, 90, and 100% IMS series.
7. Hank's Balanced Salt Solution (HBSS) at pH 7.2.
8. 0.012% Trypsin solution in HBSS (*see* Note 6).
9. HBSS acidified with a few drops of 1*M* HCl to an approximate pH of 6.0.
10. 10% Giemsa solution made up in 10 m*M* phosphate buffer, pH 6.8.

2.2. Hybridization

11. Deionized formamide: Add 5 g of monobed resin per 50 mL of analar formamide and stir for 45 min at room temperature. Filter through Whatman No. 1 paper prior to use.
12. Denionized formamide/dextran sulfate: Add 10% w/v of dextran sulfate (DS) to deionized formamide and allow it to dissolve with constant

agitation overnight. *Make sure the DS is fully dissolved!* Aliquot and store at –20°C. Replace this stock every three months (*see* Note 3).

13. 10X SSCP: 1.5*M* NaCl, 0.15*M* Na citrate, and 0.15*M* $NaH_2PO_42H_2O$.
14. Salmon sperm DNA: 10 mg/mL stock solution, sonicated and heat inactivated.
15. 2X SSC: 17.53 g of NaCl and 8.82 g of Na Citrate in 1 L of distilled water.

2.3. Posthybridization

16. Normal rabbit serum (Sigma, UK) in 100 µL aliquots: Store at –20°C.
17. 2X SSC containing 4% bovine serum albumin (BSA): Filter sterilize and store at +4°C (*see* Note 4).
18. Blocking solution: Add 900 µL of 4% BSA/2X SSC to 100 µL aliquot of normal rabbit serum; make up immediately prior to use.
19. Stringent washes: (a) For genomic DNA probes—0.5X SSC/50% formamide (not deionized) at 37°C for 40 min. (b) For repetitive (centromere specific) alphoid sequence human probes—0.2X SSC/50% formamide at 37°C for 40 min.
20. 2X SSC/0.1% BSA: Filter-sterilize and store at +4°C.
21. Monoclonal antibiotin IgG (made in mouse) (Dako, Denmark): Store at +4°C. For a working solution, dilute 1:300 in 2X SSC/0.1% BSA immediately prior to use.
22. Biotinylated antimouse IgG (Dako): store at +4°C. For a working solution dilute 1:200 in 2X SSC/0.1% BSA immediately prior to use.
23. 10X Tris-buffered saline (TBS): 80 g of NaCl, 6 g of Tris, and 44 mL of 1*M* HCl, made up to 1 L with distilled water. Add the HCl last to adjust the pH to 7.6.
24. Streptavidin–horseradish peroxidase (HP) conjugate (Dako): Store at +4°C. Working solution: dilute 1:600 in 1X TBS prior to use.
25. Hydrogen peroxide (30 vol): Store at +4°C.
26. 3'3' Diaminobenzidine (DAB): Store at –20°C (*see* Note 5).
27. Imidazole.
28. DAB substrate solution (DAB/H_2O_2): Add 10 mg of DAB and 12 mg of imidazole to 20 mL of 0.2*M* Tris-HCl buffer, pH 7.6. Mix thoroughly on a magnetic stirrer, but protect the solution from excessive light exposure. Dilute H_2O_2 1:30 in distilled water. Add 200 µL of 1% H_2O_2 to the DAB solution, stir for a further 1 min, filter through 0.2 µm Swinnex, and protect from light until used.
29. Sartorius mini-sart 0.2-µm filters.
30. 1X SSC/0.1% SDS.
31. Xam mounting medium

3. Methods

3.1. Prehybridization

1. Biotinylation of whole plasmid (probe) or genomic DNA is carried out according to the protocol provided with the Gibco-BRL nick translation kit. It is convenient to biotinylate 2 µg of DNA at each reaction in a total reaction vol of 100 µL.
2. Incubate the mix at 15°C for approx 1.5 h.
3. Add 10 µL of the stop solution (provided with kit).
4. Add 2.5 µL of 5% SDS.
5. Prepare a G50 Sephadex mini spin-column and equilibrate by spinning (1500 rpm on a bench centrifuge) through 100 µL of 1X SSC/0.1% SDS.
6. Discard the wash fraction, add biotinylated DNA to the column, and spin at 1500 rpm at room temperature for 3 min.
7. Collect and measure the vol of the fraction.
8. Aliquot and store the biotinylated DNA at -20°C.
9. Immerse the target slides in 50 mL of buffered HBSS, pH 7.2 (in a coplin jar) for a minimum of 1 min.
10. Transfer the slides into 0.012% trypsin solution in HBSS, pH 7.2, for 5–35 s (*see* Note 6).
11. Rinse the slides in acid HBSS, and stain for 5 min in 10% Giemsa. Wash in pH 6.8 phosphate buffer and mount the slides in the same buffer under a cover slip. Following photography (if required), destain the slides in analar methanol for 5 min, and air-dry.
12. Place the air-dried slides in a staining trough — add NaOH denaturing solution for 5 min at room temperature.
13. Wash the slides sequentially in a 70, 90, and 100% alcohol (IMS) (2 min in each), and air-dry.

3.2. Hybridization

Biotinylated repetitive sequence probe DNA is used at final concentration of 2 ng/µL of complete hybridization mixture (*see* Note 7). This example assumes that the probe DNA is at a concentration of 16 ng/µL, and that 14.5 µL of complete hybridization mix will be added to each slide.

1. For the hybridization mix, add the following to a 1.5 mL Eppendorf tube: probe DNA, 18.7 µL (i.e., 300 ng); formamide/dextran sulfate, 75.0 µL (i.e., 1/2 vol); 10X SSCP, 30.0 µL (i.e., 1/5 vol); salmon sperm (carrier) DNA, 7.5 µL (i.e., 1/20 vol); and distilled water, 18.8 µL (i.e., DNA + water = 1/4 vol). Total, 150 µL.

2. Mix the contents thoroughly, spin briefly in a microfuge and boil the probe/hybridization mix for 5 min; quench immediately on ice.
3. Add the hybridization mix at a ratio of 3.0 µL per cm^2 of cover slip (e.g., 14.5 µL to a 22 × 22 mm cover slip).
4. Place the slides sideways into a glass or plastic staining trough and place this into an easily sealable container (e.g., a Tupperware box with a clip-on lid). Humidify the chamber by placing some tissues soaked in 2X SSC on the bottom. Incubate overnight at 42°C. If the slides are well humidified and level, there is no need to seal the cover slips with rubber cement.

3.3. Posthybridization

1. Remove the hybridization cover slips by immersing the slides in 2X SSC at room temperature for 3–5 min; transfer the slides to a clean trough containing fresh 2X SSC. Incubate for a further 5 min at room temperature.
2. Pour off the 2X SSC and replace with a pre-heated stringent wash salt solution.
3. At the end of the incubation, pour off the stringent solution and replace with 2X SSC (room temperature). Agitate the slides gently, pour off the 2X SSC, and replace with fresh 2X SSC solution for a further 5 min at room temperature.
4. Place 60 µL of blocking solution (*see* Note 8) onto a clean 24 × 50 mm coverslip and add to each slide. Incubate at 37°C for 15 min (in a humidified 2X SSC chamber). Note: From this point on, all of the reagents are applied in the same manner.
5. Gently remove the cover slips from the slides following incubation in blocking solution and remount the slides with 60 µL of the working solution of antibiotin antibody (made in mouse). Incubate for 1.5 h at 37°C in a humidified chamber.
6. Wash off the cover slips in 2X SSC. Transfer the slides to fresh 2× SSC for 10 min at room temperature.
7. After draining excess 2X SSC from the slides (*see* Note 9), add 60 µL of a working solution of biotinylated antimouse antibody solution; incubate for 1 h at 37°C in a humidified chamber.
8. Remove the cover slips in 2X SSC and transfer slides to fresh 2X SSC. Incubate for 10 min at room temperature (the slides can be left at this stage for 1–2 h if required).
9. Pour off the 2X SSC and rinse the slides in 1X TBS; after draining excess TBS, add 60 µL of streptavidin-HP diluted 1:600 in 1X TBS. Incubate at room temperature for 20 min (in a humidified chamber).
10. During the above incubation, make up the DAB substrate.
11. Immerse the slides in 1X TBS, then transfer them to a fresh trough and

rinse 2 × 2 min in 1X TBS, then drain.

12. Add 60 µL of DAB substrate solution and leave for 5 min at room temperature. Rinse the slides in tap water to stop the reaction (*see* Note 10).
13. Counterstain for 5 min in 10% Giemsa, rinse in buffer, and air-dry (*see* Note 11). Permanent mounting without subsequent fading of the chromosome and/or hybridization signal is best achieved by mounting slides in Xam.

4. Notes

1. Biotin-11-dUTP and biotin-7-dATP have both been used successfully to label probe and genomic DNA. Biotin-11-dUTP is now only available direct from Enzo Diagnostics (NY).
2. The use of NaOH as the denaturation solution is preferred to other methods, because the quality of interphase cell and chromosome morphology is not significantly damaged by this method. It is important, however, to make the NaOH stock solutions fresh for each denaturation.
3. Use deionized formamide in the hybridization solution and ensure that the dextran sulfate is fully dissolved prior to aliquoting and storage at –20°C. Deterioration of this solution eventually leads to increased background, so replace this stock every three months.
4. The BSA solutions should be filter-sterilized to protect against bacterial contamination. It is advisable to make fresh solutions once a month.
5. Diaminobenzidine (DAB) is a possible carcinogen and should be handled with extreme caution, wearing suitable protective clothing (gloves and face mask). Batches may vary and should be tested in a control *in situ* experiment. Aliquot into 10 mg batches and store at –20°C to avoid constant freezing and thawing. Alternatively, 10 mg tablets are available from Dako and Sigma, and although relatively expensive, are easier and safer to handle than DAB in powder form.
6. Slide preparation. Chromosome banding (identification) procedures posthybridization are not possible using this *in situ* hybridization protocol. Proteinase K pretreatment may be used as an alternative to trypsin, but must be calibrated for the material used. Pretreatment of the target tissue with RNAse is not necessary when using highly repetitive sequence probes. Although acetomethanol fixation of the metaphase chromosomes removes basic proteins, further deproteinization treatment is required to optimize target interphase cells and/or chromosomes for detection of the hybridization signal. A trypsin-banding protocol allows the simultaneous identification of the target chromosomes prior to *in situ* hybridization, and for the optimum preparation of the target.

The optimum targets for biotinylated probe are conventionally made chromosome preparations, approx 1 wk old, that have been stained in Giemsa following a standard trypsin banding protocol (GTG). As a general rule, the better the chromosome morphology following trypsin treatment, the better the conditions for hybridization. However, it may be necessary to do a test series of slides with decreasing concentrations of trypsin. Excessive trypsin pretreatment will result in the cells "peeling off" the slide during the hybridization step. Trial and error at this stage will calibrate the system to the laboratory conditions being used. Cells can be prephotographed at this stage (*see* Fig. 1b) by using a cover slip with pH 6.8 phosphate buffer as mountant. Successful hybridizations are also achievable after protocols for bromodeoxyuridine (BUdR)-late labeling, centromeric (C)-, or reverse (R)-banding.

Slides stored for up to 5 yr at room temperature have been used as successful targets for human alphoid repetitive sequence probes *(15)*. When slides have been mounted in Xam or another xylene-based mountant, the cover slip is first removed by immersing the slide in xylene (or a xylene substitute) for 2–3 d. This is followed by a 10 min wash in fresh xylene (or substitute) and two 5-min washes in methanol. All GTG or Giemsa-stained slides are destained for 5 min in Analar methanol, prior to denaturation of DNA. The proportion of metaphases and interphase nuclei showing the expected signal distribution, however, was considerably less from archive material when compared with fresh, optimal slides.

7. Biotinylated probes used at a concentration of 2ng/µL in the hybridization mix give the optimal signal to background ratio as seen in Figs. 1a and c. Use thoroughly precleaned slides (e.g., washed in methanol), but coverslips can be used direct from the manufacturer's box.
8. The normal rabbit serum/BSA is a block to nonspecific binding of the primary (antibiotin, made in mouse) and secondary (biotinylated antimouse) antibodies. When removing the cover slips after this incubation, care should be taken to leave a thin film of serum to which the primary antibody is added.
9. Cover slips will float off immediately in 2X SSC after the antibody incubations, and the slides should be transferred to a clean trough containing fresh 2X SSC. Take particular care not to score or scratch the surface of the slide during these manipulations.
10. Large hybridization signals (e.g., the human X centromere specific probes) should be clearly visible under phase contrast microscopy toward the end of the 5-min developing period. However, care must be taken not to overprecipitate the DAB solution and/or not to overcounter-

stain the chromosomes with Giemsa.

11. It may be difficult to achieve satisfactory counterstaining of the chromosomes (particularly if archive material is being used), but adequate detection of the hybridization complex and the chromosome background can still be achieved by viewing the specimen using phase-contrast microscopy. In most cases, however, both hybridization signal and chromosomes should be visible using simple direct light illumination (Fig. 1c).

References

1. Bourne, J. A. (1983) Handbook of immunoperoxidase staining methods. Immunocytochemistry laboratory, Dako Corporation, CA.
2. Durnam, D. M., Gelinas, R. E., and Myerson, D. (1985) Detection of species-specific chromosomes in somatic cell hybrids. *Somat. Cell Mol. Genet.* **11,** 571–577.
3. Benham, F., Hart, K., Crolla, J., Bobrow, M., Francavilla, M., and Goodfellow, P. N. (1989) A method for generating hybrids containing nonselected fragments of human chromosomes. *Genomics* **4,** 509–517.
4. Willard, H. F. (1985) Chromosome-specific organization of human alpha satellite DNA. *Am. J. Hum. Genet.* **37,** 524–532.
5. Brewer, L. M., Gillen, M. F., and MacManus, J. P. (1990) Localization of mRNA for the oncotrophoblastic protein oncomodulin during implantation and early placentation in the rat. *Placenta* **10,** 359–375.
6. Crolla, J. A. and Llerena, J. C. (1988) A mosaic 45, X/46, X, r (?) karyotype investigated with X and Y centromere-specific probes using a nonautoradiographic *in situ* hybridization technique. *Hum. Genet.* **81,** 81–84.
7. Crolla, J. A., Smith, M., and Docherty, Z. (1989) Identification and characterization of a small marker chromosome using nonisotopic *in situ* hybridization with X and Y specific probes. *J. Med. Genet.* **26,** 192–194.
8. Fantes, J., Gosden, J., and Piper, J. (1988) Use of an alphoid satellite sequence to locate the X chromosome automatically, with particular reference to identification of the fragile X. *Cytogenet. Cell Genet.* **48,** 142–147.
9. Garson, J. A., van den Berghe, J. A., and Kemshead, J. T. (1987) Novel nonisotopic *in situ* hybridization technique detects small (1 kb) unique sequences in routinely G-banded human chromosomes: Fine mapping of N-myc and β-NGF genes. *Nucleic Acids Res.* **15,** 4761–4770.
10. Cherif, D., Bernard, O., and Berger, R. (1989) Detection of single copy genes by nonisotopic *in situ* hybridization on human chromosomes. *Hum. Genet.* **81,** 358–362.
11. Bhatt, B., Burns, J., Flannery, D., and McGee, J. O. (1988) Direct visualization of single copy genes on banded metaphase chromosomes by nonisotopic *in situ* hybridization. *Nucleic Acids Res.* **16,** 3951–3961.
12. Pinkel, D., Landegent, J., Collins, C., Fuscoe, J., Segraves, R., Lucas, J., and Gray, J. (1988) Fluorescence *in situ* hybridization with human chromosome-specific libraries: Detection of trisomy 21 and translocations of chromosome 4. *Proc. Natl. Acad. Sci. USA* **85,** 9138–9142.
13. Maniatis, T., Fritsch, E. F., and Sambrook, J. (1987) *Molecular Cloning. A Laboratory Manual.* Cold Spring Harbor Laboratory, Cold Spring Harbor, NY.

14. Wolfe, J., Darling, S. M., Erickson, R. P., Craig, I. W., Buckle, V. J., Rigby, P. N. J., Willard, H. F., and Goodfellow, P. N. (1985) Isolation and characterization of an alphoid centromeric repeat family from the human Y chromosome. *J. Mol. Biol.* **182,** 477–485.
15. Crolla, J. A., Gilgenkrantz, S., deGrouchy, J., Kajii, T., and Bobrow, M. (1989) Incontinentia pigmenti and X-autosome translocations: Nonisotopic *in situ* hybridization with an X-centromere specific probe (pSV2X5) reveals a possible X-centromere breakpoint in one of five published cases. *Hum. Genet.* **81,** 269–272.

CHAPTER 48

Biotinylated Probes in Colony Hybridization

Michael J. Haas

1. Introduction

Colony hybridization is a procedure that allows the detection of cells containing nucleic acid sequences of interest *(1)*. In this method, microbial colonies grown on, or transferred to, a supporting membrane are lysed and their nucleic acids denatured to single strands and fixed in place on the membrane. The membrane is then exposed to a similarly denatured "probe" sequence, which is identical or homologous to all or part of the target sequence, under conditions favoring reannealling. Probe sequences hybridize to complementary sequences on the membrane. Positive hybridization events are then detected by determining the presence and location of probe sequences on the membrane.

The original colony hybridization method described the use of radiolabeled probes and the detection of positive hybridization events by autoradiography *(1)*. However, because of the high waste disposal costs, short half-lives, long autoradiographic exposures, and potential health hazards associated with radioisotopes, there is interest in alternative methods to detect positive hybridizations.

Nonradioactive technology involves the attachment to the nucleic acid probe of a ligand that can subsequently be detected by chemical or enzymatic methods. The vitamin biotin is one such ligand. Biotin can be covalently

From: *Methods in Molecular Biology, Vol. 10: Immunochemical Protocols*
Ed.: M. Manson ©1992 The Humana Press, Inc., Totowa, NJ

incorporated into nucleic acids in a manner that does not interfere with their ability to hybridize with homologous sequences. This is accomplished by replacing a nucleoside triphosphate with its biotinylated analog in an in vitro DNA replication or transcription reaction, generating a biotinylated probe sequence *(2,3)*. Hybridization of such a probe to a homologous sequence immobilized on a membrane results in the retention of biotin at that site. Positive hybridization events can then be detected by assaying for biotin.

Enzymatic reaction schemes that generate insoluble colored products at sites where biotin is bound to the filters have been developed for the purpose of biotin detection in these applications. These detection reactions employ either avidin or streptavidin, two functionally identical proteins that bind to biotin with very high affinities and specificities. These proteins are retained at sites where biotinylated probes have hybridized to homologous sequences. Avidin and streptavidin have multiple biotin binding sites per molecule. They therefore retain biotin binding capability even after binding to probe sequences on the membranes. Incubation with a biotinylated form of an enzyme (e.g., alkaline phosphatase) for which there exists an assay that generates an insoluble, colored product results in the retention of signal enzyme at sites of positive hybridization. These sites are detected by applying the histochemical assay for the signal enzyme.

To facilitate our work on plasmids with no known phenotype, we have developed a method for the use and detection of biotinylated probes in colony hybridization. It is suitable both for the detection of rare positive hybridization events over a background of nonreactive colonies and for the detection of nonhybridizing colonies in a population containing sequences homologous to the probe. The latter capability could be useful in such applications as the detection of cured (i.e., plasmid-free) cells in a bacterial population containing plasmids.

2. Materials

1. Nitrocellulose filters (82-mm diameter, BA 85) are obtained from Schleicher & Schuell (Keene, NH). (Other suppliers may be acceptable.)
2. Formamide is deionized by stirring for 30 min with 10% (w/v) of a mixed-bed ion exchange resin (e.g., Bio-Rad AG 501-X8, 20-50 mesh, Bio-Rad, Richmond, CA), filtering twice through Whatman (Clifton, NJ) no. 1 paper and storing single-use aliquots at –80°C.
3. Bovine serum albumin (BSA, Fraction V, Sigma, St. Louis, MO) is used as obtained. Fatty acid-free albumin gives poor results.
4. Denatured herring sperm DNA is prepared by dissolving in water (10 mg/mL) with stirring at room temperature, shearing by 10 passages through an 18-gage needle, and immersing in boiling water for 10 min.

Aliquots are stored at –20°C. Just prior to use, these are incubated for 10 min in a boiling water bath and chilled in ice water.

5. 20X SSC buffer: 3*M* Sodium chloride, 0.3*M* sodium citrate, pH adjusted to 7 with sodium hydroxide. Sterilize by autoclaving, store at room temperature.
6. Proteinase K is obtained from Beckman, Somerset, NJ. Other sources may be acceptable. In using alternate sources, the occurrence of blue backgrounds between colonies, and oversize, blurry signals at colony sites after the final color development step indicates insufficient proteolytic activity. Prepare a solution of 200 µg/mL in 1X SSC.
7. 50X Denhardt's solution: 1% (w/v) Ficoll, 1% (w/v) polyvinyl pyrrolidone, 1% (w/v) BSA. Filter-sterilize. Store aliquots at –20°C. Do not flame the pipets used to transfer this solution. Denaturation and precipitation of the protein result from the use of hot pipets at this stage.
8. Template DNA for the production of hybridization probes must be pure. Standard methods, such as dye-buoyant density ultracentrifugation, generate acceptable products. Ethidium bromide and cesium chloride are removed prior to use of the DNA *(4)*.
9. Biotin-11-deoxyuridine-5'-triphosphate (BiodUTP) and reagents for its incorporation into DNA by nick translation are obtained commercially. The products from Bethesda Research Laboratories (BRL, Gaithersburg, MD) are acceptable. BRL now provides a prepackaged kit (BioNick) containing necessary supplies and employing biotin-14-dATP as the source of biotin. The concentration of the resulting biotinylated DNA is determined by the histochemical method for biotin *(below)*. Adequate instructions are provided with these kits.
10. Prehybridization solution: 50% formamide, 5X SSC, 5X Denhardt's solution, 25 m*M* sodium phosphate, pH 6.5, 300 µg/mL freshly denatured sheared herring sperm DNA. Filter through Whatman no. 1 paper on a Buchner funnel, then through a sterile 0.45-µm filter. Store 10-mL aliquots in glass at –20°C. Use only once.
11. Hybridization solution: 45% Formamide, 5X SSC, 5X Denhardt's solution, 20 m*M* sodium phosphate, pH 6.5, 300 µg/mL freshly denatured, sheared, herring sperm DNA, 200 ng of biotinylated DNA/mL. Before its addition, the biotinylated probe DNA is denatured by incubating for 10 min in a boiling water bath and quick-chilling in an ice bath. Shearing to reduce size is unnecessary, since the products generated by nick translation are sufficiently small. Filter and store the hybridization solution as was done for the prehybridization solution. Hybridization solution can be recovered after use and stored at –20°C. The solution can be reused at least 10 times over a time-span of at least 5 mo, without noticeable

reduction in performance. The solution is heat-denatured as described for the herring sperm DNA preparation immediately before each use.

12. Reagents for the detection of filter-bound biotin are obtained from BRL (BlueGene Nonradioactive Nucleic Acid Detection System). Comparable materials are available from Bio-Rad Laboratories.
13. Special equipment required for this protocol are a vacuum oven, a slab gel dryer, a device for the heat sealing of plastic bags (e.g., Seal-A-Meal, Sears Seal-and-Save), thin rubber sheet (such as dental sheet, A. H. Thomas, Philadelphia, PA), and a filtration device designed for the washing of nitrocellulose filters. The latter was originally described by Grunstein and Hogness *(1)* and is available from Schleicher and Schuell as the "Screen-It" colony filter hybridization device.
14. 90% (w/w) Ethanol.
15. Chloroform: Reagent grade.
16. Solutions for the posthybridization washing of filters:
    a. 0.1% (w/v) Sodium dodecylsulfate (SDS) in 2X SSC.
    b. 0.1% (w/v) SDS in 0.2X SSC.
    c. 0.1% (w/v) SDS in 0.16X SSC.
    d. 2X SSC.

3. Methods

3.1. Filter Preparation and Cell Growth

1. Use a soft lead pencil to label nitrocellulose filters with a hash mark and letter or number on one edge to allow subsequent identification and orientation (*see* Note 1).
2. Place the labeled filters between sheets of filter paper, wrap in aluminum foil, and autoclave for 10 min.
3. Seal the packets of sterile filters in an air-tight bag and store at 4°C.
4. To inoculate, place a filter on top of the solidified media in a Petri dish and spread an appropriately diluted bacterial culture over the surface.
5. Incubate the plates until the cells are approx 1–3 mm in diameter (*see* Note 2). Cell densities of approx 800/82 mm diameter filter are compatible with single colony discrimination after hybridization and color assay. If one is attempting to locate positively hybridizing sequences in a generally nonreacting population, and single colony resolution is not required in the first detection step, as many as 10^5 cells can be applied to each filter.

6. Invert a filter and gently lay it onto fresh media just prior to lysis to create a replica of the colony pattern of the filter. (Mark the plate to indicate the orientation of the filter on it.) After an appropriate incubation, this becomes a master plate from which viable analogues of desirable colonies, as identified on the filter after hybridization and processing, can be recovered.

3.2. Cell Lysis

All operations are conducted at room temperature unless otherwise noted. After Steps 1–3, gentle suction is applied to the filters (*see* Note 3). Steps 2–4 are conducted in glass Petri dishes, one filter per dish. It has not been determined if these steps can be done batchwise. It is difficult to process more than 12 filters at a time.

To achieve lysis, incubate the filters in the following fashion (*see* Note 4):

1. 7 min, colony-side up, on filter paper sheets stacked to a thickness of 4 mm and saturated with fresh 0.5*M* NaOH.
2. 5 min in 1.5*M* sodium chloride, 0.5*M* Tris-HCl, pH 7.4, 30 mL/filter.
3. 1 h in prewarmed proteinase K in 1X SSC, 30 mL/filter, 37°C.
4. 2 × 2 min in 90% (w/w) ethanol, 30 mL/filter (*see* Note 5).
5. Air-dry, 20 min.
6. Wash each filter with 100 mL of chloroform using the Screen-It colony hybridization device. A single sheet of filter paper is used as an underfilter.
7. Air-dry (approx 15 min).
8. Sandwich the filters individually between filter paper, wrap loosely in aluminum foil, and bake at 80°C *in vacuo* for 2 h.
9. Store the filters in a vacuum desiccator at room temperature.

3.3. Prehybridization, Hybridization, and Detection of Hybridization

1. For prehybridization, place pairs of filters containing lysed, fixed colonies back to back in sealable plastic bags. Add 20 mL of prehybridization solution, seal the bag, seal it within a second bag, and incubate at 42°C for 2 h. Maintain the proper temperature by submersion in a water bath.
2. After prehybridization, replace the liquid with 20 mL of hybridization solution, exclude air bubbles, reseal the bags, and immerse in the water bath. Brief incubations (1 h) are sufficient for the detection of relatively abundant sequences, such as unamplified plasmid pBR322 in *E. coli*. More extensive incubations (45 h) may be necessary to detect less abundant sequences.

3. Following hybridization, wash the filters sequentially:
   a. Twice in 250 mL of 0.1% (w/v) SDS in 2X SSC, 3 min per wash, room temperature;
   b. Twice in 250 mL of 0.1% (w/v) SDS, 0.2X SSC, 3 min per wash, room temperature;
   c. Twice in 250 mL of 0.1% (w/v) SDS, 0.16X SSC, 15 min per wash, 50°C; and
   d. Briefly in 2X SSC at room temperature.
4. Detection of the sites of hybridization-dependent binding of biotinylated probe to the filters is most readily conducted with commercially available kits. Favorable results have been obtained with the BluGene Nonradioactive Nucleic Acid Detection System from BRL. Follow the manufacturer's instructions when carrying out the following steps. After washing, sequentially expose the filters to streptavidin and biotinylated alkaline phosphatase (or to a conjugate of these two proteins). This causes the immobilization of alkaline phosphatase at sites of positive hybridization.
5. Incubate the filters with 5-bromo-4-chloro-3-indolylphosphate (BCIP) and nitroblue tetrazolium (NBT). Indoxyl generated from BCIP by the action of alkaline phosphatase condenses to form indigo (blue). Indigo then reacts with NBT to form insoluble diformazan (purple).
6. Terminate the reaction when reacting colonies are intensely purple (*see* Note 6) by replacing the dye solution with 20 m*M* Tris-HCl, 5 m*M* EDTA, pH 7.5. Nonreactive colonies should be light blue on a white background.
7. Store the moist filters in sealed bags. The elapsed time from the end of hybridization to the termination of color development is approx 3 h. Figure 1 illustrates typical results obtained with this method.

4. Notes

1. Cellulose filters give unacceptably diffuse colony patterns after lysis and should not be used. Nylon filters should be acceptable, although we have not examined their suitability.
2. The lysis of colonies larger than specified above is generally acceptable. However, with relatively mucoid strains, such as *Xanthomonas*, the lysis of oversize colonies results in smeared colony patterns. The researcher should investigate the performance of younger cells if such behavior is experienced.

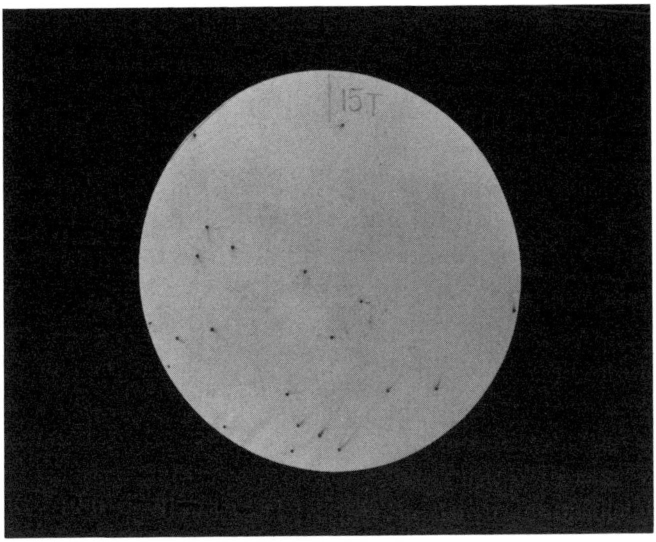

Fig. 1. Specific identification of *E. coli* containing plasmid pBR322. Approximately 225 colonies, consisting of a 10:1 mixture of plasmid-free and plasmid-containing cells, was grown on a nitrocellulose filter. The filter was subjected to the lysis protocol described here, followed by a hybridization with biotinylated pBR322. Sites of positive hybridization were detected by means of streptavidin and alkaline phosphatase. The dark sites correspond to colonies harboring pBR322. Plasmid-free cells give the faint signals present at numerous sites on the filter.

3. The application of gentle suction to the filters following Steps 1–3 of the lysis protocol reduces the dispersion of cells from their sites, promoting tighter patterns and stronger signals, and reducing the interference of signals from adjacent colonies with one another. Suction is applied by means of a slab gel dryer and a gentle vacuum source. A single sheet of filter paper serves as an underfilter. On this sheet is placed a template made by cutting into a sheet of flexible rubber holes slightly smaller than the nitrocellulose filters. The filters are placed over these holes and vacuum is applied. A brief suction suffices to remove excess moisture from the filters and to pull lysed colonies down onto them. Six filters can be treated at a time with a standard commercially available gel dryer with an 18 by 34 cm suction surface.
4. In our initial studies, the filters were swirled in the lysis solutions in an attempt to ensure lysis. After hybridization and application of the colo assay, it was found that positively reacting colonies had "tails" extendir away from them in a circular pattern across the filters. These t

obscured the signals of adjacent colonies. Tailing was eliminated by omitting the swirling action during lysis. This omission did not noticeably reduce the efficiency or sensitivity of the detection reaction.

5. The ethanol concentration in Step 4 of the lysis protocol is a w/w concentration. Ethanol solutions made up v/v, or otherwise in excess of 90% w/w, exceed the ethanol tolerance limits of some batches of nitrocellulose. Filters washed in such solutions may become brittle and be reduced nearly to powder by the end of the hybridization-color assay procedure. The appropriate solution can be made from 100% ethanol.
6. The final color development step must be conducted under dim light (i.e., incubated in a drawer) since the reagents are light sensitive. Examine the filters at frequent intervals (10 min) during this incubation. Stop the reaction when the color of positively reacting colonies is deep purple. Further incubation past this point allows the color of nonreacting colonies to darken to such a degree that they are mistaken for positives. Overdevelopment is the greatest single factor contributing to the appearance of false-positive signals.
7. The minimal probe concentration necessary for efficient detection of target sequences has not been determined. It has been noted, however, that probe concentrations of 10 to 20 ng/mL, when coupled with overnight hybridizations, are too low to give strong signals for nonreiterated target DNAs 3 Mdalton or larger in size. Maas *(5)* has reported a simple modification of the Grunstein-Hogness protocol *(1)*, which is reported to increase the sensitivity of the colony hybridization method by 100-fold. This could increase the ability to detect single copy sequences.
8. Nucleic acids can also be biotinylated by nonenzymatic methods with Photobiotin, a photoactivatable biotin analog *(6)*, which can be commercially obtained from BRL, Sigma Chemical Co., and other commercial sources. We have not compared the suitability of this method of biotin incorporation with that reported here, but expect that the method would be fully acceptable. FMC (Rockland, ME) markets an alternate nonradioactive sequence detection kit known as Chemiprobe. The basis of this system is a chemical modification of cytosine residues in the probe DNA. After hybridization, the probe is detected by means of a monoclonal antibody that specifically recognizes the sulfonated DNA. Detection of the bound monoclonal antibody is achieved by means of an alkaline phosphatase conjugated second antibody.

References

1. Grunstein, M. and Hogness, D. S. (1975) Colony hybridization: A method for the isolation of cloned DNAs that contain a specific gene. *Proc. Natl. Acad. Sci. USA* **72,** 3961–3965.
2. Langer, P. R., Waldrop, A. A., and Ward, D. C. (1981) Enzymatic synthesis of biotin-labeled polynucleotides: Novel nucleic acid affinity probes. *Proc. Natl. Acad. Sci. USA* **78,** 6633–6637.
3. Brigati, D. J., Myerson, D., Leary, J. J., Spalholz, B., Travis, S. Z., Fong, D. K. Y., Hsiung, G. D., and Ward, D. C. (1983) Detection of viral genomes in cultured cells and paraffin-embedded tissue sections using biotin-labeled hybridization. *Virology* **126,** 32–50.
4. Boffey, S. A. (1984) Plasmid DNA isolation by the cleared lysate method, in *Methods in Molecular Biology, volume 2, Nucleic Acids* (Walker, J. M., ed.) Humana, Clifton, NJ, pp. 177–183.
5. Maas, R. (1983) An improved colony hybridization method with significantly increased sensitivity for detection of single genes. *Plasmid* **10,** 296–298.
6. Forster, A. C., McInnes, J. L., Skingle, D. C., and Symons, R. H. (1985) Nonradioactive hybridization probes prepared by the chemical labeling of DNA and RNA with a novel reagent, photobiotin. *Nucleic Acids Res.* **13,** 745–761.

CHAPTER 49

Screening of λ gt11 cDNA Libraries Using Monoclonal Antibodies

Duncan F. Webster, William T. Melvin, M. Danny Burke, and Francis J. Carr

1. Introduction

The screening of a λgt11 library with monoclonal antibodies described here is a relatively uncomplicated procedure, but it requires a little introduction nevertheless. For simplicity, it is assumed that you have in your possession a library that is ready to screen, i.e., either a library bought from a company or a library that you have made. Construction of a λgt11 library is described in Volume 4, Chapter 19.

This method is basically that of de Wet et al. *(1)*, which is itself a modification of that of Young and Davis *(2,3)*. It involves transfecting the host strain of bacteria (we normally use a strain called Y1090) with bacteriophage, and mixing this with some molten agarose before plating onto media plates. The agarose embedded bacteria can still grow, and would form a continuous lawn of bacteria in a few hours if no phage were present. However, when phage are present, they lyse the cells to form clear plaques in the bacterial lawn. Even though the cells are embedded in agarose, the phage can still diffuse away from the plaque, but the rate of this diffusion is much slower than in a liquid culture. Therefore, the lysis of cells is restricted to the immediate area of the initial site of infection, i.e., each plaque arises from one phage infecting one bacterium. As can be guessed therefore, the bacteria are present in gross excess with respect to the phage for two purposes:

1. To ensure there are enough bacteria to form a lawn so that the plaques are visible (and detectable by antibodies), and

From: *Methods in Molecular Biology, Vol. 10: Immunochemical Protocols*
Ed.: M. Manson ©1992 The Humana Press, Inc., Totowa, NJ

2. To ensure a large excess of bacteria to phage, thus reducing the chance of two phage infecting the same bacterium.

The plates are incubated for several hours at a temperature that allows the phage to lyse the cells. The genetics behind this are discussed by Young and Davis *(2)* in the paper that describes the construction of the λgt11 vector.

Agar plates are then overlayed with a sheet of nitrocellulose that has been soaked in isopropyl β-D-thiogalactoside (IPTG), which is a gratuitous inducer of the β-galactosidase gene present within the λgt11. It is into this gene that the cDNA has been inserted in the unique EcoRI site, and if the cDNA is in the correct reading frame and orientation (in theory, a 1 in 6 probability), a fusion protein will be produced. The protein sequence encoded by the cDNA is an extension of the β-galactosidase protein at the carboxy terminus. Filter overlayed plates are then incubated at a slightly lower temperature that allows lysis to continue, but at a reduced rate, and hence increases the quantity of fusion protein in the plaque.

The filter is removed from the plate and washed to remove bacterial debris. The binding capacity of the nitrocellulose filters for proteins is saturated by washing in a diluted serum. Then, the first antibody is added and left to incubate overnight. The filters are washed the next day and then the second antibody is added. This is normally a horseradish peroxidase-linked second antibody, which we find is quite sensitive enough for this procedure. After an incubation of a few hours, the second antibody is removed and the filters are again washed.

Finally, the filters are stained, dried, and aligned to the plates. This alignment is used to remove the required plaques in a plug of agar. The phage are eluted overnight in a saline magnesium sulphate solution (Mg^{2+} ions are required for phage infectious activity as well as for their stability) and then screened again in a manner very similar to the one above, but with some small modifications that will be discussed in detail later. This is repeated until pure plaques are obtained as judged by 100% recognition by antibodies in a subsequent screen.

2. Materials

1. L-Broth (1 L): 10 g of bacto-tryptone, 5 g of yeast extract, 5 g of NaCl, and 2.4 g of $MgSO_4.7H_2O$. Adjust to pH 7.5.
2. Tris-buffered saline (TBS): 0.17*M* NaCl, 0.01*M* Tris base. Adjust to pH 7.5.
3. SM (λ eluant): 0.1*M* NaCl, 0.01*M* $MgSO_4.7H_2O$, 0.05*M* Tris base, 0.01% (w/v) gelatin (Swine skin, Type I). Adjust to pH 7.5.
4. Bottom agar: Add 1.5 g of agar/100 mL of L-Broth (LB).
5. Top agarose: Add 0.7 g of agarose (Type II)/100 mL of LB.
6. 2% (w/v) sodium azide (*see* Note 1 on toxicity).

7. 20% (w/v) maltose.
8. 10 m*M* isopropyl β-D-thiogalactoside (IPTG).
9. NBCS: 20% (v/v) newborn calf serum (heat-inactivated) in TBS.
10. Bacterial strain: Y1090: This is grown on an LB bottom agar plate (Step 4 above) containing 50 μg/mL of ampicillin (*see* ref. *4*).
11. Sartorius Minisart 0.2 μm sterile disposable filter.
12. 4-chloro-1-naphthol substrate solution: Dissolve 60 mg of 4-chloro-1-naphthol in 20 mL of ice-cold methanol. Make fresh for each screening.
13. Nitrocellulose (we use Schleicher and Schuell, BA85, 200 × 200 mm, cut to fit 85 mm Petri dishes with a card template).
14. Horseradish peroxidase-linked antibody (against the species of antibody that you are using).
15. Petri dishes (85 mm diameter).
16. Sterile Pasteur pipets.
17. 30% H_2O_2 solution.
18. Methanol (a general purpose grade reagent).
19. Chloroform.

All solutions should be made up with double distilled water. Solutions 1, 3, 4, and 5 should be autoclaved and can then be stored at room temperature. Since solution 7 (maltose) cannot be autoclaved, it is sterilized by filtration (Step 11). Solution 8 (IPTG) is stored in 20 mL aliquots at –20°C, this is enough for roughly 20 filters. Reagents 14, 17, and 18 are stored at 4°C. Reagent 12 is stored at –20°C. Solution 3 (SM) will probably have to be warmed for the gelatin to go into solution; this is normally done before adjusting the pH. The serum used in solution 9 is stored at –20°C in 50 mL aliquots, whereas solution 9 itself is prepared immediately before use and stored at 4°C. It is possible to use either 3% (w/v) gelatin, BSA, or ovalbumin instead of the serum. Gelatin gives reasonable results; however, the solution tends to solidify on cold days and hence serum is normally used instead.

3. Methods

3.1. Initial Screening of a λgt11 Library

3.1.1. Growth of Phage

1. Remove a single colony from the Y1090 stock plate using a plastic disposable loop and inoculate a sterile 150 mL conical flask containing 50 mL of LB. To this add 0.5 mL of a 20% maltose solution. Incubate overnight at 37°C with shaking (180 rpm).
2. Place the number of plates to be screened in a 42°C incubator to warm up (*see* Note 2).
3. For each plate to be screened, take 3 mL of the Y1090 overnight

culture and spin for 10 min, 4°C at 2000*g*, to pellet the bacteria.

4. Calculate the dilution required to give 0.2×10^5 plaques on a plate. Use SM to dilute a portion of the library to a value 10 or 100-fold more concentrated, i.e., if the library has a titer of 10^{10} PFU (plaque forming units)/mL, then dilute 1 µL of the library to 1 mL with SM. This then has a titer of 10^7 PFU/mL. If 2 µL of this is then mixed with 198 µL of cells (a further 10^2 dilution) there should be 0.2×10^5 PFU on the plate (*see* Note 3).
5. Melt the top agarose and put 3-mL aliquots in sterile capped test tubes prewarmed in a water bath at 45°C (one tube for each plate; *see* Note 4).
6. Pour off the supernatant and resuspend the pellet in 0.2 mL of SM for each plate to be screened.
7. Place the correct number of phage (e.g., 2 µL of a 10^3 dilution of the library) in a 1.5 mL Eppendorf tube and add 198 µL of the resuspended cells.
8. Incubate at 37°C with shaking (180 rpm) for 15 min. This allows the phage to adsorb to the bacteria and any great extension (to 30 min or longer) of this time could result in bacterial lysis and significantly increase the number of phage present in the mixture.
9. Remove plates from the oven and make sure each plate has a clear identifying mark or number on its base (*see* Note 5).
10. Using a sterile Pasteur pipet, transfer the phage/bacteria mixture to one of the tubes and mix it up and down once. Quickly pour the still molten agarose onto one of the agar plates and ensure that the agarose covers the entire plate surface before it sets. Put this to one side.
11. Repeat Step 10 as quickly as possible for each plate.
12. Allow the agarose to set, approx 5 min or less, and then place the plates, inverted, at 42°C for 2 h.
13. After 1 h, soak nitrocellulose filters in IPTG solution and leave to dry on some clean chromatography paper. Once dry, mark each one with the same mark as a plate. Remember, do not touch nitrocellulose with bare hands, as it will bind any proteins on the surface of your skin. This is also true of the bench-top if you happen to drop the filter onto it.
14. Remove the plates from the incubator and overlay each with a nitrocellulose filter soaked in IPTG. Make sure that the correct filter is placed on the correct plate. Care must be taken not to trap air bubbles beneath the filter as these can appear to be positives when the filters are stained later on (*see* Note 6).
15. Using a red hot needle, pierce the filter and the agar to mark the position of the filter on the plate. Make sure to pierce the filter in an asymmetric pattern, e.g., a single hole at one edge, a group of two holes at 120° to the first, and finally, a group of three holes at 240° to the first.
16. Incubate the filters inverted for a further 2 h at 37°C.

3.1.2. Recognition of Phage by Antibody

1. Just before the end of the incubation, put roughly 10 mL of TBS into Petri dishes — the same number of dishes as before.
2. Remove the filters with a pair of tweezers and place in the Petri dishes containing the TBS so that the side that was in contact with the agarose is facing upwards in the dish.
3. Incubate the filters at 30°C for 5 min with gentle rocking (roughly 60 rpm).
4. Store the plates inverted overnight at 4°C.
5. Discard the TBS and add 10 mL of the 20% NBCS to each filter. Return to the 30°C shaker for 10 min.
6. While the filters are being blocked with the NBCS, thaw the appropriate amount of the antibody to be used in the screening. This will depend on how many plates you are screening and the concentration at which the antibody is to be used, i.e., if your antibody is normally used at a 1 in 500 dilution for immunoblots, then that dilution can also be used for screening. Allow 9 mL per plate and calculate the total vol required. Place the correct amount of antibody in a beaker and also add enough sodium azide to make it to 0.02% (a 1 in 100 dilution of the stock). Now add 20% NBCS to the correct total vol (*see* Note 7).
7. Discard the NBCS on the plates and pour on the antibody solution prepared in Step 6 above.
8. Return to a shaker at 30°C and leave overnight.
9. Discard the antibody solution the next day and add 10 mL of TBS to the plates. Shake the plates as before (30°C) for 10 min.
10. Wash the filters once more as in Step 9.
11. Dilute the second antibody; we use a horseradish peroxidase-conjugated antibody in 20% NBCS as for the first antibody (Step 6), **but do not add azide to the second antibody** (*see* Note 8).
12. Discard the TBS wash and add the second antibody solution. Place the filters in the shaker for a further 2 h.
13. Pour off the second antibody solution and wash twice in 10 mL of TBS as before (Steps 9 and 10).
14. During the last wash, prepare the 4-chloro-1-naphthol substrate solution. Also, in a separate beaker, add 60 µL of 30% H_2O_2 to 100 mL of TBS.
15. Discard the TBS wash and mix the two solutions prepared in Step 14 above.
16. Add 10 mL to each filter and return to the shaker for either 10 min or until the filters have a pale blue background (*see* Note 9).
17. Discard the staining solution and wash the filters under a running cold tap.
18. Blot dry on a piece of chromatography paper.

3.1.3. Isolation of Phage Recognized by Antibody

Possible positive plaques will appear on the filter as very small blue dots. At this stage, they will be quite difficult to see because they are usually less than 1 mm in diameter. However, they should be circular and of approximately the same size as the plaques on the plate. They should not be visible on the other side of the filter since only one side of the filter bound the antigen; if they are, then they can be regarded as a false reaction, probably because of a flaw in the surface of the nitrocellulose. Also if the positive spot appears to have a pin hole in its center when viewed with a light shining from behind, it is probably a false reaction as well. These guidelines should be used to decide which of the possible positives are likely to be real. Once you have decided which you are going to rescreen, follow the procedure below (*see* Note 10).

1. Collect as many small sterile bijoux bottles as there are plaques to be picked.
2. Into each, place 2 mL of SM and three drops of chloroform from a Pasteur pipet.
3. Take the plates from the refrigerator and match each filter to its plate.
4. With a ball point pen, draw a circle round each plaque to be picked on both sides of the filter. Mark them so that they can be identified, e.g., the phage from plate 2 could be called 2.1, and so on.
5. Mark the bijoux bottles in an identical fashion, one for each plaque to be picked.
6. Align a filter and a plate using the pierce marks.
7. Take a 1-mL Gilson tip and cut it roughly 1 cm from the end. This will give an opening about 5 mm wide, which is convenient for taking agar plugs containing the positive phage.
8. Holding the plate so that it does not move, position the cut tip over one of the marked plaques and press down through the agarose and agar to cut a plug. Remove the tip, being careful not to leave the plug behind (it may be necessary to angle the tip to prevent this) and place it into the correct bijoux. The plug is removed by shaking the tip in the SM.
9. The bottle can now be closed, and when all the plugs are taken, they should be stored at 4°C until they are to be rescreened.

3.2. Rescreening of Picked Plaques

This is essentially the same as the initial screening (Sections 3.1.1. and 3.1.2.) with some differences that are outlined below.

1. Instead of using a fixed dilution of phage, as in Section 3.1.1., Step 4, a range of dilutions must be used. In general, if dilutions in the range 10^2–10^4 are used, then two of the three plates are usually suitable to

continue with into Section 3.1.2., i.e., the density of plaques is such that they have some space between them. The exact range of dilutions varies according to how densely plated the original plate was (*see* Note 11).

2. In Section 3.1.1., Step 12, the plates are incubated at 42°C for 2 h, but here they are left for 4 h. Consequently, the plaques are much larger, with a diameter of about 1 mm.
3. The final differences occur in Section 3.1.2. Instead of seeing the small dot-like stain, there should be this time a round blue spot on the filter of the same size as the plaque, i.e.,roughly 1 mm. This may have an unstained region in its center. This occurs if the phage grew fast and had lysed all the cells in the center of the infection by the time the filter was placed on it. However, there can be a variation in this even when a phage can be said to have been isolated to purity and so, a mixture of the two types does not mean that the phage is contaminated. For confidence, it may be worth isolating one of each type of plaque and rescreening.

   The pin hole seen at the center of some positives and referred to in the introduction to Section 3.1.3., should not be confused with the larger plaques of the second or later screening that have unstained centers. The plaques referred to earlier look like normal positive plaques on the filter, except that when viewed with a light shining from behind, the light can be seen shining through the filter.
4. Instead of picking all the positive plaques, this time pick the ones that are the furthest separated from the rest. If three plaques are picked, this gives a backup in case one of the isolates proves very difficult to purify. Otherwise, they are picked in exactly the same manner as before (*see* Note 12).

4. Notes

1. **Sodium azide is very toxic by ingestion and inhalation. Contact with acids liberates a highly toxic gas. It is also irritating to the skin and eyes. It reacts slowly with water at ambient temperatures but decomposes violently in hot water. It forms readily detonatable salts with many metals.** Suitable care should be taken in weighing this compound, wear rubber gloves and mask. As it decomposes slowly in water, make a fresh solution at regular intervals, every 2–3 mo.
2. Your level of confidence with the method determines how many plates to use. Initially try 6; this is a reasonable number that is still worth screening. Once you can perform the plating out part of the experiment efficiently, then choose your own limit! We recommend probably no

more than 25 at a time, because if you start at 9 AM with that number, you should finish sometime between 5 and 6 PM, i.e., to Section 3.1.2., Step 8. The reason for putting plates into an incubator before plating is to prevent the agarose from setting on contact with the plates, since the bacteria do not seem able to grow in the set "lumps" of agarose that result.

3. Although it is recommended that 10^5 phage on a plate should be aimed for, you may find that this concentration does not leave any space between the plaques. At this stage, it may be more desirable to lower the concentration by a small amount, a factor of 10 should be enough, and thereby possibly increase the speed at which phage can be purified. It also depends on whether you can see any bacterial growth after the 4 h. If you cannot, then the phage concentration should be reduced. You may be killing the bacteria before the protein in the phage is induced and so it would be impossible to find anything. To get around this, try using a range of dilutions for the first screening, e.g., start at a lower dilution than the theoretical optimum and then work to higher dilutions. Then, just screen at what looks like the densest plating that allows bacterial growth to be seen.
4. If you find that you get very few or no plaques, check the temperature of your water bath. Temperatures above 50°C seem to kill the bacteria. Also, make sure that the agarose has time to cool down to 45°C before you add cells.
5. At first, only take three plates out of the oven. This will prevent them cooling down too much before the agarose is put on. If the plates are stacked inverted on the bench, then you can mark the bottom of each plate just before you use it and keep them warmer by their combined lower heat loss.
6. To do this, hold the filter at opposite edges, bend the filter in the middle and allow the middle of the filter to touch the center of the plate. Then, using gentle pressure, push down on the filter until it is fully in contact with the plate. This should exclude virtually all air bubbles, except those already in the agarose.
7. We determine the concentration at which to use the antibodies by testing them on Western blots to detect the protein band of interest. We use the combination of first and second antibody dilutions that gives the strongest staining but the lowest background signal. If you have trouble with every phage being identified faintly, the antibody concentration could be the problem.

If you wish to screen with more than one antibody you can either do multiple lifts with filters (*see* ref. 4 for timings) and incubate each one with a different antibody, or you can add all the antibodies at the same time to one filter. If you do the latter, add antibodies at their normal individual concentrations. The purified phage can be tested with each antibody separately at a later date.

8. There are two reasons for not adding sodium azide to the second antibody. The first is that it is an inhibitor of the horseradish peroxidase-conjugated to the second antibody and its presence would greatly reduce any staining that developed later. The second reason is that the antibody is only going to be incubated for 1–2 h, during which bacterial growth should not be as great a problem as with the overnight incubation of the first antibody.
9. Each batch of filters should be incubated for the same length of time. If your second antibody reacts quickly, because of activity or the amount bound, then a blue background will appear quickly. However, some do not give a significant background and, therefore, can be left in for 10 min or more to get a strong stain.
10. In identifying positives, remember that you are looking for a very small blue dot that may look somewhat like a piece of dust that has drifted onto your filter. However, if on rubbing it with your finger it does not move or smudge, then it is probably a positive. True positives have not been known to be rubbed off under this scrutiny!
11. When rescreening, you want a large enough number of phage on the plate to give you a chance of picking up a positive. Therefore, start with a 10^2 dilution of the phage eluted from the plug and increase by a factor of 10 each time, i.e., three plates with a decreasing number of phage on them. This serial dilution increases the chance of finding a positive in a position that will make it easier to purify with few other plaques near it. The concentration of phage in the bijoux depends on the number of plaques in the plug, so a little trial and error comes into this. Once it has been done, however, it is usually possible to use the same dilution for the subsequent phage.
12. An increasing proportion of positives will be seen at each stage of the purification. To isolate all of these would be a monumental task. Also, as they came from the same initial plaque, they should be the same, and hence, after the initial purification, only three positive phages should be picked from each plate even if there is a greater number than that. Obviously, the plaques that appear the easiest to purify are the best to pick here.

References

1. de Wet, J. R., Fukushima, H., Dewji N., Wilcox, E., O'Brien, J. S., and Helinski, D. R. (1984) Chromogenic immunodetection of human serum albumin and α-L-fucosidase clones in a human hepatoma cDNA expression library. *DNA* **3,** 437–447.
2. Young, R. A. and Davis, R. W. (1983) Efficient isolation of genes by using antibody probes. *Proc. Nat. Acad. Sci. USA* **80,** 1194–1198.
3. Young, R. A. and Davis, R. W. (1983) Yeast RNA polymerase II genes: Isolation with antibody probes. *Science* **222,** 778–782.
4. Maniatis, T., Fritsch, E. F., and Sambrook, J. (1982) *Molecular Cloning, A Laboratory Manual.* Cold Spring Harbor, Cold Spring Harbor, NY.

CHAPTER 50

Expression of Foreign Genes in Mammalian Cells Using an Antibody Fusion System

Simon J. Forster, Francis J. Carr, William J. Harris, and Anita A. Hamilton

1. Introduction

Whereas the expression of foreign genes in mammalian cells usually proves successful, the purification of gene products is often a difficult and time-consuming process. The availability of monoclonal antibodies to the foreign protein can greatly assist in small scale purification, but where antibodies are not available, alternatives have to be sought. One useful approach involves the fusion of the foreign gene adjacent to a gene segment encoding an antibody heavy chain variable region *(1)*. By transfection of this construct into a cell line producing a compatible light chain or by cotransfection of the fusion product with a light chain gene, an antibody-like molecule can be produced and purified using the corresponding antigen.

This chapter will describe the use of the plasmid pSV-$V_{NP}\gamma_{2b}\delta(C_H2, C_H3)$ *(2)* which is shown in Fig. 1. This vector is derived from pSV$_2$*gpt* *(3)*, but additionally encodes a chimeric antibody heavy chain, which consists of the variable region, V_{NP}, of a mouse antibody that binds to the haptens NP (4-hydroxy-3-nitrophenacetyl) and the related NIP (5-iodo-4-hydroxy-3-nitrophenacetyl), and the C_H1, hinge and *N*-terminal part of the C_H2 domain of a mouse γ_{2b} heavy chain. The vector contains an ampicillin resistance gene for selection in *E. coli* and the *E. coli gpt* gene that codes for the enzyme XGPRT, and allows selection in mammalian cells in medium containing

From: *Methods in Molecular Biology, Vol. 10: Immunochemical Protocols*
Ed.: M. Manson

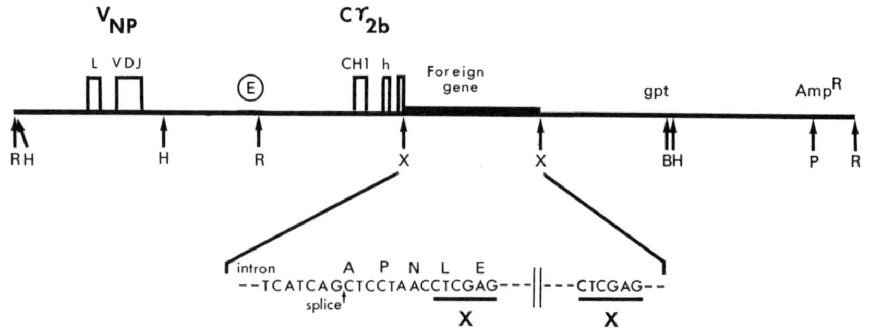

Fig. 1. Structure of plasmid pSV-$V_{NP}\gamma_{2b}\delta(C_H2, C_H3)$. The V_{NP} and γ_{2b}genes are inserted in pSV_2*gpt* between the *Eco*R1 and *Bam*H1 sites. The *Bam*H1 site is converted to *Xho*1 by the addition of linkers. B, *Bgl*II; H, *Hind*III; P, *Pvu*1; R, *Eco*R1; X, *Xho*1; E, immunoglobulin enhancer.

mycophenolic acid and xanthine. The unique *Xhol* cloning site is at the start of the heavy chain C_H2 region. The foreign gene is inserted as an *Xho*l fragment, the 5' site being in phase to continue the reading frame. Thus, the insert replaces the rest of the heavy chain constant region (*see* Note 1). The SV40 polyadenylation site is provided in the vector and the immunoglobulin enhancer is included for high-level expression in myeloma cells. Association of pSV-V_{NP} heavy chain with light chains of the λ1 subgroup gives rise to a hybrid antibody molecule that binds to NP and NIP (Fig. 2). A convenient source of light chain for this purpose is from the J558L mouse plasmacytoma, a derivative of J558 *(4)*, which expresses λ1 but no heavy chain. The plasmid is transferred to the J558L cells by spheroplast fusion *(5)*.

The advantage of this approach is that the foreign gene insert is expressed fused to the equivalent of an antibody Fab fragment. The antigen-binding capability of the fusion product is then used as a basis for screening transfectants and purification of the fusion product. We have found it most convenient to use an ELISA method for initial screening of transfectants and a Western blot method for a fuller characterization of selected transfectants. Detailed descriptions of these methods are given (Sections 3.3. and 3.4.).

For the ELISA screen, the wells of microtiter plates are coated with NIP conjugated to bovine serum albumin. The wells are then filled with culture medium conditioned by transfectants. After washing, bound Fab fusion protein is detected using commercially-available antibody to mouse lambda chain and an appropriate enzyme conjugate. Positive transfectants can be further analyzed by Western blotting, using the same antibodies, allowing the molecular weight of the fusion product to be estimated.

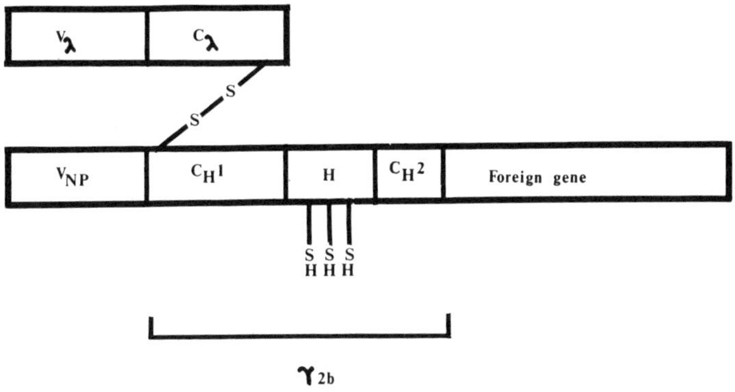

Fig. 2. Structure of the antibody fusion protein. The diagram shows the structure of the Fab-like form, consisting of a lambda light chain derived from J558L cells and the chimeric heavy chain produced by the vector. The heavy chain consists of a variable region with NP-binding activity; C_H1, hinge and part of the C_H2 domain of murine heavy chain of isotype γ_{2b}, and the foreign gene. The fusion protein may be secreted in this form or as an $F(ab')_2$-like form consisting of two Fab-like forms linked at the hinge region by three disulphide bonds.

2. Materials

2.1. Cloning and Transfection

1. Plasmid pSV-$V_{NP}\gamma_{2b}\delta(C_H2, C_H3)$: Kindly supplied by M. Neuberger, MRC Laboratory of Molecular Biology, Cambridge, UK.
2. L-broth (1 L): 10 g of tryptone, 5 g of yeast extract, 5 g of NaCl; adjust to pH 7.2 and autoclave.
3. L-agar (1 L): Add 15 g of agar to the L-broth and autoclave.
4. Ampicillin stock: 25 mg/mL in water, sterilize by filtration.
5. Enzymes: *Xho*I, other restriction enzymes, and T4 DNA ligase are supplied with appropriate buffers by most manufacturers. Follow their instructions for storage and shelf-life.
6. 7.5*M* ammonium acetate.
7. 100% ethanol (Analar grade or equivalent).
8. 70% ethanol.
9. Calf intestinal alkaline phosphatase.
10. Phosphatase buffer (10X): 0.5*M* Tris-HCl, pH 9, 10 m*M* $MgCl_2$, 1 m*M* $ZnCl_2$ and 10 m*M* spermidine.
11. Geneclean™ kit: BIO 101 Inc., La Jolla, CA.
12. TE buffer: 10 m*M* Tris-HCl, pH 8, 1 m*M* ethylenediaminetetra-acetic acid disodium salt (Na_2EDTA); autoclave.

13. Frozen competent cells: Available commercially from several manufacturers with detailed protocols for transformation.
14. Phenol: Recrystallized phenol equilibrated with several changes of 0.5*M* Tris-HCl, pH 8, and finally 0.1*M* Tris-HCl, pH 8. Phenol causes severe burns and should be handled with care.
15. Chloroform: Chloroform/isoamyl alcohol 24:1 (v/v).
16. Phenol/chloroform: 1:1 mixture of solutions 14 and 15.
17. Solution A: 50 m*M* glucose, 10 m*M* Na_2EDTA, and 25 m*M* Tris-HCl, pH 8, autoclave.
18. Lysozyme: 16 mg/mL in solution A, make fresh.
19. Solution B: 0.2*M* NaOH, 1% (v/v) SDS, make fresh.
20. Solution C: 3*M* sodium acetate, pH 5.2, autoclave.
21. RNase: RNase A, 10 mg/mL, in 10 m*M* Tris-HCl, pH 8, 15 m*M* NaCl. Heat to 100°C for 15 min and cool slowly to room temperature. Store in aliquots at –20°C.
22. J558L mouse plasmacytoma cell line: Obtained from G. Koch, MRC Laboratory of Molecular Biology, Cambridge, UK.
23. DMEM: Dulbecco's Modification of Eagle's Medium with 10% (v/v) fetal calf serum, 4 m*M* L-glutamine, 1 mg/mL streptomycin, and 2000 U/mL penicillin G.
24. Gentamicin stock: 10 mg/mL, sterilize by filtration.
25. Fungizone (Amphotericin B) stock: 250 µg/mL, sterilize by filtration.
26. Mycophenolic acid stock: 500 µg/mL in phosphate-buffered saline (*see* Step 39 below), neutralize with NaOH, sterilize by filtration. Note that this compound is toxic.
27. Xanthine stock: 5 mg/mL in 50 m*M* NaOH, autoclave.
28. Selective medium (1 L): DMEM with 10 mL of mycophenolic acid stock, 5 mL of xanthine stock, and 2.5 m*M* HCl.
29. Chloramphenicol stock: 34 mg/mL in 100% ethanol, sterilize by filtration.
30. Sucrose/Tris: 20% (v/v) sucrose, 50 m*M* Tris-HCl, pH 8, autoclave.
31. Lysozyme: 5 mg/mL in 0.25*M* Tris-HCl, pH 8. Prepare immediately before use.
32. 0.25*M* Na_2EDTA, pH 8, autoclave.
33. 50 m*M* Tris-HCl, pH 8, autoclave.
34. DMEM/sucrose/$MgCl_2$: Add 10% (w/v) sucrose and 10 m*M* $MgCl_2$ to DMEM, sterilize by filtration.
35. PEG: 50% (v/v) polyethylene glycol 1500, autoclave.
36. Protease inhibitors: Phenylmethylsulphonylfluoride (PMSF), 0.1*M* in ethanol and *N*-ethylmaleimide (NEM), 0.25*M* in ethanol. Note that these

inhibitors are toxic by inhalation and skin contact and should be handled with gloves in a fume hood.

All solutions must be made up in tissue culture grade water (Milli-Q or equivalent). Solutions 4, 5, 9, 10, 21, 24–27, 29, and 36 are stored at –20°C and 11, 14, 17, 23, 28, 30 and 34 are stored at 4°C. Solutions 18, 19, and 31 are made immediately before use. The remaining solutions are stored at room temperature. All solutions used with the cultured cells must be kept rigorously sterile and all manipulations must be performed in a sterile flow hood.

2.2. Screening, Analysis, and Purification

37. NIP-Cap-OSu (Cambridge Research Biochemicals, Northwich, Cheshire, UK): 31 mg in 300 µL of *N,N'*-dimethylformamide (*see* Note 4).
38. 5 mg/mL bovine serum albumin (BSA) in 1% (w/v) sodium bicarbonate.
39. Phosphate-buffered saline (PBS): 0.14*M* NaCl, 2.7 m*M* KCl, 1.5 m*M* KH_2PO_4, and 8.1 m*M* Na_2HPO_4, pH 7.2.
40. PBS containing 0.1% (w/v) sodium azide. Sodium azide is toxic by ingestion and inhalation, and should be handled with gloves in a fume hood.
41. PBST: PBS containing 0.05% (v/v) Tween-20.
42. PBS containing 1 m*M* NIP-cap-OH (Cambridge Research Biochemicals).
43. Coating buffer: 50 m*M* carbonate–bicarbonate pH 9.6, 0.02% sodium azide.
44. Tris/saline: 10 m*M* Tris-HCl, 0.9 (w/v) NaCl, pH 7.4.
45. Tris/saline/BSA: Tris/saline containing 3% (w/v) BSA.
46. Antibodies: Goat antibody to mouse lambda chain (Sera-Lab, Crawley Down, UK); rabbit antigoat IgG conjugated to alkaline phosphatase (Sigma) (*see also* Note 5).
47. Alkaline phosphatase substrate buffers: For ELISA, 100 m*M* glycine, 1 m*M* $MgCl_2$ and 1 m*M* $ZnCl_2$, pH 10.4; for Western blot, 0.1*M* Tris-HCl, 0.1*M* NaCl, and 5 m*M* MgCl, pH 9.5.
48. Materials for SDS-PAGE according to Laemmli (*see* Note 6).
49. 1*M* Iodoacetamide; make only 1 mL and store at –20°C. Discard when a brownish tinge appears.
50. Blot buffer: 192 m*M* glycine, 12.5 m*M* Tris-HCl, pH 8.3 containing 10% (v/v) methanol.
51. Blot blocking solution: Tris/saline containing 3% (w/v) nonfat dried milk.
52. NBT (nitro blue tetrazolium) stock: 75 mg/mL in 70% (v/v) *N,N'*-dimethylformamide. *N,N'*-dimethylformamide is toxic by inhalation and skin absorption andl should be handled with gloves in a fume hood.

53. BCIP (5-bromo-4-chloro-3-indolyl phosphate) stock: 50 mg/mL (*p*-toluidine salt) in *N,N'*-dimethylformamide.
54. 3% (w/v) sodium bicarbonate.
55. ω-Aminohexyl Sepharose (Sigma).
56. 50 mg of NIP-cap-OSu dissolved in 400 µL of 1,4-dioxan. 1,4-dioxan is toxic by ingestion and inhalation and should be handled with gloves in a fume hood.
57. Small disposable plastic syringe or column (for example, Bio-Rad).
58. Glass wool.

Solutions 37, 38, 42, and 54 should be freshly made as required. Solutions 45, 51, 52, and 53 should be stored at –20°C. For antibodies and antibody-enzyme conjugates, the manufacturer's storage recommendations should be followed. The remaining solutions may be stored at 4°C or at room temperature. Solutions 39 and 44 may be autoclaved for extended storage at room temperature. Solid NIP-cap-OSu and NIP-cap-OH should be stored desiccated at –20°C.

3. Methods

3.1. Cloning into pSV-$V_{NP}\gamma_{2b}\delta(C_H2, C_H3)$

1. Digest 10 µg of vector DNA made up to 40 µL in the appropriate buffer (supplied by the manufacturer) with 20–50 U of *Xho*1. Digest 10–20 µg of DNA containing the gene of interest in a similar manner with *Xho*1. Check that the DNAs are fully digested by running 1 µL samples on a 0.8% agarose minigel.
2. Precipitate one-half the digested vector DNA with ethanol. Add 0.5 vol of 7.5*M* ammonium acetate and 2 vol of 100% ethanol. Put at –20°C for 30 min, then spin at full speed in a microfuge for 10 min, wash the pellet with 70% ethanol, dry, and resuspend in 10 µL of TE. Add 5 µL of 10X phosphatase buffer, 35 µL of water, and 0.02 U of calf intestinal alkaline phosphatase. Incubate at 37°C for 30 min, then add another 0.02 U and continue the incubation for a further 30 min. Extract with phenol/chloroform, followed by chloroform, and precipitate the DNA with ethanol as previously.
3. Load the digested vector, the digested and phosphatase-treated vector, and the digested foreign gene DNA onto preparative agarose minigels, i.e., with one or several wide slots, and electrophorese until the fragments are well separated. Excise the linear vector bands and the band containing the gene of interest. Purify these DNAs from the gels using Geneclean™ according to the manufacturer's instructions. Run a small sample of each on a 0.8% agarose minigel alongside standard DNA samples and estimate the concentrations. Precipitate the DNA fragments with ethanol.
4. Resuspend the DNA fragments in TE at approx 200 µg/mL. Set up a 25

μL ligation reaction, in the buffer provided by the manufacturer, containing equimolar amounts of phosphatase-treated vector and DNA to be inserted, at a total DNA concentration of 40 μg/mL. The vector is 10 kb, so if the insert is 2 kb, 5 times as much insert as vector DNA should be added; 0.2 μg and 1 μg in 25 μL. Add 3 U of T4 DNA ligase and incubate at 16°C overnight. Also, set up control ligations containing digested vector DNA and digested, phosphatase-treated vector DNA only.

5. Transform 10 μL each of the ligated DNAs into competent HB101 cells. Plate the transformation mixes onto L-agar plates containing 50 μg/mL ampicillin (2 mL of stock/L).
6. Pick ampicillin-resistant colonies to 10 mL of L-broth containing ampicillin and grow overnight. To prepare DNA minipreps, spin down one microcentrifuge tubeful of culture. Remove the supernatant and resuspend the pellet in 80 μL of cold solution A by vortexing. Add 20 μL of lysozyme in solution A, mix and incubate for 5 min at room temperature. Add 200 μL of solution B, invert 3 times to mix, and incubate on ice for 5 min. Add 150 μL of solution C, vortex for 10 s, and incubate on ice for 10 min. Spin at full speed in a microfuge for 5 min. Remove the supernatants to fresh tubes containing 200 μL of phenol/chloroform. Mix and spin briefly, then remove the aqueous layers to fresh tubes and add 1 mL of ethanol. Leave at room temperature for 10 min. Spin for 10 min at full speed in a microfuge, wash the pellet in 70% ethanol, and dry. Resuspend the DNA in 50 μL of TE containing RNase (100 μg/mL).
7. Use 10 μL portions of these DNAs for restriction analysis to determine the presence and orientation of inserts. The gene of interest must be inserted in the correct orientation for expression. By digesting with a restriction enzyme that cuts at one end of the inserted gene and a second enzyme that cuts in the vector (e.g., *Bgl* II), the orientation of the insert can be deduced.
8. Streak an L-agar plate containing ampicillin with the correct bacterial clone and prepare a frozen stock.

3.2. Transfection into J558L

3.2.1. Preparation of Bacterial Spheroblasts

1. Add 0.2 mL of ampicillin stock to 100 mL of L-broth in a 500 mL conical flask and inoculate with *E. coli* HB101 harboring pSV-V_{NP} (one colony or a loopful of frozen stock). Incubate with shaking at 37°C until late log phase (an A_{650} of 1.0). Add 0.5 mL of chloramphenicol stock (170 μg/mL final concentration) and continue shaking overnight.
2. Remove a small sample (0.5 mL) and put aside, then spin down the cells at 1000*g* at 4°C for 15 min and discard the supernatant. Spin again briefly

and remove any remaining liquid. Resuspend the cells in 2.5 mL of ice-cold sucrose/Tris.

3. Add 0.5 mL of lysozyme and swirl on ice for 5 min.
4. Add 1 mL of cold EDTA and swirl on ice for 5 min.
5. Add 1 mL of 50 m*M* Tris-HCl and incubate, swirling at 37°C for 15 min. This stage is critical; if the original culture was too thin, the cells may start to lyse. If this occurs, move on to Step 6 immediately.
6. Slowly dilute the spheroplasts into 20 mL of DMEM/sucrose/$MgCl_2$ while swirling at room temperature.
7. Leave for 10 min at room temperature. Examine a sample of the cells with a phase contrast microscope (40–100× objectives) to check for good conversion of *E. coli* rods to round spheroplasts. Compare to the original cells (*see* Note 2). If the spheroplast suspension is good, only one-half (5 mL) is required for the fusion.

3.2.2. J558L Plasmacytoma Cells

1. Twenty-four hours prior to transfer, set up 2 × 75 cm^2 flasks of J558L cells at 1–2 × 10^5 cells/mL at 25 mL per flask in DMEM from an actively growing culture.
2. Pool and count the cells to check that they are at the required density of 4–5 × 10^5 mL. Spin at 150*g* for 5 min at room temperature and discard the supernatant. Progressively resuspend the pellet in 10 mL of DMEM/sucrose/$MgCl_2$.

3.2.3. Fusion

1. Prepare a tube containing 10 mL of DMEM at 37°C. Keep the bottle of DMEM at 37°C and warm the PEG to 40°C.
2. Spin the diluted spheroplast suspension (5 or 10 mL as required) at 2500 rpm for 10 min at room temperature. Remove the supernatant carefully.
3. Gently layer the J558L cell suspension on top of the spheroplasts without disturbing the pellet and spin at 1000*g* for 7 min at room temperature.
4. Carefully take off the supernatant and break up the pellet by gently tapping the bottom of the tube. Place this in a beaker of water at 40°C for the duration of the fusion.
5. Using a 1-mL pipet and stirring the cells with the tip, add 0.8 mL of PEG, prewarmed to 40°C, over a period of 30 s, and continue stirring for a further 30 s.
6. With the same pipet, take 1 mL of DMEM from the 10 mL and add to the fusion, stirring continuously as before. Repeat such that the 10 mL of DMEM is added over 30 s. This step must not be delayed as PEG is toxic to the cells.

7. Slowly add 12–13 mL of prewarmed DMEM. Spin at 150*g* for 8 min at room temperature.
8. Resuspend the cells in DMEM at 10^5 cells/mL. Distribute 1 mL per well to 24-well dishes as required and place in a 37°C, 5% CO_2 humidified incubator.
9. Allow the cells to settle for about 3 h, then remove the medium and replace with DMEM containing gentamicin (0.5 mL/100 mL) and fungizone (10 mL/100 mL). This removes bacterial debris that may interfere with cell growth.
10. After 24 h, remove the medium and replace with selective medium 2 mL/well). *Gpt*$^+$ clones, probably several per well, should appear after 8–10 d.
11. Supernatants from positive clones can be analyzed directly by ELISA, or individual clones can be isolated and expanded for further analysis and purification. The J558L cells should be grown to a density of 2×10^6/mL or greater for maximal production of hybrid antibody. Conditioned medium is stored at –20°C after addition of protease inhibitors NEM and PMSF (20 mL/L of each) and Na_2EDTA (20 mL/L).

3.3. Screening by ELISA

3.3.1. Preparation of NIP-cap-BSA

1. Add 75 µL of NIP-Cap-OSu in DMF to 5 mL of BSA-bicarbonate solution.
2. Leave stirring overnight at 4°C.
3. Dialyze for 48 h against three changes of 2 L of PBS.
4. Determine the protein concentration and coupling ratio by reading A_{280} and A_{430}. Protein concentration is:

$$\frac{A_{280} - (0.59 \times A_{430})}{1.4} \quad \text{mg/mL}$$

and the coupling ratio (hapten:BSA) is:

$$\frac{68 \times A_{430}}{4.85 \times [\text{BSA concentration in mg/mL}]}$$

The dialyzed material will need dilution before the absorbances can be read. Try 1 in 5 (in PBS) initially. The concentration to be used in the second equation above is that of the diluted sample (in mg/mL).

3.3.2. Screening (see Notes 5 and 7)

1. Coat wells of a suitable microtiter plate with 1 µg of NIP-cap-BSA per well as follows: Calculate the total vol needed at 200 µL per well; dilute NIP-cap-BSA to 5 µg/mL (i.e., 1 µg/200 µL) in coating buffer. Dispense

200 µL per well and incubate for at least 2 h at 37°C or overnight at 4°C.
2. Empty wells. Do not wash. Block wells by filling with 250 µL Tris/saline/BSA and incubating for at least 2 h at 37°C or overnight at 4°C.
3. Empty wells and wash 3× with PBST.
4. Fill wells with 200 µL of conditioned medium and incubate at 37°C for 1 h.
5. Wash 3× with PBST.
6. Fill (200 µL) wells with goat antimouse lambda chain diluted 1:1000 in PBST and incubate for 1 h at 37°C. Empty and wash three times with PBST.
7. Fill wells with alkaline phosphatase conjugate diluted 1:1000 in PBST. Incubate for 1 h at 37°C. Empty and wash three times with PBST.
8. Develop color: Fill wells with substrate buffer containing 2.2 mg/mL *p*-nitrophenyl phosphate. Incubate in the dark for an appropriate time. Positive wells should give a yellow color, which can be read at 405 nm, within 30 min. If necessary, the reaction can be stopped by adding 50 µL of 3*M* NaOH to each well.

3.4. Western Blot Analysis

1. Run samples on SDS-PAGE according to the method of Laemmli, except that 2-mercaptoethanol should be replaced by iodoacetamide (final concentration 0.1*M*) (*see* Notes 5 and 6).
2. Transfer the separated proteins to nitrocellulose paper. We routinely use a semidry blotting apparatus and transfer at 0.8 mA/cm^2 for 1 h, constant current. Any commercial blotting apparatus may be used and the manufacturer's instructions should be followed (*see* Note 8 and this vol., Chapter 24).
3. Block the blot in a known volume of blocking solution for 1 h at room temperature, or overnight at 4°C.
4. Add antilambda chain antibody to an appropriate dilution (normally 1:1000) and continue incubation for at least 1 h at room temperature, or overnight at 4°C.
5. Wash the blot with multiple changes of Tris/saline, over one-half hour.
6. Add the conjugate diluted 1:1000 in Tris/saline, and incubate for 1 h at room temperature.
7. Wash over one-half hour as before, first with Tris/saline, then with alkaline phosphatase substrate buffer.
8. Make up the substrate solution: For each 10 mL of substrate buffer, add 40 µL of NBT stock and 40 µL of BCIP stock, in that order, mixing between. Add the substrate solution to the blot and allow color development to proceed. Stop by washing several times with water. Dry between sheets of Whatman 3MM under a weight.

3.5. Affinity Purification on NIP Columns

3.5.1. Preparation of NIP-cap-Sepharose

1. Swell 10 g of ω-aminohexyl sepharose by stirring for about 5 h at room temperature in 100 mL of 3% sodium bicarbonate.
2. Cool to 4°C and add 50 mg of NIP-cap-OSu dissolved in 400 µL of 1,4-dioxan.
3. Stir overnight at 4°C.
4. Wash exhaustively with water, then PBS containing 0.1% azide.
5. Store at 4°C in PBS/azide.

3.5.2. Purification (see *Note 9*)

1. Block the end of a suitably sized syringe (e.g., 2 mL) with a small wad of glass wool (wear gloves). Alternatively, suitable small disposable columns are available commercially.
2. Add an appropriate amount of NIP-cap-Sepharose slurry and allow to settle.
3. Block nonspecific binding sites by filling the column with Tris/saline/BSA and leaving for 30 min at room temperature.
4. Wash with PBS.
5. Pass the sample through the column.
6. Wash the column extensively with PBS.
7. Elute with 1 m*M* NIP-cap-OH in PBS.
8. Analyze the eluted material by SDS-PAGE or by Western blot.
9. Dialyze against a large volume of PBS (no azide) overnight at 4°C.
10. Read the absorbance of the dialyzed material at 280 nm and calculate the approximate "antibody" concentration using A_{280} of 1.4 = 1 mg/mL.

4. Notes

1. A gene cut out with *Sal*I can be cloned into the vector *Xho*I site as the cohesive ends are the same. However, it is likely that genes will require modification to introduce *Xho*I sites at the correct locations. *Xho*I sites can be positioned precisely by oligonucleotide-directed in vitro mutagenesis or synthetic oligonucleotides (either commercially available linkers or adaptors or specially made sequences) can be added at convenient sites. Consideration must be given to the effect on the protein of the amino acid changes resulting from these manipulations. Proteins with *N*-terminal signal peptides and *C*-terminal membrane anchor regions require special attention. The 3' *Xho*I site must be positioned at the

signal peptide cleavage site or a fused protein will not be produced. Hydrophobic regions, normally located within the cell membrane, must be removed to allow secretion of the product.

2. It may be necessary to try variations of the incubation times given here to obtain good spheroplast preparations.
3. DMEM, and particularly fetal calf serum, must be carefully selected to ensure optimal growth of the cells.
4. NIP-cap-OSu comprises the NIP moiety with a 5-carbon (caproic acid) spacer arm and an *N*-hydroxysuccinimidyl group that is reactive with amino groups, thus allowing attachment to bovine serum albumin and ω-aminohexyl Sepharose. NIP-cap-OH is the free acid form.
5. If antibodies to the foreign gene product are available, these can also be used in the ELISA and Western blot with an appropriate conjugate depending on the species in which the antibodies were raised. For a discussion of the reason for using antilambda rather than anti-γ_{2b} chain antibodies, *see* Note 6 below.
6. A description of the materials and apparatus for SDS-PAGE according to the method of Laemmli may be obtained from a number of sources, for example *(6)* and Volume 1, Chapter 7. We have found a gradient of 6–12.5% polyacrylamide suitable to cover the range of molecular weights from free lambda chains to large fusion products. If an antibody to the fused foreign gene product is available, this can be used to probe a blot made from a gel run under reducing conditions. Otherwise, the substitution of iodoacetamide for mercaptoethanol maintains the integrity of the disulphide bonds, and hence, the attachment of the lambda chain to the heavy chain. Although the use of an antibody to the γ_{2b} chain would also allow the use of reducing conditions, we have not been able to find any commercially available antibody that performs adequately. This may be as a result of the major determinants being in the deleted C_H2 and C_H3 regions, the C_H1 domain being masked in the intact folded antibody used as immunogen, the denaturation of determinants during SDS-PAGE, or a combination of these. We have in fact been able to raise an antiserum that does react with the denatured truncated γ_{2b} chain by transfecting cells with the pSV-$V_{NP}\gamma_{2b}\delta(C_H2, C_H3)$ plasmid (lacking any insert in the cloning site), purifying the Fab-like fragment, separating the truncated γ_{2b} chain from the lambda chain by preparative SDS-PAGE in reducing conditions, and immunizing rabbits with gel slices containing this chain.
7. It is good practice to do a transfection and an ELISA screen using plasmid with no insert in the cloning site. About 50% of transfectants will be positive in the ELISA screen, and these can be used as positive and

negative controls in subsequent ELISA screens and Western blots. We have, however, had one case of a formerly negative clone beginning to secrete Fab-like fragment after several weeks in culture.

8. Some means of checking the efficiency of transfer should be included. We use prestained marker proteins (Rainbow markers, Amersham International), which are visible on the blot after successful transfer. Alternatively, duplicate tracks that are cut off and stained with Amido Black can be included. (Soak in 0.1% (w/v)Amido Black in methanol/acetic acid/water 45:10:45 for a few minutes and destain in ethanol/acetic acid/water 90:2:8.)
9. Although this expression system is designed to allow the Fab fusion protein to be secreted as an antibody-like molecule, it is quite possible that a particular insert will give expression but not secretion, perhaps because of incorrect folding, or the presence of a hydrophobic domain in the foreign protein (*see* Note 1 above). Provided that the Fab fusion is not too rapidly degraded intracellularly, it may still be possible to characterize and purify it using the techniques described in this chapter, with slight modifications. If expression but not secretion is suspected, the cells should be harvested by centrifugation (150*g* for 10 min), washed several times in PBS, then lysed in a small volume of 50 m*M* Tris, 150 m*M* NaCl, 5 m*M* EDTA, pH 8 containing 1% (v/v) Nonidet-P40, 2 m*M* PMSF, and 5 m*M* NEM. Lysates should be cleared by centrifugation if necessary, and can then be used in the methods given above, with the following modifications. For the ELISA screen, dilute lysates 1 in 5 with PBST before use. For purification on a NIP column, after blocking the column with Tris/saline/BSA, wash it extensively with the lysis buffer just described, omitting PMSF and NEM. After loading the sample, wash again with this buffer and elute with 1 m*M* NIP-cap-OH in this buffer. No modification is necessary to the Western blot procedure.

References

1. Neuberger, M. S., Williams, G. T., and Fox, R. O. (1984) Recombinant antibodies possessing novel effector functions. *Nature* **312,** 604–608.
2. Williams, G. T. and Neuberger, M. S. (1986) Production of antibody-tagged enzymes by myeloma cells: Application to DNA polymerase I Klenow fragment. *Gene* **43,** 319–324.
3. Mulligan, R. C. and Berg, P. (1981) Selection for animal cells that express the *Escherichia coli* gene coding for xanthine–guanine phosphoribosyltransferase. *Proc. Natl. Acad. Sci. USA* **78,** 2072–2076.
4. Lundblad, A., Steller, R., Kabat, E. A., Hirst, J. W., Weigert, M.G., and Cohn, M. (1972) Immunochemical studies on mouse myeloma proteins with specificity for dextran or for levan. *Immunochemistry* **9,** 535–544.

5. Oi, V. T., Morrison, S.L., Herzenberg, L. A., and Berg, P. (1983) Immunoglobulin gene expression in transformed lymphoid cells. *Proc. Nat. Acad. Sci. USA* **80,** 825–829.
6. Hames, B. D. (1981) An introduction to polyacrylamide gel electrophoresis, in *Gel Electrophoresis of Proteins: A Practical Approach* (Hames, B. D. and Rickwood, D. J., eds.) IRL, Oxford, pp. 1–91.

Index

A

B

P

R

S